HIGH SCHOOL ALGEBRA

EMBRACING A COMPLETE COURSE FOR HIGH SCHOOLS
AND ACADEMIES

BY

WILLIAM J. MILNE, Ph.D., LL.D.,
PRESIDENT OF NEW YORK STATE NORMAL COLLEGE, ALBANY, N. Y.

NEW YORK ∴ CINCINNATI ∴ CHICAGO
AMERICAN BOOK COMPANY

MILNE'S MATHEMATICS

MILNE'S ELEMENTS OF ARITHMETIC
For Primary and Intermediate Schools

MILNE'S STANDARD ARITHMETIC
A Complete Course for Schools and Academies

MILNE'S ELEMENTS OF ALGEBRA
A Course for Beginners

MILNE'S HIGH SCHOOL ALGEBRA
A Course for High Schools and Academies

Copyright, 1892, by AMERICAN BOOK COMPANY.

H. S. Alg.

E.-P 27

PREFACE.

THIS work is prepared to meet the demands of our best High Schools and Academies. The plan pursued in the development of the subjects is substantially the same as that adopted in the author's Inductive Algebra, of which this is a revision, but the scope of the treatise has been considerably extended, so that it may more fully meet the demands of institutions that are preparing students for our higher scientific schools, and for advanced standing in our colleges.

It is believed that the treatment of the subject will commend itself to teachers on account of its simplicity, its clearness, and its thoroughness. The student is led by natural and properly graduated exercises to a thorough comprehension of the principles of the science, and then he is given such abundant practice in applying them, that they become fixed in the memory, and the most rapid progress is secured.

Improvement in methods of teaching has been very great in recent years, and it is necessary, therefore, that new text-books should keep pace with the very noticeable advance in educational science. This work is prepared for the purpose of meeting this demand, and it is confidently believed that it is constructed upon a plan which exemplifies the methods of the best teachers of the subject.

In no instance has the theoretical treatment of subjects been subordinated to the practical, and yet the theoretical is at all times illustrated and enforced by numerous practical exercises in which the book abounds.

The author desires to express his obligations to others who have preceded him in preparing text-books upon this subject, particularly to the authors of the many excellent works which have recently appeared in England. He is also specially indebted to Professor Oliver S. Westcott, A.M., of Chicago, Ill. His distinguished success as a teacher, his keen insight into the depths of mathematical science, and his eminently practical views of truth have peculiarly fitted him to give wise and valuable assistance in the preparation of the book.

The High School Algebra is submitted to the public with the hope that its scientific arrangement, its progressive development of principles, its accuracy of statements, its precision in definition, its clearness in discussion, its abundant examples, and its special adaptation for its purposes, may meet what is believed to be a popular demand.

WILLIAM J. MILNE.

State Normal College,
    Albany, N.Y.

CONTENTS.

	PAGE
ALGEBRAIC PROCESSES	9
Preliminary Definitions	10
ALGEBRAIC EXPRESSIONS	17
ADDITION	20
Equations and Problems	26
SUBTRACTION	28
Signs of Aggregation	33
Transposition in Equations	35
Equations and Problems	37
MULTIPLICATION	40
Equations and Problems	47
Special Cases in Multiplication	50
DIVISION	57
Zero and Negative Exponents	64
Equations and Problems	65
FACTORING	72
COMMON DIVISORS	88
COMMON MULTIPLES	95
FRACTIONS	99
Reduction of Fractions	100
Clearing Equations of Fractions	109
Addition and Subtraction of Fractions	114

FRACTIONS. — *Continued.* PAGE

    Multiplication of Fractions 119

    Division of Fractions... 122

    Review of Fractions...................... 125

SIMPLE EQUATIONS... 129

SIMULTANEOUS EQUATIONS..................................... 144

    Two Unknown Quantities 144

    Three or More Unknown Quantities 158

INVOLUTION 166

EVOLUTION... 175

    Square Root of Numbers.................................... 180

    Cube Root of Numbers..................................... 187

    Theory of Exponents....................................... 191

RADICAL QUANTITIES.. 199

    Reduction of Radicals...................................... 200

    Addition and Subtraction of Radicals 205

    Multiplication of Radicals 206

    Division of Radicals.. 209

    Involution and Evolution of Radicals........................ 210

    Rationalization 211

RADICAL EQUATIONS 219

QUADRATIC EQUATIONS.. 223

    Pure Quadratics... 224

    Affected Quadratics.. 227

    Equations in the Quadratic Form............................ 236

    Simultaneous Quadratic Equations 240

    Properties of Quadratics................................... 250

RATIO 256

PROPORTION.. 258

    Principles of Proportion.................................... 259

    Fractional Equations solved by the Principles of Proportion.... 268

CONTENTS.

	PAGE
PROGRESSIONS	270
Arithmetical Progression	271
Special Applications	277
Geometrical Progression	279
Special Applications	286
IMAGINARY QUANTITIES	299
Zero and Infinity	302
Interpretation of Negative Results	304
Indeterminate Equations	306
Inequalities	310
LOGARITHMS	314
Tables of Logarithms	316
Permutations and Combinations	328
THE BINOMIAL THEOREM	332
Positive Integral Exponents	332
Undetermined Coëfficients	336
Reversion of Series	342
Recurring Series	344
Any Exponent	347
Theory of Equations	352
Transformation of Equations	358
Commensurable Roots	365

HIGH SCHOOL ALGEBRA.

ALGEBRAIC PROCESSES.

1. EXAMPLE 1. Two boys had together $21. If the elder had twice as much as the younger, how much had each?

ARITHMETICAL PROCESS.

A certain sum = the money the younger had.
2 times that sum = the money the elder had.
3 times that sum = the money both had.
Therefore, 3 times that sum = $21.
The sum = $7, what the younger had.
2 times $7 = $14, what the elder had.

The above process may be abridged by using the letter s for the expressions, *a certain sum* and *that sum*. In Algebra it is common to use the letter x, or some other one of the last letters of the alphabet, for a number whose value is to be found. Therefore, the following is the

ALGEBRAIC PROCESS.

Let x = the money the younger had.
Then $2x$ = the money the elder had.
And $3x$ = the money both had.
Therefore, $3x$ = $21.
x = $7, what the younger had.
$2x$ = $14, what the elder had.

2. An Equation is an expression of equality between two numbers or quantities.

Thus, $4 + 7 = 11$, and $2x = 16$, are equations.

3. A Problem is a question requiring solution.

4. A Solution of a problem is a process of finding the result sought.

5. A Statement of a problem is an expression of the conditions of the problem, in algebraic language.

Solve algebraically the following:

2. A man paid $30 for a coat and a vest. If the coat cost 4 times as much as the vest, what was the cost of each?

3. Two boys earned together $36. If James earned 3 times as much as Henry, how much did each earn?

4. A farmer picked 24 bushels of apples from two trees. If one tree bore twice as many bushels as the other, how many bushels did each bear?

5. A and B together furnish $800 capital, of which A furnishes 3 times as much as B. How much does each furnish?

6. A man had 450 sheep in three fields. In the second he had twice as many as in the first, and in the third 3 times as many as in the second. How many were there in each field?

7. Two boys together solved 350 problems, of which William solved 4 times as many as Charles. How many did each solve?

8. A certain number added to itself is equal to 260. What is the number?

9. A farmer sold a horse and a cow for $250, receiving 4 times as much for the horse as for the cow. How much did he receive for each?

10. A has 3 times as many sheep as B, and both have 4? How many has each?

ALGEBRAIC PROCESSES.

11. A farm of 480 acres was divided between a brother and a sister, the brother having 3 times as many acres as the sister. How many acres had each?

12. The greater of two numbers is 5 times the less, and their sum is 540. What are the numbers?

13. A and B had a joint capital of $1750. A furnished 4 times as much as B. How much did each furnish?

14. A farmer raised 1320 bushels of grain. If he raised 5 times as much corn as wheat, how many bushels of each did he raise?

15. A farmer raised 1350 bushels of wheat, corn, and rye. If he raised twice as much corn as rye, and 3 times as much wheat as corn, how many bushels of each did he raise?

16. A, B, and C contributed $560 for the relief of the sick. A gave a certain sum, B gave twice as much as A, and C gave twice as much as B. How much did each give?

17. The number 169 can be divided into three integral parts such that the second part is 3 times the first, and the third 9 times the first. What are the parts?

18. The profits of a business for 3 years were $10,890. The second year the gain was twice the gain of the first year, and the gain the third year was twice as much as that of both previous years. What was the gain the third year?

19. The expenses of a manufactory doubled each year for three years. The third year they were $13,800. What were the expenses for each of the other years?

20. A lecturer received $300 for 2 lectures. For the second lecture he received 3 times as much as he did for the first. How much did he receive for each?

21. A, B, and C own 10,000 head of cattle. B owns 3 times as many as A, and C owns $\frac{1}{4}$ as many as are owned by A and B. How many does each own?

22. A number plus twice itself, plus 3 times itself, plus 4 times itself, equals 30. What is the number?

23. John has 5 times as many hens as ducks. He has in all 12 fowls. How many ducks has he?

24. A man has two daughters and one son. He wishes to divide $6000 among them so as to give the elder daughter twice as much as the younger, and the son as much as both the daughters. How much must he give each?

25. Walter has 3 times as many slate-pencils as Albert has lead-pencils. The lead-pencils cost 3 cents apiece, and the slate-pencils 1 cent apiece, and together they cost 30 cents. How many slate-pencils has Walter?

26. Divide 36 into 4 parts so that the second shall be 8 times the first, the third shall be $\frac{1}{3}$ of the first and second, and the fourth shall be $\frac{1}{2}$ of the other three.

27. What number added to 5 times itself equals 90?

28. What number added to twice itself, and that sum added to 4 times the number, equals 28?

29. What number added to 7 times itself equals 104?

30. A and B enter into partnership to do business. A furnishes 4 times as much of the capital as B, and both together furnish $15,500. How much does each furnish?

31. A gentleman dying, bequeathed his property of $14,400 as follows: to his son 3 times as much as to his daughter, and to his widow twice as much as to both son and daughter. What was the share of each?

32. A farmer bought some grain for seed — in all, 32 bushels. He purchased 3 times as many bushels of oats as of barley, and as many bushels of wheat as of oats and barley. How many bushels of each kind did he purchase?

33. A merchant bought 3 pieces of cloth which together measured 144 yards. The second was 3 times as long as the first, and the third was 8 times as long as the first. What was the length of each piece?

ALGEBRAIC PROCESSES.

34. A farmer had an orchard containing 560 trees. The number of peach trees was 3 times the number of cherry trees, and the number of apple trees 8 times the number of peach trees. How many were there of each kind?

35. James has 6 times as much money as John. He finds also that he has 30 cents more than John. How much has each?

36. A library contains 10,000 volumes. The books of fiction are 9 times as many as the scientific works, the books of travel and biography each one third as many as the books of fiction, and all the other works 4 times as many as the scientific works. How many books of fiction are there in the library?

37. Mary had 40 cents more than Sarah, and Mary's money is 5 times as much as Sarah's. How much has each?

38. A farmer had 217 cattle in three fields. The first field contained twice as many as the third, and the second twice as many as the first. How many were there in each field?

39. The earnings of a manufactory doubled each year. If, at the end of four years, they amounted to $15,000, what were the earnings the first year and the fourth year?

40. Three men engaged in business with a joint capital of $6000. A furnished three times as much as C, and B furnished $\frac{1}{2}$ as much as A and C. How much did each furnish?

41. In a certain school there are 600 pupils. The pupils in the second class are twice as many as the pupils in the first class, the pupils in the third class are as many as in both the first and second classes, while the number in the fourth class is double the number in the third. How many pupils are there in each class?

6. Quantity is the amount or extent of any thing.

Numbers are used to express quantity. In Algebra, however, the word *quantity* is frequently used for the word *number*.

Thus 35 gallons expresses a quantity.

In Algebra $2a$, $(x+y)$, $4ax$ are called quantities.

14   HIGH SCHOOL ALGEBRA.

7. Known Numbers, or **Quantities,** are such as have definite values, or those whose values are given, or to which any value can be assigned. They are represented by *figures* and the *first letters* of the alphabet.

Thus, 6, 8, 215, representing given numbers, and a, b, c, etc., representing any numbers, are known numbers, or quantities.

8. Unknown Numbers, or **Quantities,** are those whose values are to be found. They are represented by the *last letters* of the alphabet.

Thus, x, y, z, v, w, etc., are used to represent unknown numbers, or quantities.

9. Algebra is that branch of mathematics which treats of general numbers, or quantities, and of the nature, transformation, and use of equations.

The *Signs in Algebra* are, for the most part, the same as those used in Arithmetic.

10. The Sign of Addition is an upright cross : $+$. It is called *Plus*. Placed between quantities, it shows that they are to be added.

Thus, $a + b$ is read a plus b, and means that a and b are to be added.

11. The Sign of Subtraction is a short horizontal line: $-$. It is called *Minus*. Placed between two quantities it shows that the second is to be subtracted from the first.

Thus, $a - b$ is read a minus b, and means that b is to be subtracted from a.

12. The Ambiguous or **Double Sign** is $\pm$, a combination of the sign of Addition and the sign of Subtraction.

Thus, $a \pm b$ shows that b may be added to or subtracted from a.

The signs $+$ and $-$ are sometimes used for other purposes besides indicating operations to be performed. Thus, if distances *east* from a given meridian are indicated by the sign $+$ prefixed to the number of degrees, distances *west* will be indicated by placing the sign $-$ before the number of degrees.

ALGEBRAIC PROCESSES.

If the sign + indicates *gain*, the sign − will indicate *loss*. If degrees of temperature *above* the zero point are marked with the sign +, those *below* the zero point will be marked with the sign −. The signs + and − are, therefore, sometimes, *signs of opposition*, as well as of *operation*.

|−4 −3 −2 −1 0 +1 +2 +3 +4|

13. The **Sign of Multiplication** is an oblique cross: ×. It is read *multiplied by* or *times*. Placed between two quantities, it shows that the one is to be multiplied by the other.

Multiplication may also be indicated by a dot (·), or by writing the literal factors side by side.

Thus, each of the expressions $a \times b$, $a \cdot b$, and ab, shows that a is to be multiplied by b.

14. The **Sign of Division** is a short horizontal line between two dots: ÷. It is read *divided by*. Placed between two quantities, it shows that the one at the left is to be divided by the one at the right.

Division may also be indicated by writing the dividend above the divisor, with a line between them.

Thus, each of the expressions $a \div b$ and $\dfrac{a}{b}$ shows that a is to be divided by b.

15. The **Sign of Equality** is two short horizontal lines placed one above the other: =. It is read *equals*, or *is equal to*. When it is placed between two expressions an *Equation* is formed.

Thus, $x + y = 4$ is an equation.

16. The **Signs of Aggregation** are: The *Parenthesis*, ( ); the *Vinculum*, ——; the *Bracket*, [ ]; and the *Brace*, { }. They show that the quantities included by them are to be subjected to the same process.

Thus, each of the expressions $(a+b)c$, $\overline{a+b} \times c$, $[a+b]c$, and $\{a+b\}c$, shows that the sum of a and b is to be multiplied by c.

When quantities are under the vinculum, or are included within any of the other signs of aggregation, they are commonly said to be *in parenthesis*.

17. The **Sign of Involution** is a small figure or letter, called an *Exponent*, written a little above and to the right of a quantity to indicate how many times the quantity is used as a factor.

Thus, a^5 shows that a is to be used as a factor 5 times.

When no exponent is written, the exponent is **1**.

Thus, a is regarded as a^1; b as b^1.

An exponent is also called an **Index.**

18. A **Power** of a quantity is the product arising from using the quantity a certain number of times as a factor.

Thus, 4 is the second power of 2; a^3, the third power of a.

Powers are *named* from the number of times the quantity is used as a factor.

Thus, a^5 is read the *fifth* power of a, or *a fifth*.
The *second* power of a quantity is also called the *square*, and the *third* power the *cube* of the quantity.

19. A **Root** of a quantity is one of the equal factors of the quantity.

Thus, 2 is a root of 4; a is a root of a^3.

20. Roots are *named* from the number of equal factors into which the quantity is separated.

Thus, one of *two* equal factors is the *second* root, one of *three* equal factors the *third* root, etc.
The *second* root of a quantity is also called the *square* root, and the *third* root the *cube* root of the quantity.

21. The **Sign of Evolution** is $\sqrt{\ }$, called the *Radical Sign.* When it is placed before a quantity it shows that a root of the quantity is required.

The quantity or number written at the opening of the radical sign is called the *Index*. It shows what root is sought.

When no quantity or *Index* is written at the opening of the radical sign, the *square root* is indicated; if 3, as $\sqrt[3]{\ }$, the *third root*; if 4, as $\sqrt[4]{\ }$, the *fourth root*, etc.

ALGEBRAIC EXPRESSIONS.

22. The Sign ∴ is called the Sign of Deduction. It means *therefore* or *hence*.

23. A **Coëfficient** is a figure or letter placed before a quantity to show how many times the quantity is taken additively.

Thus, in the expression $7\,b$, 7 is the coëfficient of b, and it shows that $7\,b$ is equal to $b + b + b + b + b + b + b$.

Although the first figure or letter of an expression is usually regarded as the coëfficient, strictly speaking, any factor or the product of any number of the factors may be considered as the coëfficients of the product of the remaining factors.

Thus, in $3\,ax$, 3 may be regarded as the coëfficient ax, $3\,a$ as the coëfficient of x, or $3\,x$ as the coëfficient of a.

24. Coëfficients expressed by numbers are called **Numeral Coëfficients**; those expressed by letters, **Literal Coëfficients**; those expressed by figures and letters, **Mixed Coëfficients**.

When no coëfficient is expressed, the coëfficient is **1**.

ALGEBRAIC EXPRESSIONS.

25. An **Algebraic Expression** is the expression of a quantity in algebraic language.

EXERCISES.

1. Interpret in ordinary language $a^2 + 3\sqrt{a^2 - x^2}$.

INTERPRETATION. — The algebraic expression interpreted or **read** is, the sum of a square and 3 times the square root of the remainder when x square is subtracted from a square. Or, the sum of a square and 3 times the square root of the quantity a square minus x square.

Copy and read the following expressions:

2. $a + b$.
3. $3\,b - a$.
4. $a^2 + b$.
5. $a - \sqrt{b}$.
6. $x^2 + b + c^2$.
7. $4(a + b) - c$.
8. $x^3 + \sqrt{x^5 - y}$.

9. $\sqrt{a + \sqrt{2\,b} + c}$.
10. $\dfrac{x + 4(x - 3\,y)}{2 - \sqrt{4\,x - z}}$.
11. $\sqrt{a + b} + x^2 - 4$.
12. $\dfrac{\sqrt{a} + x + y^2}{\sqrt{a} - (x + y)}$.
13. $\dfrac{3\,x + y^2 - \sqrt{xy}}{4\,y^2 - z + 2\,xy^2}$.

ALGEBRA.—2.

HIGH SCHOOL ALGEBRA.

When $a=1$, $b=2$, $c=3$, $d=4$, $e=5$, find the numerical value of each of the following expressions by using the number for the letter which represents it:

Thus, $a+b+3d-e = 1+2+12-5 = 10$.

1. $3a+b$.
2. $2c-b$.
3. $3d+a-b$.
4. $2c^2-a-b$.
5. $d+c-2a$.
6. $d-(a+b)$.
7. a^2+b^2-d.
8. $(a+b)d-c$.
9. $(a+b)(d-c)$.
10. $(a^2+b^2) \div (a+b)$.
11. $4(3a-b)$.
12. $7a(3d-2a)$.
13. $abcd(a+b+c+d)$.
14. $(a+b+c)(a+b+c)$.
15. $(d+e-b)-(c-b)$.
16. $(d-a+c)(e-b)$.
17. $(bc+d)-b(a+c)$.
18. $(b^2-a+c)+(c-a)$.
19. $(c+d^2-a) \div (d-b)$.
20. $2b(a+c)(e-c)$.
21. $3ad(c-a+d)-e^2$.
22. $\dfrac{cd}{b}(a+b+c) - \sqrt{d}$.
23. $\sqrt{5e} + a + \dfrac{3b}{2}$.
24. $\left(\dfrac{a}{2}+\dfrac{c}{2}\right)d$.
25. $\dfrac{(a^2+b^2)\,3b}{2e+a}$.
26. $a^2+b^2+c^2+d^2-e^2$.
27. $\sqrt{d}+(a+b)^2-e$.
28. $(a+b)(b-a)4a$.
29. $a+3\sqrt{2e+\sqrt{4e+3d+2b}}$.
30. $\left(\dfrac{ae}{a+c}+\dfrac{3d}{c}\right)^2$.
31. $\dfrac{3c(d^3-c^3)-a}{2d+b}$.
32. $\dfrac{b^2c^2}{a^2} - 5\sqrt{(a+c)^2}$.
33. $2b(a^2+c^2)-\sqrt{(d+e)}$.
34. $(a+d)^2+ab\sqrt{(bd+a)}$.
35. $\left(\dfrac{4bc}{a+e}\right)^2 \div \dfrac{d^2}{b}$.
36. $\dfrac{a+c^2}{e}+\dfrac{de}{b+c}-\dfrac{(a+e)^2}{6}$.
37. $4(abc-bc) + \dfrac{[\overline{a+b^2}+(c+d)^2]}{(b+d)}$.
38. $\{[\overline{a+b} \times c(d+e)-a]b\} \div bd\sqrt{(a+d)^2}$.

ALGEBRAIC EXPRESSIONS. 19

26. The **Terms** of an algebraic expression are the parts connected by $+$ or $-$.

Thus, in the expression $2a + 3x - 2cd$, there are three terms.

Several terms in parenthesis are considered but one quantity.

Thus, $a + (b+c-d)x$ consists of but *two* terms, viz.: a and $(b+c-d)x$, although the quantity in parenthesis consists of three terms itself.

27. A **Positive Term** is one that has the sign $+$ before it.

When the first term of an expression is positive, the sign $+$ is usually omitted before it.

Thus, in the expression $a + 3c - 2d + 5e$, the first, second, and fourth terms are *positive*.

28. A **Negative Term** is one that has the sign $-$ before it.

Thus, in the expression $3a - 2d - 3c + 2b - e$, the second, third, and fifth terms are *negative*.

29. **Similar Terms** are such as contain the same letters with the same exponents.

Thus, $3x^2$ and $12x^2$ are similar terms; also $2(x+y)^2$ and $4(x+y)^2$. ax^2 and bx^2 are similar terms when a and b are regarded as coëfficients.

30. **Dissimilar Terms** are such as contain different letters, or the same letters with different exponents.

Thus, $3xy$ and $2yz$ are dissimilar terms, as are also $3xy$ and $3xy^2$.

31. A **Monomial** is an algebraic expression consisting of but one term.

Thus, xy, $3ab$, and $2y$ are monomials.

32. A **Polynomial** is an algebraic expression consisting of more than one term.

Thus, $x + y + z$ and $3a + 2b$ are polynomials.

33. A **Binomial** is a name applied to a polynomial of two terms.

Thus, $2a + 3b$ and $x - y$ are binomials.

34. A **Trinomial** is a name applied to a polynomial of three terms.

Thus, $x + y + z$ and $2a + 3b - 2c$ are trinomials.

ADDITION.

35. 1. How many books are 5 books, 3 books, and 7 books?

2. How many b's are $4b$, $3b$, $5b$, and $2b$?

3. How many x's are $3x$, $5x$, $9x$, $13x$, and $10x$?

4. How many ab's are $2ab$, $3ab$, $4ab$, $6ab$, and $9ab$?

5. James has no money, and owes one person 5 cents, another 3 cents, and another 2 cents. What is his financial condition?

6. If the sign − is placed before each sum which he owes, what sign should be placed before the entire amount?

7. What sign will the sum of negative quantities have?

8. How many $-a$'s are $-9a$, $-3a$, $-7a$, $-8a$?

9. Asa owes one person 10 cents, another 12 cents, and another 15 cents. If James owes him 5 cents and Henry owes him 9 cents, what is Asa's financial condition? What is the value of -10, -12, -15, $+5$, and $+9$?

10. How much is the debt in excess in the following: -8 dollars, -7 dollars, -9 dollars, 5 dollars, and 12 dollars?

11. Which is in excess, and how much, in the following $3a$, $-5a$, $-2a$, $7a$, $-6a$, $9a$, $-2a$?

12. When no sign is prefixed to a number, or quantity, what sign is it assumed to have?

36. Addition is the process of uniting several quantities so as to express their value in the simplest form.

37. The **Sum** is the result obtained by adding.

ADDITION.

38. Principles.—1. *Only similar quantities can be united by addition into one term.*

2. *Dissimilar quantities are added by writing them one after the other with their proper signs.*

In Algebra an *indicated* operation is often regarded as an operation performed, as in Principle 2.

39. To add similar monomials.

1. What is the sum of $3a$, a, $4a$, and $5a$?

PROCESS.

$3a$
a
$4a$
$5a$
―――
$13a$

EXPLANATION.—The sum of $5a$, $4a$, a, and $3a$ is determined by adding the coëfficients, or numbers, which tell how many a's there are. Hence, the *sum* is $13a$.

2. What is the value of $2a + 4a - 2a + 3a - a - 3a$?

PROCESS.

$2a \quad -2a$
$4a \quad - a$
$3a \quad -3a$
――― ―――
$9a \quad -6a$

$9a - 6a = 3a$

EXPLANATION.—Since the quantities are similar, they are written in columns.

The sum of the *positive* quantities is $9a$, and the sum of the *negative* quantities is $-6a$.

$9a - 6a = 3a$. Hence, the value is $3a$.

Find the sum of each of the following:

3.	4.	5.	6.	7.
$4b$	$3ax$	$4x^2y$	$-4z^2y^2$	$-2cx^3$
b	$2ax$	$7x^2y$	$-3z^2y^2$	$-cx^3$
$7b$	ax	$3x^2y$	$-z^2y^2$	$-8cx^3$
$9b$	$4ax$	$2x^2y$	$-8z^2y^2$	$-cx^3$
$5b$	$9ax$	$9x^2y$	$-7z^2y^2$	$-cx^3$

8. Find the sum of ax, $3ax$, $7ax$, $9ax$, $8ax$, and $2ax$.

9. Find the sum of $7mn$, mn, $2mn$, $8mn$, $3mn$, and $5mn$.

10. Find the sum of $-3x^2y^2$, $-x^2y^2$, $-5x^2y^2$, $-7x^2y^2$, $-9x^2y^2$, and $-x^2y^2$.

11. Find the sum of $3x^3y^3$, $4x^3y^3$, $3x^3y^3$, x^3y^3, $7x^3y^3$, and x^3y^3.

Express in the simplest form:

12. $3a + 4a - 2a + 7a - 3a - 6a + a$.

13. $9a^3x - 3a^3x + a^3x + 2a^3x - 7a^3x - a^3x$.

14. $4\sqrt{xy} + 2\sqrt{xy} - 3\sqrt{xy} + \sqrt{xy} + 4\sqrt{xy} - 2\sqrt{xy}$.

15. $3(xy)^3 + 4(xy)^3 - 3(xy)^3 - (xy)^3 - 7(xy)^3$.

16. $2(x+y)^4 + 6(x+y)^4 - 7(x+y)^4 - 3(x+y)^4 - 4(x+y)^4 + 9(x+y)^4 - 9(x+y)^4 + 8(x+y)^4 - (x+y)^4$.

17. $3(x-y) + 5(x-y) - 2(x-y) + 7(x-y) - (x-y) + 9(x-y) - 6(x-y) - 8(x-y) + 4(x-y)$.

18. $7(a+b) + 3(a+b) - 5(a+b) + 6(a+b) - 4(a+b) - 2(a+b) + (a+b) - 3(a+b)$.

19. $(a-x) + 6(a-x) - 2(a-x) + 4(a-x) + 5(a-x) - 7(a-x) + 5(a-x) - 3(a-x)$.

20. $5(a-b)^2 - 5(a-b)^2 + 5(a-b)^2 + 7(a-b)^2 - 8(a-b)^2 + 2(a-b)^2 - (a-b)^2$.

21. $3(x+y)^3 + 5(x+y)^3 + 7(x+y)^3 - 4(x+y)^3 - 3(x+y)^3 + 6(x+y)^3 - 7(x+y)^3 + 9(x+y)^3$.

22. $6\sqrt{a^2-x^2} + 2\sqrt{a^2-x^2} - 5\sqrt{a^2-x^2} + 3\sqrt{a^2-x^2} - 4\sqrt{a^2-x^2} + 7\sqrt{a^2-x^2} + \sqrt{a^2-x^2}$.

40. To add when some terms are dissimilar.

1. Find the sum of $x + 2y + z$, $x - y$, and $x + 3y + 2z$.

PROCESS.

$x + 2y + z$
$x - y$
$x + 3y + 2z$
_____
$3x + 4y + 3z$

EXPLANATION.—For convenience in adding, similar terms are written in the same column. Since there are three different sets of similar quantities, their sum, or the simplest expression, is the sums of the different sets of quantities connected by their proper signs, for only similar quantities can be united into one term.

ADDITION.

2. Express in its simplest form the following: $3x + 2xy + z - 3xy + 2x - 3z + 4x - 3xy - 2xy + 6z - 7x + 2w$.

PROCESS.

$3x + 2xy + z$
$2x - 3xy - 3z$
$4x - 3xy + 6z$
$-7x - 2xy \qquad + 2w$
———————————————
$2x - 6xy + 4z + 2w$

EXPLANATION. — The quantities are arranged so that similar terms are written in the same column. Beginning at either hand, each column is added separately, and the dissimilar terms of the result are connected by their proper signs, for the dissimilar terms can not be united in one term. (Prin. 2.)

RULE. — *Write similar terms in the same column. Add each column separately by finding the difference of the sums of the positive and negative terms. Connect the results with their proper signs.*

EXAMPLES.

3.

$3a + 2b$
$-2a + 3b - c$
$2a \qquad + 2c$
$\qquad 3b - 7c$
———————
$3a - 4b$

4.

$5x + 3xy$
$2x - 7xy$
$-3x - 6xy$
$\qquad 4xy - 3z$
$3x \qquad + 4z$
———————

5.

$3x + 4z - xz$
$2x - 4z$
$3z - 4xz$
$3x + 6z - 4xz$
———————
$\qquad\qquad 7xz$

Express in their simplest form the following:

6. $3x + 2y - 3z - 2y + 3z - 6x + 4y + 3z + 3x + 3z - 6y$.

7. $4xy + z - y + 3z - y - 3xy + xy - y + z + 4x - 3y + z$.

8. $3ac + 4ay + 2ac - 3ay + 2ay + 2ac - 3ac + ay$.

9. $9b + 2cd - 3e - 3cd + 9b + 3cd - 6e - 2b - 4c + 3cd$.

10. $3x^2y + 3xy - 3z + 6xy - 6x^2y + 2z - 3xy + 6z - 4z$.

11. $a + 6b + 3c - 4a + 3c + 3a - 6b + d + 2c - 3a + 7d$.

12. $x^2y + y + w - 3y + 2w + 2x^2y + z - 3x^2y - 3y + 2w$.

13. $9a^2b^2 - 3c^3y^3 + 2d^2 - 4c^3y^3 + 4a^2b^2 - 3d^2 + 2d^2 - 3a^2b^2$.

HIGH SCHOOL ALGEBRA.

14. Add $3ab + 3\sqrt{xy} + 4$, $4\sqrt{xy} - 2ab + 7$, $7ab + 3 + 2\sqrt{xy}$, $2\sqrt{xy} + 4 - 4ab$, and $3ab - 2\sqrt{xy} + 7$.

15. Add $3x^3 - 4x^2 - x + 7$, $2x^3 - x^2 + 3x - 10$, $2x^2 - 7x^3 - 2x + 4$, $3x^3 - 2x^2 + 12 - 3x$, $11x^3 + 5x^2 + 6x - 7$.

16. Add $\frac{3}{4}ax^2 + \frac{2}{3}a^2 + 2x^3y + b^3$, $3ax^2 + \frac{1}{4}x^3y + 3a^2 - 2b^3$, $2ax^2 + 3x^3y - a^2 - \frac{2}{3}b^3$, and $\frac{1}{16}ax^2 + \frac{1}{8}x^3y + 3a^2 - \frac{2}{5}b^3$.

17. Add $ac^2 + ab^2 + \frac{1}{4}a^3 - a^2b + \frac{3}{4}abc + \frac{1}{3}a^2c$, $a^2b + b^3 + ab^2 + bc^2 + 2abc + \frac{1}{2}b^2c$, and $a^2c - ac^2 + b^2c - bc^2 + c^3 + abc$.

18. Add $2(a-x) + 4x^2$, $(a-x) - 3x^2$, $6x^2 - 3(a-x)$, $7(a-x) - 5x^2$, and $x^2 - (a-x)$.

19. Add $7(a+b)^2 + b^2c^2$, $6b^2c^2 - 5(a+b)^2$, $3(a+b)^2 - 4b^2c^2$, $6b^2c^2 + 8(a+b)^2$, and $7(a+b)^2 - 8b^2c^2$.

20. Add $6(ab+c) + 7(a-x) + ax$, $5ax - 8(ab+c) - 5(a-x)$, $3(a-x) - 2(ab+c) - 4ax$, and $3(ab+c) + 2ax - (a-x)$.

21. Add $(a+c)^2 - 3a(x+y)$, $5a(x+y) - 5(a+c)^2$, $7(a+c)^2 - 7a(x+y)$, and $a(x+y) - 9(a+c)^2$.

22. Add $a(x+1) - 4(y-2) + a^2$, $3(y-2) - 5a^2 - 2a(x+1)$, $7a^2 + 4a(x+1) - 2(y-2)$, and $3a(x+1) + (y-2) + 3a^2$.

23. Add $\sqrt{a-x} + 5x^2$, $7x^2 - 3\sqrt{a-x}$, $7\sqrt{a-x} - 6x^2$, and $3x^2 - 4\sqrt{a-x}$.

24. Add $5a - 6(b+c) + 7$, $5(b+c) - 6a - 4$, $8a - 9(b+c) - 9$, and $3a - 5(b+c) + 2$.

25. Add $7(x+3) - 4y^4 + ab^2$, $3y^4 - 2ab^2 - 6(x+3)$, $5ab^2 - 5y^4 + 3(x+3)$, and $7y^4 - 2ab^2 + (x+3)$.

26. Add $ax(a-1) + (b^2-2) + y^2$, $2(b^2-2) - 3y^2 + 3ax(a-1)$, $5y^2 - 6ax(a-1) - 6(b^2-2)$, and $4ax(a-1) + (b^2-2) - 7y^2$.

ADDITION. 25

27. Add $7\sqrt{a+b} + x\sqrt[3]{a-3b} + abc$, $7x\sqrt[3]{a-3b} - 5\sqrt{a+b} - 3abc$, $4\sqrt{a+b} + 2abc - 5x\sqrt[3]{a-3b}$, and $6\sqrt{a+b} - 5x\sqrt[3]{a-3b} - 7abc$.

28. Add $4a\sqrt[5]{b-c} - 3\sqrt{x} + y$, $8\sqrt{x} - 5a\sqrt[5]{b-c} - 5y$, $6y + 7\sqrt{x} + 3a\sqrt[5]{b-c}$, and $a\sqrt[5]{b-c} - 7\sqrt{x} + 2y$.

29. What is the sum of $4x^3 + ax^3 - bx^3 + 2x^3$?

PROCESS.

$+4x^3$
$+ax^3$
$-bx^3$
$+2x^3$
───────
$(6+a-b)x^3$

EXPLANATION. — The quantities 4, a, $-b$, and 2 may be regarded as the coëfficients of x^3, and the sum obtained by adding these quantities and prefixing it to x^3. The sum of the coëfficients is $6 + a - b$.
∴ the sum of the quantities is $(6 + a - b)x^3$.

30. What is the sum of $2ax - 3bx + 4cx + 3dx$?

31. What is the sum of $2ax^2 + 4bx^2 + 3cx^2 + 4x^2$?

32. Add $2(a+b)$, $3a(a+b)$, $4(a+b)$, $2a(a+b)$.

33. Add $5(a+3)$, $2(a+3)$, $3a(a+3)$, $2b(a+3)$.

34. Add $3a\sqrt{x+y}$, $2\sqrt{x+y}$, $2a\sqrt{x+y}$, $3\sqrt{x+y}$.

35. Add $5(x+y)$, $a(x+y)$, $b(x+y)$, $-4(x+y)$.

36. Add $b(x-y)$, $b^2(x-y)$, $b^3(x-y)$.

37. Add $a\sqrt{a-b}$, $4c\sqrt{a-b}$, $3a\sqrt{a-b}$, $2c\sqrt{a-b}$.

38. Add $6b(x^2+y^2)$, $a(x^2+y^2)$, $-c(x^2+y^2)$, $-5b(x^2+y^2)$, $2c(x^2+y^2)$.

39. Add $5\sqrt{a^2-c^2}$, $3x\sqrt{a^2-c^2}$, $6\sqrt{a^2-c^2}$, $2x\sqrt{a^2-c^2}$.

40. Add $7(x+y+1)$, $2b(x+y+1)$, $-5(x+y+1)$, $3b(x+y+1)$.

41. Add $a\sqrt[3]{x-y}$, $b\sqrt[3]{x-y}$, $c\sqrt[3]{x-y}$, $(a+b+c)\sqrt[3]{x-y}$.

EQUATIONS AND PROBLEMS.

41. Simplify the following and find the value of x:

1. $3x + 4x + 2x - 3x - 2x + 4x = 16.$

SOLUTION.

$$3x + 4x + 2x - 3x - 2x + 4x = 16$$
Uniting terms, $\qquad 8x = 16$
Whence, $\qquad\qquad x = 2$

2. $5x + 2x - 3x + 4x - 6x + 7x = 18.$

3. $5x + 6x - 9x - 3x + 2x + 4x = 20.$

4. $3x - 2x + 5x + 7x + 4x - 3x = 26 + 2.$

5. $3x - 4x + 2x + 6x - 4x + x = 15 + 3 - 2.$

6. $x + 4x + 6x - 3x + 7x - 9x = 21 + 7 - 4.$

7. $9x - 2x - 3x + 7x - 5x + 4x = 35 + 9 - 4.$

8. $8x - 4x + 7x + 3x - 6x - 4x = 37 - 3 + 2.$

9. $11x - 3x + 7x - 4x + 6x - 3x = 23 + 7 - 2.$

10. $10x - 4x + 2x + 7x - 6x + 2x = 35 + 6 + 3.$

Solve the following problems:

11. James solved twice as many problems as Henry, and Henry solved 3 times as many as Harvey. If they all solved 70 problems, how many did each solve?

12. A had twice as much money as B, and B had twice as much as C. If they all had $140, how much had each?

13. William had twice as many marbles as Henry, and Henry had 3 times as many as Samuel. How many had each, if they all had 50 marbles?

14. A merchant owes B a certain sum of money, and C twice as much. Various persons owe him in all 10 times as

ADDITION.

much as he owes B. After paying all his debts he will have $1400 left. How much does he owe B and C?

15. After taking 5 times a number from 13 times a number and adding to the remainder 8 times the number, the result was 5 more than 155. What was the number?

16. A circulating library contained 10 times as many books of reference and 3 times as many historical books as works of fiction. The works of reference exceeded the works of fiction and history by 12,000 volumes. How many volumes were there of each?

17. A merchant failed in business, owing A 10 times as much as B, C three times as much as B, and D twice the difference of his indebtedness to B and C. The entire debt to these persons was $36,000. How much did he owe each?

18. At a local election there were three candidates for an office who polled the following votes respectively: A received twice as many as B, and B $1\frac{1}{2}$ times as many as C. The vote for all lacked 3 votes of being 1125. How large a vote did each receive?

19. A man earned daily for 5 days 3 times as much as he paid for his board, after which he was obliged to be idle 4 days. Upon counting his money after paying for his board he found that he had 2 ten-dollar bills and 4 dollars. How much did he pay for his board, and what were his wages?

20. A man loaned the same sum of money to each of 4 men. One man had the money for 2 years, another for 3 years, another for 4 years, and another for 5 years. If the entire interest money received was $420, how much did each man pay?

21. In a company of 77 persons it was found that there were twice as many women as men, and twice as many children as women. How many were there of each?

22. A man gave to a hospital a sum of money equal to twice what he gave to a library, and to a school four times what he gave to the library. If he gave to all $70,000, how much did he give to each?

SUBTRACTION.

42. 1. What is the difference between 7 miles and 9 miles?

2. What is the difference between $9m$ and $3m$?

3. What is the remainder when $8a$ is taken from $12a$?

4. What is left when $3a^2b$ is taken from $12a^2b$? What is the sum of $12a^2b$ and $-3a^2b$?

5. What is left when $5pq^2$ is taken from $13pq^2$? What is the sum of $13pq^2$ and $-5pq^2$?

6. Instead of subtracting a *positive* quantity, what may be done to secure the same result?

7. What is the remainder when 7 is subtracted from 13? When $7-3$ is subtracted from 13?

8. How does the result when $7-3$ is subtracted from 13 compare with the result when 7 is subtracted from 13?

9. What is the remainder when $8a$ is subtracted from $11a$? When $8a-5a$ is subtracted from $11a$?

10. How does the result when $8a-5a$ is subtracted from $11a$ compare with the result when $8a$ is subtracted from $11a$?

11. Instead of subtracting a *negative* quantity what may be done to secure the same result?

43. Subtraction is the process of finding the difference between two quantities; or
The process of finding a quantity which, added to one given quantity, will produce another.

44. The Minuend is the quantity from which another is to be subtracted.

SUBTRACTION.

45. The **Subtrahend** is the quantity to be subtracted.

46. The **Difference**, or **Remainder**, is the result obtained by subtracting.

47. PRINCIPLES. — 1. *The difference between similar quantities, only, can be expressed in one term.*
2. *Subtracting a positive quantity is the same as adding a numerically equal negative quantity.*
3. *Subtracting a negative quantity is the same as adding a numerically equal positive quantity.*

48. To subtract when the terms are positive.

1. From $9a$ subtract $3a$.

PROCESS.
$9a$
$3a$
―――
$6a$

EXPLANATION. — When 3 times any number is subtracted from 9 times that number, the remainder is 6 times the number; therefore, when $3a$ is subtracted from $9a$, the remainder is $6a$. Or, since subtracting a positive number or quantity is the same as adding an equal negative quantity (Prin. 2), $3a$ may be subtracted from $9a$ by changing the sign of $3a$ and adding the quantities. Therefore, to subtract $3a$ from $9a$, we find the sum of $9a$ and $-3a$, which is $6a$.

2. From $13a$ take $15a$.

PROCESS.
$13a$
$15a$
―――
$-2a$

EXPLANATION. — After subtracting from $13a$ as much as we can of $15a$, there will be $2a$ yet to be subtracted, and the result will be $-2a$. Or, since subtracting a positive quantity is the same as adding an equal negative quantity (Prin. 2), $15a$ may be subtracted from $13a$ by finding the sum of $13a$, and $-15a$, which is $-2a$. Therefore, when $15a$ is taken from $13a$, the result is $-2a$.

	3.	4.	5.	6.	7.	8.
From	$15a$	$13xy$	$15x^3y^2$	$19xyz$	$3x^2y^3z$	$10a^2b^3c$
Take	$6a$	$8xy$	$17x^3y^2$	$22xyz$	$15x^2y^3z$	$13a^2b^3c$

Subtract the following:

9. $8x + 2y$ from $12x + 6y$.

10. $9a + 3b$ from $10a + 2b$.

11. $7xy + 2z$ from $5xy + 4z$.

12. $3x^2y^2 + 6z$ from $5x^2y^2 + 3z$.

13. $8xy^3z + 3xy$ from $6xy^3z + 2xy$.

14. $4p^2qs + 3pq^2s$ from $5p^2qs + 6pq^2s$.

15. $5m^2nx + 3mnx$ from $7m^2nx + 2mnx$.

16. $5x^2y + 2y^2$ from $9x^2y + 7y^2$.

17. $3xy^2 + 4z$ from $xy^2 + z$.

18. $5p^2q^2 + 5pq$ from $p^2q^2 + 4pq$.

19. $x^2y^2z^2 + 4y^2$ from $15x^2y^2z^2 + 2y^2$.

20. $8yz^4 + y^4z$ from $3yz^4 + 3y^4z$.

21. $3p^2q^2 + 4qs$ from $9p^2q^2 + 2qs$.

22. $10xyz^3 + 4xyz$ from $xyz^3 + xyz$.

49. To subtract when some terms are negative.

23. From $6a - 2b$ subtract $3a - 4b$.

PROCESS.
$$\begin{array}{r} 6a - 2b \\ 3a - 4b \\ -+ \\ \hline 3a + 2b \end{array}$$

EXPLANATION. — Since the subtrahend is composed of two terms, each term must be subtracted separately. Subtracting $3a$ from $6a - 2b$ leaves $3a - 2b$, or the result may be obtained by adding $-3a$ to $6a - 2b$. But since the subtrahend was $4b$ less than $3a$, to obtain the true remainder, $4b$ must be added to $3a - 2b$, which gives $3a + 2b$. Therefore, the subtraction may be performed by changing the sign of each term of the subtrahend and adding the quantities.

RULE. — *Write similar terms in the same column. Change the sign of each term of the subtrahend from + to —, or from — to +, or conceive it to be changed, and proceed as in Addition.*

	24.	25.	26.	27.	28.
From	$4a^3x$	$3x^3y^3$	$2x + y$	$6y - 2z$	$7ax - 4by$
Take	$-2a^3x$	$-5x^3y^3$	$-2x - 2y$	$3y + 4z$	$3ax - 9by$

	29.	30.	31.
From	$3a + 2b - 3c$	$4x + 3y - 3z$	$4xy + 3z + x^2$
Take	$2a - 4b + 5c$	$2x - 4y - 5z$	$2xy - 3z + 4x^2 - y$

SUBTRACTION.

32. From $a+b+c$ subtract $a+2b-c$.
33. From $3x+2y-3z$ subtract $2x-3y+4z$.
34. From $6a^2+2b^2+3c^2$ subtract $3a^2-3b^2-2c^2$.
35. From $3a^3-2c^3-4d^3$ subtract $4c^3-3a^3+2d^3$.
36. From $8x^4-3y^2+2z^3$ subtract $4y^2-3x^4+2z^3$.
37. From $9p^2+4q^2+r^3$ subtract $3r^3-4p^2-2q^2$.
38. From $ax+2ay+z$ subtract $2ax-2ay+z$.
39. From $2xy+5yz+3xz$ subtract $2xy-3yz-4xz$.
40. From $8x^3y^2+16xy^3+10xy$ subtract $14x^3y^2-8xy^3-4xy$.
41. From $5x^3y^3+10x^4y-6yz^3$ subtract $10x^4y-4x^3y^3+5yz^3$.
42. From $3x^2+2xy+z^2+w$ subtract $2x^2-3xy-4z^2$.
43. From $15x^3+10y^3+8z^3-r^3$ subtract $5y^3+4z^3+6r^3$.
44. From $4xy^2+3x^5y+4x-3$ subtract $4xy^2-3x-7$.
45. From $4bx^3+3ay^2+4-cy$ subtract $cy-5-bx^3$.
46. From $3x^2y^4+3xy-5x$ subtract $2x^2y^4-2xy+4x-5$.
47. From $4x^5y^2-3xy^5-7z^4$ subtract $2x^5y^2+6xy^5+2z^4+9$.
48. From $7ar^2-4bs^3+3rs$ subtract $3ar^2+p+2bs^3+7$.
49. From $15x^5-24x^3y^3-16y^4$ subtract $15x^3y^3+4z-5y^4+x^5$.
50. From $3x^m-4x^ny^m+4y^m$ subtract $4x^m+2x^ny^m-4x^{2m}$.
51. From $3x^{2n}-2x^{3n}y^m-y^{m-1}$ subtract $3y^{m-1}+2x^{3n}y^m-4x^{2n}$.
52. From $3\sqrt{xy}+2z-\sqrt[3]{y^2}$ subtract $2\sqrt{xy}-3z-2\sqrt[3]{y^2}$.
53. From $4(a+b)^2-3a+4c$ subtract $a-2(a+b)^2-2c$.
54. From $5\sqrt{a+b^2}-3\sqrt[3]{x+y}$ subtract $6\sqrt[3]{x+y}-7\sqrt{x+y}$.
55. From $5\sqrt{a+b^2}-3\sqrt[3]{c+d}$ subtract $4\sqrt{a+b^2}+2\sqrt[3]{c+d}$.

HIGH SCHOOL ALGEBRA.

56. From $ax + by$ subtract $cx - dy$.

PROCESS.
$$ax + by$$
$$cx - dy$$
$$\overline{(a-c)x + (b+d)y}$$

Since a and c may be regarded as the coëfficients of x, and b and $-d$ the coëfficients of y, the difference between the quantities may be found by writing the difference between the coëfficients as coëfficients of x and y respectively. Since c cannot be subtracted from a, the subtraction is indicated by $(a-c)$, and since $-d$ cannot be subtracted from b, the subtraction is indicated by $(b+d)$, consequently the remainder may be written $(a-c)x + (b+d)y$.

57. From $ay + 2x$ subtract $cy - dx$.

58. From $cx + 2xy - 3dz$ subtract $3ax + 2xy - 2az$.

59. From $2cd - 3ab + ce^2$ subtract $2ad + 3ab - 2de^2$.

60. From $ax + by - z$ subtract $bx - ay - cz$.

61. From $5ay + 2cz - 6x$ subtract $cy - az - dx$.

62. From $ax^2 + 2cy + 3x^3y$ subtract $2bx^2 - 3ay - cx^3y$.

63. From $2px^2 + ry^2 - 3qxy$ subtract $rx^2 + sxy - py^2$.

64. From $cx - 14aby + 7a^2b^2$ subtract $9x - 14aby + 15b^2$.

65. From $3x + 7y - 8z$ subtract $bcx - ay + qz$.

66. From $(a-b)x + (a+b)y$ subtract $(a-c)x + (b-c)y$.

67. From $(a+b-c)x + (a-b+c)y$ subtract $(a+b-d)x - (a+b-c)y$.

68. From $4b(x-y) + 4cdx$ subtract $4c(x-y) + ax$.

69. From $5\sqrt{x} + 3\sqrt{y}$ subtract $3\sqrt{x} + 2a\sqrt{y}$.

70. From $(a^2+b^2)x - (a^2-b^2)y$ subtract $(b^2-c^2)x - (a^2+c^2)y$.

71. From $a\sqrt{x+y} + b\sqrt{x-y} + dx$ subtract $b\sqrt{x+y} - a\sqrt{x-y} - cx$.

72. From $n\sqrt[m]{x^2y^2} - (m+n)\sqrt[n]{xy}$ subtract $(m-n)\sqrt[n]{xy} - (n+3)\sqrt[m]{x^2y^2} + y^2$.

SIGNS OF AGGREGATION.

50. The subtrahend is sometimes expressed with a sign of aggregation, and written after the minuend with the sign — between them.

Thus, when $b + c - d$ is subtracted from $a + b$, the result is sometimes indicated as follows: $a + b - (b + c - d)$.

1. What change must be made in the signs of the terms of the subtrahend when it is subtracted from the minuend?

2. When a quantity in parenthesis is preceded by the sign —, what change must be made in the signs of the terms when the subtraction is performed or when the parenthesis or other similar sign is removed?

The term *parenthesis* is commonly used to include all signs of aggregation.

See § 16, note.

51. PRINCIPLES.—1. *A parenthesis, preceded by the minus sign, may be removed from an expression if the signs of all the terms in parenthesis are changed.*

2. *A parenthesis, preceded by the minus sign, may be used to inclose an expression if the signs of all the terms to be inclosed in parenthesis are changed.*

When quantities are inclosed in a parenthesis preceded by the *plus* sign, the parenthesis may be removed without any change of signs, and consequently, any number of terms may be inclosed in a parenthesis with the *plus* sign without any change of signs.

The student should remember that in expressions like $-(x^2 - y + z)$ the sign of x^2 is plus, and the expression is the same as if written $-(+ x^2 - y + z)$.

Simplify the following:

1. $a - (a + b)$.
2. $x - (x - y)$.
3. $a + b - (- a)$.
4. $a - (- a - b)$.
5. $a - (a - b)$.
6. $y - (- x - y)$.
7. $4a - (2a + y)$.
8. $3x + 2y - (2x - 2y)$.

9. $5x - 3y - (-2x + 4y)$.

10. $7x + 3z - (x + y + z)$.

11. $2x - 3y^2z - (x + z^2 - 3y^2z)$.

12. $3xy + 2x^3y - (4xy - x^3y + x^2)$.

13. $3x^2 + 2y^2 - (-4x^2 - 2y^2 - z^2)$.

14. $3ab^2 - 2ac^2 - (-3ab^2 - 6ac^2)$.

15. $(a+b) + (a-b) - (2a - 2b)$.

16. $(a+b-c-x) - (b-c-x+a) + (x-a)$.

17. $(3x - 4c) + (x - 3c) - (4x - 7c - 4)$.

18. $(3a^2 - 2a^2x - 7) - (7 + 3a^2 - 4a^2x + 2) - 3$.

19. $(a^2 + 2ab + b^2) - (a^2 - 2ab + b^2) - (-4ab)$.

20. $1 - (1-x) + (2+x) - (1 + x^2 - x)$.

21. $\{(a+b)x + 4\} - (a-b)x + 7$.

When the expression contains two or more parentheses, they may be removed *in succession* by beginning with the outside or the inside one.

Thus, $\quad a + b - (c - a + [d+b] - c + 2b - d)$
$= a + b - c + a - [d+b] + c - 2b + d$
$= a + b - c + a - d - b + c - 2b + d$
$= 2a - 2b$.

Simplify the following:

22. $2a - (2b - d) - \{a - b - (2c - 2d)\}$.

23. $2a - [3b + (2b - c) - 4c + \{2a - (3b - \overline{c - 2b})\}]$.

24. $2a - [2a - \{2a - (2a - \overline{2a - a})\}]$.

25. $x^2 - \{5mc^2 - [x - (3c - 3mc^2) + 3c - (x^2 - 2mc^2 - c)]\}$.

26. $a - b - c - (d + 2a + [3b - 2c + d] - 4a - 2b)$.

27. $x^2 + 2y - (x^3 + [2y + 3x^2 - 4x^3] - 6y + 3x^2) + 4x^2$.

28. $-(x^2y + 2y - 3) - (x^2y - [6y + 7 - 3x^2y] + 9)$.

29. $ab + bc - (3ab + [3bc + 2bd - 3ab] + 2bd) - 6c$.

SUBTRACTION.

30. $-\{3ax-[2xy+3z]+z-(4xy+[3ax+6z]+3z)\}.$

31. $x-[-\{-(-x)+x\}-2x].$

32. $y^2-(c^2+x^2)-[y^2-\{-(-\overline{x^2+y^2})\}].$

33. $(a-b)-\{-a-(b-a)+(a-b)\}.$

34. $3a-(2a+1)+\{a-(3-\overline{4-a})\}.$

35. $-7-[-\{-a-(-a-\overline{a-3})\}].$

36. $(x^2+1)-[ax-\{-(-2ax+7)-\overline{ax-x^2-7}\}+2x^2].$

TRANSPOSITION IN EQUATIONS.

52. 1. If $x-5=20$, what is the value of x?

2. If $x+5=20$, what is the value of x?

3. In the equation $x-5=20$, what is done with the 5 in obtaining the value of x? In the equation $x=20+5$, how does the sign of the 5 compare with its sign in the previous equation?

4. In the equation $x+5=20$, what is done with the 5 in obtaining the value of x? In the equation $x=20-5$, how does the sign of the 5 compare with its sign in the previous equation?

5. In changing the 5's from one side, or member, of the equation to the other, what change was made in the sign?

6. When a number, or quantity, is changed from one member of an equation to the other, what change must be made in its sign?

7. If 5 is added to one member of the equation $2+3=5$, what must be done to the other member so as to preserve the equality?

8. If 5 is subtracted from one member of the equation $2+3=5$, what must be done to the other member so as to preserve the equality?

9. If one member of the equation $2 + 3 = 5$ is multiplied by 5, what must be done to the other member to preserve the equality?

10. If one member of the equation $2 + 3 = 5$ is divided by 5, what must be done to the other member to preserve the equality?

11. If one member of the equation $7 + 9 = 16$ is raised to the second power, or if the second root of one member is found, what must be done to the other member to preserve the equality?

12. What, then, may be done to the members of an equation without destroying the equality?

53. The **Members of an Equation** are the parts on each side of the sign of equality.

54. The **First Member** of an equation is the part on the left of the sign of equality.

55. The **Second Member** of an equation is the part on the right of the sign of equality.

56. **Transposition** is the process of moving a term from one member of an equation to the other.

57. An **Axiom** is a truth that does not need demonstration.

Axioms. — 1. *Things that are equal to the same thing are equal to each other.*

2. *If equals be added to equals, the sums will be equal.*

3. *If equals be subtracted from equals, the remainders will be equal.*

4. *If equals be multiplied by equals, the products will be equal.*

5. *If equals be divided by equals, the quotients will be equal.*

6. *Equal powers of equal quantities are equal.*

7. *Equal roots of equal quantities are equal.*

SUBTRACTION.

58. PRINCIPLE. — *A term may be transposed from one member of an equation to the other if its sign is changed from $+$ to $-$, or from $-$ to $+$.*

EQUATIONS AND PROBLEMS.

59. 1. $2x - 3 = x + 6$. Find the value of x.

PROCESS.

$$2x - 3 = x + 6$$
$$+3 = +3$$
$$\overline{2x = x + 9}$$
$$x = x$$
$$\overline{x = 9}$$

OR,

$$2x - 3 = x + 6$$
$$2x - x = 6 + 3$$
$$x = 9$$

EXPLANATION. — Since the known and unknown quantities are found in both members of the equation, in order to find the value of x, the known quantities must be collected in one member and the unknown in the other.

Since -3 is found in the first member, it may be caused to disappear by adding 3 to both members (Axiom 2), which gives the equation $2x = x + 9$.

Since x is found in the second member, it may be caused to disappear by subtracting x from both members (Axiom 3), which gives, as a resulting equation, $x = 9$.

Or, since a term may be moved from one member of an equation to the other, if its sign is changed (Prin.), -3 may be transposed to the second member if it is changed to $+3$, and x may be transposed to the first member if it is changed to $-x$. Therefore, the resulting equation will be $2x - x = 6 + 3$. By uniting the terms the result is $x = 9$.

The result may be *verified* by substituting the *value* of x for x in the original equation. If both members are then identical, the value of the unknown quantity is correct. Thus, if 9 be substituted for x in the original equation, it becomes $18 - 3 = 9 + 6$, or $15 = 15$. Therefore, the value of x is 9.

RULE. — *Transpose the terms so that the unknown quantities stand in the first member of the equation, and the known quantities in the second.*

Unite similar terms, and divide each member of the equation by the coëfficient of the unknown quantity.

VERIFICATION. — *Substitute the value of the unknown quantity in the original equation. If both members are then identical in value, the value of the unknown quantity found is correct.*

1. The same quantity, with the same sign upon opposite sides of an equation, may be cancelled from both.
2. The equality will not be destroyed if the signs of all the terms of an equation are changed at the same time.

Transpose, and find the value of x in the following:

2. $x + 3 = 7$.
3. $2x - 4 = 12$.
4. $2x - 10 = 14$.
5. $3x + 7 = 28$.
6. $3x - 5 = 25$.
7. $7x - 3 = 25$.
8. $9x + 6 = 24$.
9. $8x - 13 = 27$.
10. $7x + 5 = 26$.
11. $10x - 5 = 35$.
12. $12x + 6 = 30$.
13. $13x - 4 = 35$.
14. $2x + 2 = 6 + x$.
15. $3x - 4 = 6 + x$.
16. $3x + 5 = 11 - x$.
17. $4x + 2 = 3x + 8$.
18. $4x - 11 = 9 - x$.
19. $4x + 3 = 3x + 10$.
20. $7x - 5 = 19 + 4x$.
21. $9x - 3 = 30 - 2x$.
22. $2x + 35 = 5x + 2$.
23. $3x - 15 + 24 = 25 - 10$.
24. $4x + 13 + 38 = 10x - 3x$.
25. $3x - 6 = x + 14 - 4$.

Solve the following problems:

26. What number increased by 9 is equal to 34?

27. What number diminished by 15 equals 31?

28. What number increased by 9 equals 27?

29. What number diminished by 10 equals 33?

30. What number added to twice itself gives a sum equal to 45?

31. What number added to three times itself gives a sum equal to 72?

32. What number is there whose double exceeds the number by 10?

33. What number is there such that, if 10 be added to it, twice the sum will be 44?

34. Twice a certain number increased by 4 is equal to the number increased by 15. What is the number?

SUBTRACTION.

35. Three times a certain number diminished by 5 is equal to the number plus 21. What is the number?

36. A man walked 71 miles in three days, walking 3 miles more the second day than the first, and 5 miles more the third day than the second. How far did he travel each day?

SOLUTION. — Let $x =$ the number of miles he traveled the 1st day.
Then, $x + 3 =$ the number of miles he traveled the 2d day.
And, $x + 8 =$ the number of miles he traveled the 3d day.
Therefore, $x + x + 3 + x + 8 = 71$
Transposing, $x + x + x = 71 - 3 - 8$
Uniting similar terms, $3x = 60$
Whence, $x = 20$, the number of miles he traveled the 1st day.
$x + 3 = 23$, the number of miles he traveled the 2d day.
$x + 8 = 28$, the number of miles he traveled the 3d day.

37. Three boys had together 85 cents. James had 10 cents more than John, and Henry had 5 cents more than James. How much had each?

38. A farmer remembered that he had 395 sheep distributed in three fields, so that there were 20 more in the second than in the first, and 25 more in the third than in the second, but he could not tell how many there were in each field. Find the number in each field.

39. A drover being asked if he had 100 head of cattle, replied that if he had twice as many as he then had and 4 more he would have 100. How many had he?

40. A gentleman left his estate, amounting to $6900, to be divided among his four sons, so that each should have $150 more than his next younger brother. How much was the share of each?

41. The expenses of a manufacturer for 4 years were $9500. An examination showed an annual increase of $250. What were his yearly expenses?

MULTIPLICATION.

60. 1. If a man walks 4 miles per hour, how far will he walk in 3 hours?

2. How many m's are 3 times $4m$? 2 times $4m$? 5 times $6m$?

3. If a boy can gather 3 quarts of chestnuts per hour, how many quarts can he gather in 4 hours? How many q's are 4 times $3q$?

4. A vessel sails 6 miles north per hour, indicated by $+6$. How far will she sail in 3 hours? What sign should be placed before the product to indicate the direction sailed?

5. How many are 3 times $+6$? 3 times $+6a$? 2 times $+5b$? 3 times $+7x$? 4 times $+3a^2$?

6. When a positive quantity is multiplied by a positive quantity, what is the sign of the product?

7. If a vessel sails 5 miles south per hour, indicated by -5, how far will she sail in 4 hours? What sign should be placed before the product to indicate the direction sailed?

8. How many $-m$'s are 4 times $-5m$? 3 times $-6m$? How many are 5 times $-4b$? 6 times $-3x$?

9. When a negative quantity is multiplied by a positive quantity, what is the sign of the product?

10. How does the product of 4×5 compare with the product of 5×4? What effect has it upon a product to change the order of the factors when they are abstract numbers? What, then, is the product of $-4 \times +3$? Of $+3 \times -4$? Of $-5x \times +7$? Of $+7 \times -5x$?

MULTIPLICATION. 41

11. When a positive quantity is multiplied by a negative quantity, what is the sign of the product?

12. What is the product of -3×6?

13. Since -3×6 is -18, if -3 is multiplied by $6-2$, how many times -3 must be subtracted from -18 to obtain the true result?

14. If the subtraction is indicated, what are the signs of the remainder when -6 is subtracted from -18?

15. What is the product of -5×4?

16. Since -5×4 is -20, if -5 is multiplied by $4-3$, how many times must -5 be subtracted from -20? If the subtraction is indicated, what are the signs of the remainder when -15 is subtracted from -20?

17. Since, in the results just obtained, -3 multiplied by -2 gives $+6$ and -5×-3 gives $+15$, what may be inferred as to the sign of the product when a negative quantity is multiplied by a negative quantity?

18. What is an exponent? What does it show? In the expression 5^3, what does the 3 show? In the expression a^5, what does the 5 show?

19. When a^3 is multiplied by a^2, how many times is a used as a factor? How many times is a used as a factor when a^2 is multiplied by a^5?

20. How, then, may the number of times a quantity is used as a factor in multiplication, be determined from the exponents of the quantity in the expressions which are multiplied?

21. How is the exponent of a quantity in the product determined?

22. Multiply $3a^2$ by $2a$. How is the coëfficient in the product obtained from the coëfficients in the factors?

61. Multiplication is the process of taking one quantity as many times as there are units in another.

HIGH SCHOOL ALGEBRA.

62. The **Multiplicand** is the quantity to be taken or multiplied.

63. The **Multiplier** is the quantity showing how many times the multiplicand is to be taken.

64. The **Product** is the result obtained by multiplying.

65. The multiplicand and multiplier are called the *factors* of the *product*.

66. The **Signs of Multiplication.** (See Art. 13.)

67. Principles. — 1. *Either factor may be used as multiplier or multiplicand when both are abstract.*

2. *The sign of any term of the product is $+$ when its factors have* LIKE *signs, and $-$ when they have* UNLIKE *signs.*

3. *The coëfficient of a quantity in the product is equal to the product of the coëfficients of its factors.*

4. *The exponent of a quantity in the product is equal to the sum of its exponents in the factors.*

68. The principle relating to the signs of the terms of the product is illustrated as follows:

$$+a \text{ multiplied by } +b = +ab$$
$$-a \text{ multiplied by } +b = -ab$$
$$+a \text{ multiplied by } -b = -ab$$
$$-a \text{ multiplied by } -b = +ab$$

69. To multiply when the multiplier is a monomial.

1. What is the product of $3a^2x$ multiplied by $2a^3x^2y$?

PROCESS.
$3a^2x$
$2a^3x^2y$
―――
$6a^5x^3y$

EXPLANATION. — Since the multiplier is composed of the factors 2, a^3, x^2, and y, the multiplicand may be multiplied by each successively. 2 times $3a^2x = 6a^2x$; a^3 times $6a^2x = 6a^5x$ (Prin. 4); x^2 times $6a^5x = 6a^5x^3$ (Prin. 4); y times $6a^5x^3 = 6a^5x^3y$, since literal quantities when multiplied may be written one after another without the sign of multiplication. Or,

The coëfficient of the product is obtained by multiplying 3 by 2 (Prin. 3). The literal quantities are multiplied by adding their exponents (Prin. 4). Hence, the product is $6a^5x^3y$.

MULTIPLICATION.

2. What is the product of $2a - b^2$ multiplied by $-3b$?

PROCESS.
$$2a - b^2$$
$$-3b$$
$$\overline{-6ab + 3b^3}$$

EXPLANATION. — Since $2a$ multiplied by $-3b$ is the same as $2a$ times $-3b$ (Prin. 1), the product of $2a$ multiplied by $-3b$ is $-6ab$. But, since the *entire* multiplicand is $2a - b^2$, the product of b^2 multiplied by $-3b$ must be *subtracted* from $-6ab$. b^2 multiplied by $-3b$ gives as a product $-3b^3$, which subtracted from $-6ab$ gives the entire product $-6ab + 3b^3$; or,

Since $2a$ and $-3b$ have *unlike* signs, the *sign* of their product is $-$ (Prin. 2); and, since $-b^2$ and $-3b$ have *like* signs, the *sign* of their product is $+$ (Prin. 2). Hence, the product is $-6ab + 3b^3$.

RULE. — *Multiply each term of the multiplicand by the multiplier, as follows:*

To the product of the numerical coëfficients, annex each literal factor with an exponent equal to the sum of the exponents of that letter in both factors.

Write the sign $+$ before each term of the product when its factors have like signs, and $-$ when they have unlike signs.

	3.	4.	5.	6.	7.	8.
Multiply	-8	4	$7a$	$-3x$	$4x$	$3x^2$
By	3	-3	3	4	-5	$2x^4$

	9.	10.	11.	12.	13.
Multiply	$3x^3$	$4x^3y$	$3x^2y^2$	$-4xmy$	$-10xyz^2$
By	$2x^4$	$2x^2y$	$2x^3y^4$	$3x^2my^2$	$4x^2y^4z$

	14.	15.	16.	17.	18.
Multiply	$-2x^3y^2z^2$	$-4a^2b$	$6abx^3$	$-3x^2y^2$	$-5a^3b^2xy$
By	$-8xyz$	$5ab^2$	$-4a^2b^2x$	$-2x^3y$	$-3a^3b^2xy$

	19.	20.	21.	22.	23.
Multiply	$-3c^2dy$	$4ax^3y$	$5a^2x^2y^3$	$-4x^2y$	$-6x^2y^2z^2$
By	$4c^2d$	$-5x^2y^3$	$-3a^2x^4z$	$5xy^3z$	$-4x^2y^2$

	24.	25.	26.	27.
Multiply	$4a^2x^2y^2$	$5ax^2y$	$(x+y)$	$4(a+b)$
By	$-3y^2z^2$	$-3bx^2z$	2	-3

HIGH SCHOOL ALGEBRA.

	28.	29.	30.
Multiply	$-5(y+z)^2$	$(a-b)^3$	$2(c+d)^2$
By	$-3(y+z)^3$	$4(a-b)^3$	$3(c+d)^3$

	31.	32.	33.	34.
Multiply	$2(x+y+z)^4$	$3x^n$	$4a^n$	$-5ax$
By	$-5(x+y+z)^3$	$4x^n$	$-5a^{2n}$	$3a^2x^{3n}$

	35.	36.	37.	38.
Multiply	$2ax^n$	$3a^nx^n$	$-4x^ny^m$	$3x^{n-1}$
By	$4ax^m$	$-5a^{2n}x^{3n}$	$-5x^my^n$	$-5x^{n+3}$

Multiply:

39. $x^2 - 2y$ by $3y$.

40. $x^2y - 2z$ by $2z$.

41. $4x^2 - 2xy$ by $3xy$.

42. $-3x^2 - 2y^2$ by $2x^2y$.

43. $4x^2y^2 + 2z^2$ by $-4x^2z^2$.

44. $3x^2y^2 - 2yz$ by $3xyz$.

45. $4x^3 + 2y + 3z$ by xy.

46. $3x^2y + y - 3xz$ by $2xz$.

47. $6x^2y^2 + 4y^2$ by $3xy^2$.

48. $4ab - 3ac$ by $3acd$.

49. $5ac - 6ax$ by $-5acx$.

50. $5abc - 3acd$ by $-4abcd$.

51. $3a^2xy - 2a^2bc$ by $-2ax^2$.

52. $3xy + 7z$ by $2x^2yz$.

53. $4ab^2 - 3c^2$ by $3bc^2$.

54. $5a^2x^3 - 4ax^8$ by $3a^2x^2$.

55. $4x^4y + 3y^2$ by $5x^ny^2$.

56. $6a^2x - 9ax^2$ by $3a^2x$.

57. $2a^5x^2 - 3ax^n$ by $3a^{2n}x$.

58. $7a^2bc^2 + 4dx$ by $5dy^2$.

59. $\frac{1}{3}a^2x^4 - \frac{3}{5}ax$ by $\frac{3}{4}a^3x^3$.

60. $2a^nb + 3ab^2$ by $7a^nb^m$.

70. To multiply when the multiplier is a polynomial.

1. Multiply $x - 2y$ by $2x + y$.

PROCESS.

$$\begin{array}{r} x - 2y \\ 2x + y \\ \hline \end{array}$$

$2x$ times $(x-2y) = 2x^2 - 4xy$

y times $(x-2y) = xy - 2y^2$

$(2x+y)$ times $(x-2y) = 2x^2 - 3xy - 2y^2$

MULTIPLICATION.

RULE. — *Multiply each term of the multiplicand by each term of the multiplier, and add the partial products.*

              2.   3.

Multiply $ab + 2c$ $3x^2 - axy$
By $2ab - 3c$ $2x^2 + 3axy$

        $2a^2b^2 + 4abc$ $6x^4 - 2ax^3y$
             $- 3abc - 6c^2$ $+ 9ax^3y - 3a^2x^2y^2$

Product, $2a^2b^2 + abc - 6c^2$ $6x^4 + 7ax^3y - 3a^2x^2y^2$

Multiply:

4. $x + y$ by $x - y$.

5. $3a + c$ by $a + 3c$.

6. $4a - 2b$ by $3a - 3b$.

7. $2y + 3z$ by $3y - 4z$.

8. $2x + y$ by $2x + 2y$.

9. $3x - 4y$ by $3x - 4y$.

10. $5a + 2c$ by $3a - 7c$.

11. $ax + by$ by $ax + by$.

12. $2ac + 3bc$ by $2ac - 3bc$.

13. $3bd - 4bc$ by $2bd + 3bc$.

14. $3x^2y^2 - 4z^2$ by $2x^2y^2 + 3z^2$.

15. $3x^2z^2 + 2y$ by $2xy^2 + z$.

16. $4ab^2 + 3bc^2$ by $2ab + 2bc^2$.

17. $5x^3y - 3ax$ by $5xy^3 - 2ax$.

18. $a^2 + 2ab + b^2$ by $a + b$.

19. $x^2 + 4x + 4$ by $x + 2$.

20. $a^2 + ay - y^2$ by $a - y$.

21. $2a^2 + ab - 2b^2$ by $3a - 3b$.

22. $a^6 + a^4 + a^2$ by $a^2 - 1$.

23. $x^4 + x^2y^2 + y^4$ by $x^2 - y^2$.

24. $2x - 3y + 4z$ by $3x + 2y - 5z$.

25. $2a^2 + 5ab - 3c^2$ by $3a^2 - 2ab + 5c^2$.

26. $3x^2 - 4xy + 5y^2$ by $7x^2 - 2xy - 3y^2$.

27. $1 - 3x + 3x^2$ by $1 - 2x + 2x^2$.

28. $a^2 + ax + x^2$ by $a^2 - ax + x^2$.

29. $x + 2y - z$ by $x + y - 2z$.   31. $x^n + y^n$ by $x^n + y^n$.

30. $a^m + b^m$ by $a^m - b^m$.   32. $x^m + y^m$ by $x^n + y^n$.

33. $x^{m+n} + y^{m+n}$ by $x^{m+n} + y^{m+n}$.

34. $a^{m-n} + b^{m-n}$ by $a^{m-n} - b^{m-n}$.

The multiplication of polynomials is sometimes *indicated* by placing them in parenthesis. When the multiplication is performed, they are said to be *expanded*

Expand:

35. $(x+y)(x+y)$.
36. $(2x-y)(2x-y)$.
37. $(3x-4y)(3x+4y)$.
38. $(4x+6y)(4x-6y)$.
39. $(3ax+2y)(3ax+2z)$.
40. $(2x-4xy)(2x-2z)$.
41. $(3a^2-2bc)(3a^2+2bc)$.
42. $(a^2+b)(a+b^2)$.
43. $(a+b+c)(a-b-c)$.
44. $(a+b)(a+b)(a+b)$
45. $(a-b)(a+b)(a-b)(a+b)$.
46. $(x^2+2x+1)(x^2-2x+1)$.
47. $(a^2-2ab+b^2)(a^2+2ab+b^2)$.
48. $(1+a)(1-a)(1+a)(1+a)$.
49. $(x^2-y^2)(x^2-y^2)(x^2-y^2)(x^2-y^2)$.
50. $(a^2+b^2)(a^2-b^2)(a^4-b^4)(a^8-b^8)$.
51. $(a^2b^2-b^2)(a^2b^2+b^2)(a^4b^4-b^4)(a^8b^8-b^8)$.
52. $(x+y-z)(x-y+z)(y-x+z)(x+y+z)$.
53. $(1-x)(1+x)(1+x^2)(1+x^4)(1+x^8)$.
54. $(2x-3)(2x+3)(4x^2+9)$.
55. $(8y^p-3y^{2p})(8y^p+3y^{2p})$.
56. $(m^5-m^2n+m^3n^2-m^2n^3+mn^2+n^5)(m^2+n^2)$.

Multiply:

57. $x^3+3x^2-5x+20$ by $x^2-4x-10$.
58. $x^2-2x^4+2-x^3+3x$ by $4-3x+2x^2$.
59. $a^2-ab+b^2+a+b+1$ by $a+b-1$.
60. $x^4-2x^3-x^2+2x+1$ by x^2-x+1.
61. $3m^3-2m^2p+3mp^2-p^3$ by m^2-mp+p^2.
62. $a-2b+3c$ by $a+2b-3c$.
63. a^2+b-2c by a^2-b-c.
64. $m^2+2mn+n^2+p^2$ by $m^2-2mn+n^2-p^2$.

MULTIPLICATION.

65. $r^2 - 2rt - t^2$ by $r^2 + 2rt + t^2$.
66. $m^{p-1} - n^{p-1}$ by $m - n$.
67. $x^{p+2-2c} - x^{p-2c}y^{p-3c+2} + y^{2p+4-3c}$ by $x^{p+2c} - y^{p+3c-2}$.
68. $x^{a-1} - x^a y^b + y^{b-1}$ by $x^{1-a} - y^{1-b}$.
69. $c^{2p} + c^p d^q + d^{2q}$ by $c^p - d^q$.
70. $a^2 + b^2 + c^2 + d^2$ by $a^2 - b^2 + c^2 - d^2$.
71. $a^m + b^m$ by $a^n + b^n$.
72. $x^2 - 2xy + y^2$ by $x^2 - y^2$.
73. $a^4 - a^3 y + a^2 y^2 - a y^3 + y^4$ by $a + y$.

EQUATIONS AND PROBLEMS.

71. 1. $5(x - 3) = 2(x + 3) + 3$. Find the value of x.

PROCESS.

$$5(x-3) = 2(x+3) + 3$$

Multiplying, $\quad 5x - 15 = 2x + 6 + 3$

Transposing, $\quad 5x - 2x = 15 + 6 + 3$

Uniting, $\quad\quad\quad 3x = 24$

$$x = 8$$

EXPLANATION.—Since the multiplication is indicated in one term on each side of the equation, in finding the value of x, the multiplication must be performed.

The known quantities are then transposed to the second member and the unknown quantities to the first member, similar terms are united, and the value of x is found.

Find the value of x and verify the result in the following:

2. $3(2x - 5) = 21$.
3. $4 + 3(3x - 7) = 19$.
4. $3(4x + 7) + 5 = 50$.
5. $5x + 3(2 - x) = 40$.
6. $6x + 3(4x + 3) = 41$.
7. $5(x + 6) = 2(x + 3) + 30$.
8. $3(2x - 4) = 4(x - 5) + 32$.
9. $3(x + 2) = 4(x - 2) + 15$.
10. $3x - 2(x + 1) = 13 - 7$.
11. $5x - 3(x - 4) = 4x + 7$.

12. $4(x-5) - 3(x+6) = 0$.

13. $(2+x)(x+3) = x^2 + 2x + 18$.

14. $5(2x-2) = 27 + 3(2x+1)$.

15. $10(x-5) = (x+1) + 5(x+1)$.

16. $5(x+3) - 2(2x-7) = 3(x-7)$.

17. $3 + 7(x-2) - 4(2x-7) = 16 + (x-2)$.

18. $6x = 15 + 3(x-3) - 3(x-10)$.

19. $19 = 2(4-x) + 5(7+2x) - 48$.

20. $2x + 3(6x-5) - 5 = x - 1$.

21. $3(x-7) = 14 + 2(x-10) + 2$.

Solve the following problems, and verify the results:

22. There are two numbers whose sum is 40. One is twice the other increased by 5. What are they?

SOLUTION.—Let x represent the first number.

Then, $2(x+5)$ will represent the second number.

And, $\quad x + 2(x+5) = 40$

$\qquad x + 2x + 10 = 40$

$\qquad\quad 3x = 40 - 10$

$\qquad\quad 3x = 30$

$\qquad\quad x = 10$, the first number.

$\qquad 2(10+5) = 30$, the second number.

23. What number is that to which, if 3 times the sum of the number and 2 is added, the result will be 22?

24. If B were 5 years younger, A's age would be twice B's. The sum of their ages is 20. How old is each?

25. Two boys find that they have together 21 cents. They discover that if Henry had 5 cents less, John's money would be just 3 times Henry's. How much has each?

26. Two pedestrians travel toward each other at the rate of 5 miles per hour until they meet. When they meet they dis

MULTIPLICATION.

cover that one has traveled 3 hours longer than the other, and that the entire distance traveled by both is 55 miles. How far does each travel?

27. Three men, A, B, and C, each had a sum of money. A had twice as much as B, and B twice as much as C. A and B each lost 10 dollars and C gained 5 dollars, when the difference between what A and B had was equal to what C then had. How much had each?

28. A farmer plowed two fields containing together 50 acres. If the smaller field had contained 10 acres more, it would have been half the size of the larger. How many acres were there in each field?

29. A commenced business with twice as much capital as B. During the first year A gained $500 and B lost $300, when A had 3 times as much money as B. What was the original capital of each?

30. A man wishing to buy a quantity of butter found two firkins, one of which lacked 6 pounds of containing enough, and the other weighed 14 pounds more than he wanted. If three times the quantity in the first firkin was equal to twice the quantity in the second, how many pounds did he wish to purchase? How many pounds were there in each firkin?

31. Ten persons bought a bicycle, but four of them being unable to pay their share, the others had each to pay $8 more. What was the cost of the bicycle?

32. Two men start at the same time from two towns 49 miles apart and travel toward each other. One travels 4 miles per hour, but rests 2 hours; the other travels 5 miles per hour and rests 3 hours. How many miles has each traveled when they meet?

33. A man bought 13 dozen bananas for $3.50. For a part of them he paid 25 cents per dozen, and for the rest he paid 30 cents per dozen. How many did he buy of each kind?

34. A is three times as old as B, and 8 years ago he was seven times as old as B. How old is each now?

ALGEBRA. — 4.

SPECIAL CASES IN MULTIPLICATION.

72. The square of the sum of two quantities.

$$(a+b)(a+b) = a^2 + 2ab + b^2$$
$$(x+y)(x+y) = x^2 + 2xy + y^2$$

1. When a quantity is multiplied by itself, what power is obtained?

2. How are the terms of the second power, or square, of the quantities, obtained from the quantities?

3. What signs have the terms?

73. PRINCIPLE. — *The square of the sum of two quantities is equal to the square of the first quantity, plus twice the product of the first and second, plus the square of the second.*

Since $a^2 \times a^2 = a^4$; $a^3 \times a^3 = a^6$; $a^4 \times a^4 = a^8$; it is evident that the exponent of the second power or square of a quantity is equal to twice the exponent of the given quantity.

EXAMPLES.

Write out the products or powers of the following:

1. $(c+d)(c+d)$.
2. $(m+n)(m+n)$.
3. $(r+s)(r+s)$.
4. $(x+2)(x+2)$.
5. $(a+3)(a+3)$.
6. $(3a+x)(3a+x)$.
7. $(2b+c)(2b+c)$.
8. $(2y+1)(2y+1)$.
9. $(m+2n)(m+2n)$.
10. $(2c+2d)(2c+2d)$.
11. $(2x+3)(2x+3)$.
12. Square $2x+4y$.
13. Square $3a+2b$.
14. Square x^2+y^2.
15. Square $4x+3y$.
16. Square $3p+2q$.
17. Square $2x^2+5y^2$.
18. Square $2y^3+3x^3$.
19. Square x^n+y^n.
20. Square $x^{3n}+y^{2n}$.
21. Square $x^{p-q}+y^p$.
22. Square $x^{m+n}+y^{2n}$.

MULTIPLICATION.

23. Find the square of 31.

SOLUTION. $\qquad 31 = 30 + 1.$

$$31^2 = (30 + 1)^2 = 30^2 + 2 \times 30 \times 1 + 1^2 = 961.$$

Square:

24. 22.	**27.** 52.	**30.** 91.	**33.** 202.
25. 23.	**28.** 71.	**31.** 101.	**34.** 207.
26. 41.	**29.** 82.	**32.** 103.	**35.** 303.

36. Find the square of $4\frac{1}{2}$.

SOLUTION. $\qquad\qquad 4\frac{1}{2} = 4 + \frac{1}{2}$

$$(4 + \tfrac{1}{2})^2 = 4^2 + 2 \times \tfrac{1}{2} \times 4 + (\tfrac{1}{2})^2 = 16 + 4 + \tfrac{1}{4} = 20\tfrac{1}{4}$$

Observe that the *middle term* of the square of any number expressed by an integer and the fraction $\frac{1}{2}$ is equal to the integer. Hence the square of such a number will be the integer multiplied by itself for the first term + the integer for the middle term + the square of $\frac{1}{2}$ for the third term; the sum of the first two terms of the square will be the integer multiplied by the integer increased by 1; and the third term will be $\frac{1}{4}$.

Thus, $\qquad\qquad (7\tfrac{1}{2})^2 = 8 \times 7 + \tfrac{1}{4} = 56\tfrac{1}{4}$

Find the square of:

37. $5\frac{1}{2}$.	**40.** $12\frac{1}{2}$.	**43.** 2.5.	**46.** 45.	**49.** 8.5.
38. $8\frac{1}{2}$.	**41.** $10\frac{1}{2}$.	**44.** 3.5.	**47.** 7.5.	**50.** 85.
39. $7\frac{1}{2}$.	**42.** $9\frac{1}{2}$.	**45.** 4.5.	**48.** 75.	**51.** 35.

74. The square of the difference of two quantities.

$$(a - b)(a - b) = a^2 - 2ab + b^2$$
$$(x - y)(x - y) = x^2 - 2xy + y^2$$

1. How are the terms of the power obtained from the terms of the quantity squared?

2. What signs connect the terms of the power?

3. How does the square of $(a - b)$ differ from the square of $(a + b)$?

75. PRINCIPLE. — *The square of the difference of two quantities is equal to the square of the first quantity, minus twice the product of the first and second, plus the square of the second.*

EXAMPLES.

Write out the products of:

1. $(a-c)(a-c)$.
2. $(y-z)(y-z)$.
3. $(r-s)(r-s)$.
4. $(b-c)(b-c)$.
5. $(x-1)(x-1)$.
6. $(x-2y)(x-2y)$.
7. $(x-2yz)(x-2yz)$.
8. $(2x-3z)(2x-3z)$.
9. $(2a-c)(2a-c)$.
10. $(3y-2z)(3y-2z)$.
11. $(3x-4y)(3x-4y)$.

12. Square $2a+2d$.
13. Square $2r-3s$.
14. Square $2s-q$.
15. Square $3m-4n$.
16. Square $2v-w$.
17. Square $2x^2-2y^2$.
18. Square $2x-3$.
19. Square $3ax-2x^2$.
20. Square x^m-y^n.
21. Square $x^{m-n}-y^{m+n}$.
22. Square $2x^2-3x^n y^n$.

23. Find the square of 19.

SOLUTION. $\qquad 19 = 20 - 1$
$$19^2 = (20-1)^2 = 20^2 - 2 \times 20 \times 1 + 1^2 = 361$$

Find the square of:

24. 18.	27. 38.	30. 59.	33. 78.	36. 997.
25. 29.	28. 49.	31. 58.	34. 99.	37. 998.
26. 39.	29. 48.	32. 79.	35. 98.	38. 999.

76. The product of the sum and difference of two quantities.

$$(a+b)(a-b) = a^2 - b^2$$
$$(x-y)(x+y) = x^2 - y^2$$

1. How are the terms of the product obtained from the quantities?

2. What sign connects the terms?

77. PRINCIPLE. — *The product of the sum and the difference of two quantities is equal to the difference of their squares.*

MULTIPLICATION.

EXAMPLES.

1. $(c+d)(c-d)$.
2. $(r+s)(r-s)$.
3. $(m+n)(m-n)$.
4. $(c+a)(c-a)$.
5. $(x-1)(x+1)$.
6. $(2-x)(2+x)$.
7. $(c+2d)(c-2d)$.
8. $(2x+3)(2x-3)$.
9. $(3m+4n)(3m-4n)$.
10. $(2x+5y)(2x-5y)$.
11. $(ab+cd)(ab-cd)$.
12. $(2x+4)(2x-4)$.
13. $(2x^2+y)(2x^2-y)$.
14. $(x^2+y^2)(x^2-y^2)$.
15. $(x^4-y^4)(x^4+y^4)$.
16. $(3v+2w)(3v-2w)$.
17. $(5xy-3)(5xy+3)$.
18. $(2a^2+3b^2)(2a^2-3b^2)$.
19. $(3ab^2+5bc^2)(3ab^2-5bc^2)$.
20. $(4x^3y^3+5)(4x^3y^3-5)$.
21. $(5x^m+4y^n)(5x^m-4y^n)$.
22. $(7a^2yx^3+6z^n)(7a^2yx^3-6z^n)$.

23. Find the product of $(x+y+z)(x+y-z)$.

SOLUTION.
$$x+y+z = (x+y)+z$$
$$x+y-z = (x+y)-z$$

By Prin. 77, $[(x+y)+z][(x+y)-z] = (x+y)^2 - z^2$
$$= x^2 + 2xy + y^2 - z^2$$

24. Find the product of $(x-y+z)(x+y-z)$.

SUGGESTION. $x-y+z = x-(y-z)$ and $x+y-z = x+(y-z)$

Find the product of:

25. $(a+b+c)(a-b-c)$.
26. $(a-b+c)(a-b-c)$.
27. $(m+n+r)(m-n-r)$.
28. $(x^2+2x+1)(x^2-2x+1)$.
29. $(x-y+4)(4-x+y)$.
30. $(x^2+xy+y^2)(x^2-xy+y^2)$.
31. $(4a+3b-c)(4a-3b+c)$.
32. $(2x+3z-4)(2x+3z+4)$.
33. $(2m^2+mn+3n)(2m^2-mn+3n)$.

34. What is the product of 32 times 28?

SOLUTION. $32 = 30+2$; $28 = 30-2$
$$32 \times 28 = (30+2) \times (30-2) = 900 - 4 = 896$$

Find the product of:

35. 19×21.	39. 74×66.	43. 34×26.	47. 99×101.
36. 29×31.	40. 89×91.	44. 38×42.	48. 98×102.
37. 33×27.	41. 78×82.	45. 57×63.	49. 96×104.
38. 56×64.	42. 97×103.	46. 45×55.	50. 94×106.

51. What is the square of 97?

SOLUTION. $\qquad (a+b)(a-b) = a^2 - b^2 \qquad (1)$

Transposing (1), $\quad (a+b)(a-b) + b^2 = a^2 \qquad (2)$

Let $\qquad a = 97$ and $b = 3$

Equation (2) becomes $(97 + 3)(97 - 3) + 9 = 97^2$

$$\therefore 97^2 = 100 \times 94 + 9 = 9409$$

52. What is the square of 38?

SOLUTION. — Let $\qquad a = 38$ and $b = 2$

Equation (2) becomes $(38 + 2)(38 - 2) + 4 = 38^2$

$$\therefore 38^2 = 40 \times 36 + 4 = 1444$$

Square by a similar process:

53. 19.	57. 59.	61. 94.	65. 103.
54. 29.	58. 49.	62. 78.	66. 107.
55. 31.	59. 98.	63. 79.	67. 112.
56. 39.	60. 96.	64. 68.	68. 997.

Form a rule for squaring numbers like those given above, and then square each number without writing the results.

78. The product of two binomials.

$$(x + 2)(x + 3) = x^2 + 5x + 6$$
$$(x + 2)(x - 3) = x^2 - x - 6$$
$$(x - 2)(x - 3) = x^2 - 5x + 6$$

1. How many terms are alike in each factor?

2. How is the first term of each product obtained from the factors?

MULTIPLICATION.

3. How is the second term of the product in the first example obtained from the factors? In the second example? In the third example?

4. How is the third term of the product in each example obtained from the factors?

5. How are the signs determined which connect the terms?

79. PRINCIPLE. — *The product of two binomial quantities having a common term is equal to the square of the common term, the algebraic sum of the other two multiplied by the common term, and the algebraic product of the unlike terms.*

EXAMPLES.

Write out the products of the following:

1. $(x+4)(x+3)$.
2. $(x-5)(x+3)$.
3. $(x+3)(x-4)$.
4. $(x-4)(x-6)$.
5. $(a+c)(a+b)$.
6. $(a+m)(a+n)$.
7. $(2x+4)(2x-5)$.
8. $(3x-5)(3x+2)$.
9. $(x^2-3)(x^2+7)$.
10. $(x^3-a)(x^3+2a)$.
11. $(3x-5)(3x-6)$.
12. $(2a+y)(2a+x)$.
13. $(5b-c)(5b+3c)$.
14. $(3a^2+2)(3a^2-3)$.
15. $(4d+5)(4d+2)$.
16. $(7y-3)(7y-4)$.
17. $(3x-7)(3x+5)$.
18. $(2y-3)(2y-4)$.
19. $(4a+b)(4a+c)$.
20. $(5a+2b)(5a-2c)$.
21. $(3ax+4)(3ax-7)$.
22. $(2a^2x+2)(2a^2x-6)$.
23. $(2x^2y^3+4)(2x^2y^3+7)$.
24. $(3ac^2+3)(3ac^2-5)$.
25. $(5c^2d^3+x)(5c^2d^3-y)$.
26. $(3ax^3+4)(3ax^3+7)$.
27. $(5cdx+1)(5cdx-5)$.
28. $(4c^3+ab)(4c^3-d)$.
29. $(ax-9)(ax+5)$.
30. $(2ax+4)(2ax-b)$.
31. $(3x^n+m^n)(3x^n+n^m)$.
32. $(ad^2x^3-10)(ad^2x^3-3)$.

80. To square any polynomial.

$$(a+b+c)^2 = a^2 + b^2 + c^2 + 2ab + 2ac + 2bc$$
$$(a+b-c+d)^2 = a^2 + b^2 + c^2 + d^2 + 2ab - 2ac + 2ad - 2bc + 2bd - 2cd$$

1. In the square of the polynomials what terms are squares?

2. How are the other terms formed?

3. How are the signs of the terms determined?

81. Principle. — *The square of a polynomial is equal to the square of each of the terms and twice the product obtained by multiplying each term by all the terms that follow it.*

Find the square of:

1. $x + y - z.$
2. $x - y + z.$
3. $a - b - c.$
4. $a + b - c.$
5. $x + y + 3.$
6. $2x + y - 7.$
7. $2x - y - z.$
8. $3x + y - 4.$
9. $2x + 3y - 6.$
10. $x - 6y - 5.$
11. $3x - 2y + 3z.$
12. $a + b + c + d.$
13. $a - b - c - d.$
14. $x + y - z - d.$
15. $x + y + z + 4.$
16. $3x + 2y + 3z + 3.$
17. $2x - 3y - 2z + 5.$
18. $2x + 5y + z + w.$
19. $3x + y + 2z + 5.$
20. $2x + 3y - 5 + 2z.$
21. $3x - 7 + 2y - 5z.$
22. $4x - 2y - 2z + 6.$
23. $2a - 3b - 2c.$
24. $4an + 3ab + 6.$
25. $3ax^2 - 2by^2 + 7.$
26. $2x + 3y - 2z + 4.$
27. $x^n + y^n + z^n + w^n.$
28. $x^{2n} + y^n - z^{3n} - 8.$

DIVISION.

82. 1. What is the product of $a^2 \times a^3$?

2. Since the product of $a^2 \times a^3$ is a^5, if a^5 is divided by a^2 what will be the quotient? What will be the quotient if a^5 is divided by a^3?

3. What is the product when x^2 is multiplied by x^4?

4. What is the exponent of the quotient when x^6 is divided by x^2? By x^4? x^9 by x^3? x^9 by x^6? x^5 by x^4?

5. How is the exponent of a quantity in the quotient determined?

6. How many times is $5x$ contained in $10x^2$? $6y$ in $18y^2$? $8z$ in $40z^2$?

7. How is the coëfficient of the quotient determined?

8. When $+5$ is multiplied by $+3$, what is the product?

9. Since $+15$ is the product of $+5 \times +3$, if $+15$ is divided by $+3$ what is the sign of the quotient?

10. What is the sign of the quotient when a positive quantity is divided by a positive quantity?

11. When $+5$ is multiplied by -3, what is the product?

12. Since -15 is the product of $+5 \times -3$, if -15 is divided by $+5$ what is the sign of the quotient? What when it is divided by -3?

13. What is the sign of the quotient when a negative quantity is divided by a positive quantity?

14. What is the sign of the quotient when a negative quantity is divided by a negative quantity?

15. What is the product of -4 by -3?

16. Since $+12$ is the product of -4×-3, if $+12$ is divided by -3 what is the sign of the quotient? What when it is divided by -4?

17. What is the sign of the quotient when a positive quantity is divided by a negative quantity?

83. Division is the process of finding how many times one quantity is contained in another. Or,

The process of finding, from a product and one of its factors, the other factor.

Division is therefore the inverse of multiplication.

84. The **Dividend** is the quantity to be divided.

85. The **Divisor** is the quantity by which we divide. It shows also into how many equal parts the dividend is to be divided.

86. The **Quotient** is the result obtained by division. The part of the dividend remaining when the division is incomplete is called the **Remainder.**

87. The **Signs of Division.** (See Art. 14.)

88. PRINCIPLES. — 1. *The sign of any term of the quotient is $+$ when the dividend and divisor have like signs, and $-$ when they have unlike signs.*

2. *The coëfficient of the quotient is equal to the coëfficient of the dividend divided by that of the divisor.*

3. *The exponent of any quantity in the quotient is equal to its exponent in the dividend diminished by its exponent in the divisor.*

89. The principle relating to the signs in division may be illustrated as follows:

$$\left.\begin{array}{l} +a \times +b = +ab \\ -a \times +b = -ab \\ +a \times -b = -ab \\ -a \times -b = +ab \end{array}\right\} \quad \text{Hence,} \quad \left\{\begin{array}{l} +ab \div +b = +a \\ -ab \div +b = -a \\ -ab \div -b = +a \\ +ab \div -b = -a \end{array}\right.$$

DIVISION.

90. To divide when the divisor is a monomial.

1. Divide $-15x^2y^3z^4$ by $3xy^2z^2$.

PROCESS.
$$3xy^2z^2 \overline{)\,-15x^2y^3z^4}$$
$$-5xyz^2$$

EXPLANATION.—Since the dividend and divisor have unlike signs, the sign of the quotient is $-$. (Prin. 1.)

Then -15 divided by 3 is -5; x^2 divided by x is x; y^3 divided by y^2 is y; and z^4 divided by z^2 is z^2 (Prin. 3). Therefore, the quotient is $-5xyz^2$.

2. Divide $12a^2x^2y^3$ by $5a^2x^2z^2$.

PROCESS.
$$\frac{12\,a^2x^2y^3}{5\,a^2x^2z^2} = \frac{12\,y^3}{5\,z^2}$$

EXPLANATION. — Since division may be indicated by writing the divisor under the dividend with a line between them, the division may be expressed as in the margin. And since the same factors are found in both dividend and divisor, they may be cancelled without changing the quotient. Hence, the quotient is $\dfrac{12\,y^3}{5\,z^2}$.

3. Divide $9a^2x^3 - 12a^3x^5 + 6ax^4$ by $3ax^2$.

PROCESS.
$$3ax^2 \overline{)\,9a^2x^3 - 12a^3x^5 + 6ax^4}$$
$$3ax \;-\; 4a^2x^3 + 2x^2$$

EXPLANATION. — Dividing $9a^2x^3$ by $3ax^2$, the result is $3ax$; dividing $-12a^3x^5$ by $3ax^2$, the result is $-4a^2x^3$; dividing $6ax^4$ by $3ax^2$, the result is $2x^2$.

Therefore, the quotient is $3ax - 4a^2x^3 + 2x^2$.

RULE. — *Divide each term of the dividend by the divisor, as follows:*

To the numerical coëfficient of the dividend divided by that of the divisor, annex each literal factor with an exponent equal to the exponent of that letter in the dividend minus its exponent in the divisor.

Write the sign $+$ before each term of the quotient when the terms of both dividend and divisor have like signs, and $-$ when they have unlike signs.

1. An equal factor in both dividend and divisor may be omitted, since it forms no part of the quotient.

2. When the quotient is not integral it should be expressed as a fraction.

	4.	5.	6.	7.	8.
Divide	$6a$	$-12a^2x$	$15a^2y^2$	$-20x^3y^2$	$24y^2z^3$
By	$3a$	$3a^2x$	$-5ay$	$-5x^2y$	$-8y^2z$

HIGH SCHOOL ALGEBRA.

Find the quotient in the following:

9. $-25 x^2y^2z^3 \div 5xyz^2$.
10. $20 a^5b^5c \div 10 abc$.
11. $30 cd^2f \div 15 cd^2$.
12. $36 ax^2y \div 18 ay$.
13. $-18 x^2yz \div 9xy$.
14. $-21 vwz^2 \div 7 vz^2$.
15. $-33 r^2sz^2 \div 11 rs$.
16. $35 m^2nx \div 5 m^2x$.
17. $20 x^3y^3z^3 \div 10 x^3yz^3$.
18. $-14 a^2x^3y^4 \div 7 axy^2$.
19. $32 r^2s^2q \div 8 r^2sq$.
20. $-18 v^2x^2y \div v^2xy$.
21. $24 a^{2n}bc^n \div -a^{2n}bc^n$.
22. $36 n^{2n}xy^{3n} \div -4 n^nxy^{2n}$.
23. $25 xyz^2 \div -5 x^2y^2z$.
24. $-28 y^2z^{2m} \div 4 y^2z^3x$.
25. $-30 n^2x^2 \div 6 m^2x^2$.
26. $28 x^2y^2z^2 \div 7 xyz$.

Divide:

27. $ax^2y - 2xy^2$ by xy.
28. $3xy^2 - 3x^2y$ by xy.
29. $4x^3y^2 + 2x^2y^3$ by $2x^2y^2$.
30. $3a^2b^2 - 6ab^3$ by $3ab$.
31. $abc^2 - a^2b^2c$ by $-abc$.
32. $9x^2y^2z + 3xyz^2$ by $3xyz$.
33. $a^2 - 3ab + ac^2$ by a.
34. $x^2y - xy^2 + x^2y^3$ by xy.
35. $x^2 - 2xy + y^2$ by x.
36. $z^2 - 3xz + 3z^2$ by z.
37. $m^2n + 2mn - 3m^2$ by mn.
38. $c^2d - 3cd^2 + 4d^3$ by cd.
39. $a(b+c)^2 + b(b+c)^3$ by $-(b+c)$.
40. $9(a-c) - 6(a-c)^3$ by $3(a-c)$.
41. $6a^3x^2 - 15a^4x^2 + 30a^3x^3$ by $-3a^3x^2$.
42. $20x^2y^4 - 14xy^3 + 8x^2y^2$ by $2x^2y^2$.
43. $28ab^3x^5 + 36a^2b^4x^4 - 32a^3b^5x^6$ by $4a^2b^3$.
44. $18c^2dx^3 + 24c^3d^2x^2 - 30c^4d^3x^4$ by $6c^2d^2x$.
45. $xy^2 + cx^2y^3 + dx^3y^4$ by xy^3.
46. $a(b-c)^3 + b(b-c)^4 - c(b-c)^2$ by $(b-c)$.

DIVISION.

47. $3(x+y) - 9(x+y)^2 + 6(x+y)^5$ by $-(x+y)^2$.

48. $x^4 - \frac{1}{2}x^3 - \frac{1}{3}x^2 - 2x - 1$ by $2x$.

49. $x^{m+1} + x^{m+2} + x^{m+3} + x^{m+4}$ by x^4.

50. $y^{n+1} - y^{n+2} - y^{n+3} - y^{n+4}$ by y^{n+1}.

91. To divide when the divisor is a polynomial.

1. Divide $x^3 + 3x^2y + 3xy^2 + y^3$ by $x + y$.

PROCESS.

$$\begin{array}{l|l} x^3 + 3x^2y + 3xy^2 + y^3 & \underline{x + y} \\ \underline{x^3 + x^2y} & x^2 + 2xy + y^2 \\ 2x^2y + 3xy^2 \\ \underline{2x^2y + 2xy^2} \\ xy^2 + y^3 \\ \underline{xy^2 + y^3} \end{array}$$

EXPLANATION. — For convenience, the divisor is written at the right of the dividend, and both dividend and divisor are arranged according to the descending powers of x.

When the first term of the dividend is divided by the first term of the divisor, the result will be the first term of the quotient. x is contained in x^3, x^2 times; therefore, x^2 is the first term of the quotient. x^2 times the divisor equals $x^3 + x^2y$. Subtracting, there is a remainder of $2x^2y$, to which the next term of the dividend is annexed for a new dividend.

When the first term of the new dividend is divided by the first term of the divisor, the result will be the second term of the quotient. x is contained in $2x^2y$, $2xy$ times; therefore, $2xy$ is the second term of the quotient. $2xy$ times the divisor equals $2x^2y + 2xy^2$. Subtracting, there is a remainder of xy^2, to which the next term of the dividend is annexed for a new dividend.

The third term of the quotient is found by dividing the first term of the new dividend by the first term of the divisor. x is contained in xy^2, y^2 times. y^2 times the divisor is $xy^2 + y^3$. Subtracting, there is no remainder. Hence, the quotient is $x^2 + 2xy + y^2$.

RULE. — *Write the divisor at the right of the dividend, arranging the terms of each according to the ascending or descending powers of one of the literal quantities.*

Divide the first term of the dividend by the first term of the divisor, and write the result for the first term of the quotient.

Multiply the divisor by this term of the quotient, subtract the product from the dividend, and to the remainder annex as many terms of the dividend as are necessary to form a new dividend.

Divide the new dividend as before, and continue to divide in this way until the first term of the divisor is not contained in the first term of the dividend.

If there be a remainder after the last division, write it over the divisor in the form of a fraction, and annex it with its proper sign to the part of the quotient previously obtained.

2. $\quad\begin{array}{l} x^4 - a^2x^2 + 2a^3x - a^4 \\ \underline{x^4 + ax^3 - a^2x^2} \\ \quad -ax^3 + 2a^3x - a^4 \\ \quad \underline{-ax^3 - a^2x^2 + a^3x} \\ \qquad\qquad a^2x^2 + a^3x - a^4 \\ \qquad\qquad \underline{a^2x^2 + a^3x - a^4} \end{array}\;\bigg|\;\begin{array}{l} x^2 + ax - a^2 \\ \hline x^2 - ax + a^2 \end{array}$

3. $\quad\begin{array}{l} x^4 - 1 \\ \underline{x^4 - x^3} \\ \quad x^3 - 1 \\ \quad \underline{x^3 - x^2} \\ \qquad x^2 - 1 \\ \qquad \underline{x^2 - x} \\ \qquad\quad x - 1 \\ \qquad\quad \underline{x - 1} \end{array}\;\bigg|\;\begin{array}{l} x - 1 \\ \hline x^3 + x^2 + x + 1 \end{array}$

4. $\quad\begin{array}{l} a^3 - x^3 \\ \underline{a^3 - a^2x} \\ \quad a^2x - x^3 \\ \quad \underline{a^2x - ax^2} \\ \qquad ax^2 - x^3 \\ \qquad \underline{ax^2 - x^3} \end{array}\;\bigg|\;\begin{array}{l} a - x \\ \hline a^2 + ax + x^2 \end{array}$

5. $\quad\begin{array}{l} 48x^3 - 76ax^2 - 64a^2x + 105a^3 \\ \underline{48x^3 - 72ax^2} \\ \quad -4ax^2 - 64a^2x \\ \quad \underline{-4ax^2 + 6a^2x} \\ \qquad\quad -70a^2x + 105a^3 \\ \qquad\quad \underline{-70a^2x + 105a^3} \end{array}\;\bigg|\;\begin{array}{l} 2x - 3a \\ \hline 24x^2 - 2ax - 35a^2 \end{array}$

DIVISION.

6. $$\begin{array}{l|l} x^4 + 4a^4 & x^2 + 2ax + 2a^2 \\ \underline{x^4 + 2ax^3 + 2a^2x^2} & \overline{x^2 - 2ax + 2a^2} \\ -2ax^3 - 2a^2x^2 + 4a^4 \\ \underline{-2ax^3 - 4a^2x^2 - 4a^3x} \\ 2a^2x^2 + 4a^3x + 4a^4 \\ \underline{2a^2x^2 + 4a^3x + 4a^4} \end{array}$$

Divide:

7. $a^2 - 2ab + b^2$ by $a - b$.

8. $x^2 + 4x + 4$ by $x + 2$.

9. $9 + 6x + x^2$ by $3 + x$.

10. $x^3 + x^2y + xy^2 + y^3$ by $x + y$.

11. $a^4 + a^3y + ay^3 + y^4$ by $a + y$.

12. $x^3 + 3x^2y + 3xy^2 + y^3$ by $x + y$.

13. $r^3 + 3r^2s + 3rs^2 + s^3$ by $r^2 + 2rs + s^2$.

14. $x^4 + 4x^3y + 6x^2y^2 + 4xy^3 + y^4$ by $x + y$.

15. $c^4 + 4c^3d + 6c^2d^2 + 4cd^3 + d^4$ by $c^2 + 2cd + d^2$.

16. $x^4 - 3x^3 - 36x^2 - 71x - 21$ by $x^2 - 8x - 3$.

17. $a^3 + 5a^2x + 5ax^2 + x^3$ by $a^2 + 4ax + x^2$.

18. $a^2 + 2bc - b^2 - c^2$ by $a - b + c$.

19. $a^4 - 4a^3y + 6a^2y^2 - 4ay^3 + y^4$ by $a^2 - 2ay + y^2$.

20. $ax^3 - a^2x^2 - bx^2 + b^2$ by $ax - b$.

21. $20a^2b - 25a^3 - 18b^3 + 27ab^2$ by $6b - 5a$.

22. $3x^4 - 8x^2y^2 + 3x^2z^2 + 5y^4 - 3y^2z^2$ by $x^2 - y^2$.

23. $4a^4 - 9a^2 + 6a - 1$ by $2a^2 + 3a - 1$.

24. $2ay + 3by + 10ab + 15b^2$ by $y + 5b$.

25. $b - 6b^3 - 2a + 54a^3 - 3a^2b$ by $2a - b$.

26. $25a^5 - a^3 - 8a - 2a^2$ by $5a^2 - 4a$.

27. $x^3 + y^3 + z^3 - 3xyz$ by $x + y + z$.

28. $18x^4 - 45x^3 + 82x^2 - 67x + 40$ by $3x^2 - 4x + 5$.

29. $16x^4 - 72a^2x^2 + 81a^4$ by $2x - 3a$.

30. $a^4 + 4a^2x^2 + 16x^4$ by $a^2 + 2ax + 4x^2$.

31. $x^4 + x^2z^2 + z^4$ by $x^2 - xz + z^2$.

32. $x^4 - y^4$ by $x - y$.

33. $x^5 + y^5$ by $x + y$.

34. $x^7 + 1$ by $x + 1$.

35. $x^4 - 81y^4$ by $x - 3y$.

36. $81a^4 - 16b^4$ by $3a + 2b$.

37. $x^n - y^n$ by $x + y$ to three terms.

38. $a^4 + 2a^2b^2 + 9b^4$ by $a^2 - 2ab + 3b^2$.

39. $8x^3 + 27y^3$ by $2x + 3y$.

40. $x^4 - a^2x^2 + 2a^3x - a^4$ by $x^2 - ax + a^2$.

41. $a^4 - 4a^3x + 6a^2x^2 - 4ax^3 + x^4$ by $a^2 - 2ax + x^2$.

42. $x^4 + 11x^2 - 12x - 5x^3 + 6$ by $3 - 3x + x^2$.

43. $8m^9 - 27n^6$ by $2m^3 - 3n^2$.

44. $6x^2y^2 - 4xy^3 - 4x^3y + y^4 + x^4$ by $x - y$.

45. $x^6 + x^3 - x^2 + 2x - x^4 - 1$ by $x^2 + x - 1$.

46. $a^5 - b^5$ by $(a^3 + b^3)(a + b) + a^2b^2$.

47. $a^6 - b^6$ by $a^3 - 2a^2b + 2ab^2 - b^3$.

ZERO AND NEGATIVE EXPONENTS.

92. 1. How many times is a^2 contained in a^2? a^3 in a^3?

2. When similar quantities have exponents, how may the division be performed?

3. What, then, will be the quotient when a^2 is divided by a^2 by subtracting exponents? a^3 by a^3? a^m by a^m?

4. Since $a^2 \div a^2$, $a^3 \div a^3$, and $a^m \div a^m$ are each equal to a^0 and to 1, what is the value of a^0, or any finite quantity with 0 for an exponent?

DIVISION.

5. What will be the quotient when a^3 is divided by a^5 by subtracting exponents? a^5 by a^7? a^6 by a^9?

6. What will be the quotient when a^2 is divided by a^5 without subtracting exponents? a^6 by a^9?

7. Since $a^3 \div a^5 = a^{-2}$ and $\frac{1}{a^2}$, $a^6 \div a^9 = a^{-3}$ and $\frac{1}{a^3}$, to what is any quantity with a *negative* exponent equal?

93. The Reciprocal of a quantity is 1 divided by the quantity. Thus the reciprocal of a is $\frac{1}{a}$, of $x+y$ is $\frac{1}{x+y}$.

94. PRINCIPLES.—1. *Any finite quantity having 0 for an exponent is equal to 1.*

2. *Any quantity having a negative exponent is equal to the reciprocal of the quantity with an equal positive exponent.*

EQUATIONS AND PROBLEMS.

95. 1. Find the value of x in the equation $ax + 4 = a^2 - 2x$.

SOLUTION. $\qquad ax + 4 = a^2 - 2x$
Transposing, $\qquad ax + 2x = a^2 - 4$
Then, $\qquad (a+2)x = a^2 - 4$
Dividing by $(a+2)$, $\qquad x = a - 2$

2. Find the value of x in the equation $bx - b^2 = 4x - 9b + 20$.

SOLUTION. $\qquad bx - b^2 = 4x - 9b + 20$
Transposing, $\qquad bx - 4x = b^2 - 9b + 20$
Then, $\qquad (b-4)x = b^2 - 9b + 20$
Dividing by $(b-4)$, $\qquad x = b - 5$

Find the value of x in the following:

3. $cx - 9 = c^2 + 6c - 3x$.

4. $ax + 16 = a^2 - 4x$.

5. $3x - 12a = 4a^2 - 2ax + 9$.

6. $dx + 9a^2 = d^2 - 3ax$.

ALGEBRA.—5.

7. $ax - a^2 = 2ab + b^2 - bx$.

8. $ax - 5ab = 2a^2 + 3b^2 - bx$.

9. $ax - c^2 = a^3 + ac + a^2c - cx$.

10. $2ax - 6a^2 = 13ab + 6b^2 - 3bx$.

11. $2ax - 10ab - 15b = 14a + 21 - 3x$.

12. $ax + bx = 5a^2 + 7ab + 2b^2 + 5ac + 2bc - cx$.

13. $2cx - 4c^3 + d^2 = 2c^2d - 2cd - dx$.

14. $b^2x + 3b^2c + 6c^3 = b^3 + 2bc^2 - 2c^2x$.

15. $4m^4 - 2m^2x - 3mx = 1 - 6m + 9m^2 - x$.

16. $a^3 + 3x - 9a^2 = ax - 27a + 27$.

17. $2m^2x + 3mn^3 + 7m^2n^2 - 4m^4 = 3mnx$.

18. $5ax = 15a^3 - 5ab + 5ab^2 + 2bx - 6a^2b + 2b^2 - 2b^3$.

19. A man being asked how much money he had, replied that if he had $25 more than 3 times what he then had, he would have $355. How much money had he?

20. A gentleman divided $10,500 among four sons, giving to the second twice as much as to the first, to the third twice as much as to the second, and to the fourth one half as much as to the other three. How much did he give to each?

21. A man who met some beggars gave 3 cents to each and had 4 cents left, but found that he lacked 6 cents of having enough to give them 5 cents each. How many beggars were there? How much money did he have?

22. A man has six sons, each 4 years older than the one next to him. The eldest is 3 times as old as the youngest. What is the age of each?

23. A vessel containing some water was filled by pouring into it 42 gallons, and there was then in the vessel 7 times as much as at first. How many gallons did the vessel contain at first?

REVIEW.

24. A man borrowed as much money as he had and spent a dollar; he then borrowed as much as he had left and spent a dollar; again he borrowed as much as he then had and spent a dollar; he then had nothing left. How much had he at first?

25. Divide $3400 among A, B, and C, giving A $100 more than B, and $200 less than C. What is the share of each?

26. There are 360 sheep in two flocks. If 40 are taken from the second and added to the first flock, the first flock will then contain twice as many as the second. How many sheep are there in each flock?

27. Two persons invest equal sums of money in trade. One gains $252, and the other loses $174. The one has now twice as much as the other. What did each invest?

28. Three persons have $152. The first has $20 less than the second, and the third as much as the first and second. How much has each?

29. Divide the number 37 into three parts, such that the second shall be 3 more than the first, and the third 5 less than the second.

30. A manufacturer shipped a certain number of harvesters to Denver, 10 more than that number to Omaha, 5 more than 3 times as many to St. Paul as to Omaha, and twice as many to Chicago as to St. Paul? How many did he ship to each place, if the whole number shipped was 1215?

REVIEW EXERCISES.

96. 1. Add $6ax - 140 + 3\sqrt{x}$, $5x^2 + 4ax + 9x^2$, $7ax + 4\sqrt{x} + 160$, and $\sqrt{x} + 3ax - 4x^2$.

2. Add $3am + 2x - 3\sqrt{y} - z$, $2\sqrt{y} + 3z - 2x^2 + 3am$, $4x^2 - 3z + 2\sqrt{y} + 3x$, and $2\sqrt{y} - 4am + 2z - 3x^2$.

3. From $\sqrt{a^2 - b^2} - 2(x + y) - 6$ subtract $4(x + y) - 3\sqrt{a^2 - b^2}$.

HIGH SCHOOL ALGEBRA.

4. From $\sqrt{x} + 2\sqrt{y} - z + 6$ subtract $3\sqrt{y} - 2\sqrt{x} - y + 2z - 16$.

5. From $a^2x^2 + 2ay - 3y^5 + z^6$ subtract $b^2x^2 + 3ay - cy^5 + 4z^6$.

6. From the sum of $x^{2n} + 3x^2y^n - 3yz + az$ and $4x^m - 3yz + 2z + 3x^2y^n$ subtract $5x^{2n} - 4z + 6x^2y^2 - 3az$.

7. Multiply $x^4 + 2x^2y + xy^3$ by $x^2 + 2xy - y^2$.

8. Multiply $x^n + 2x^ny^n + y^n$ by $x^n + 2x^ny^n + y^n$.

9. Multiply $3x^n + 2x^{2n}y^{2n} - y^n$ by $x^n - y^{2n} + x^ny^n$.

10. Multiply $3x^{n+2} + 2y^{n+m} + z^m$ by $3x^2 - 2y^{n-m} + z^{2m}$.

Expand:

11. $(x-1)(x-2)$.
12. $(x-2)(x+3)$.
13. $(x+3)(x-6)$.
14. $(x-10)(x+9)$.
15. $(a-7)(a+4)$.
16. $(x-3)(x-2)$.
17. $(x-5)(x+6)$.
18. $(a-1)(a+2)$.
19. $(y^2-7)(y^2+4)$.
20. $(c+11)(c-4)$.
21. $(m-1)(m+4)$.

22. $(p+12)(p+11)$.
23. $(y-2)(y+9)$.
24. $(d+20)(d+15)$.
25. $(18-4x)(18+4x)$.
26. $(10+1)(10-1)$.
27. $(x+2y)(x+2y)$.
28. $(a^n-b^n)(a^n+b^n)$.
29. $(3m^2-2m)(3m^2+2m)$.
30. $(x^n+x)(x^n-x)$.
31. $(a+b+c)(a+b-c)$.
32. $(x-y)(x+y)(x^2-y^2)$.

33. $(1+x)(1-x)(1+x^2)(1+x^4)(1+x^8)$.

34. $(x+y)(x+y)(x+y)(x+y)(x+y)$.

35. $(a+2)(a+2)(a-2)(a-2)$.

36. $(3a-6)(3a-6)(3a+6)(3a+6)$.

REVIEW. 69

Square:

37. $p^2 - q$.
38. $x^3 + y^2$.
39. $20 + 5$.
40. $60 + 4$.
41. $15 + 3$.
42. $2x + 5y$.
43. $3x^2 - 2y^2$.
44. $3a - 2b$.
45. $2ax + 3ay$.
46. $4a^2 - 3ay$.
47. $2a^m + b^{2n}$.
48. $x^{2n} + 2y^{2n}$.
49. $x^{2+n} - y^2$.
50. $a + b + c$.
51. $a - b + c$.
52. $a + b - c$.

Write out the product of:

53. $(2x + y)(2x - y)$.
54. $(3x + 7y)(3x - 7y)$.
55. $(4x^2 - 2y^2)(4x^2 + 2y^2)$.
56. $(3a^2b + 2b^2)(3a^2b - 2b^2)$.
57. $(7a^2 + 2b)(7a^2 - 2b)$.
58. $(5b^2 + 6c^2)(5b^2 - 6c^2)$.
59. $(8xy + 3x^2)(8xy - 3x^2)$.
60. $(9am + 8xy)(9am - 8xy)$.
61. $(ax^n + y^n)(ax^n - y^n)$.
62. $(ax^{2n} + ay^{2n})(ax^{2n} - ay^{2n})$.
63. $(a + x)(a - x)(a^2 + x^2)(a^4 + x^4)$.
64. $(x^2 + y^2)(x^2 - y^2)(x^4 + y^4)(x^8 + y^8)$.
65. $(2x + 3)(2x - 3)(4x^2 + 9)(16x^4 + 81)$.

Divide:

66. $4a^4 - 5a^2b^2 + b^4$ by $2a^2 - 3ab + b^2$.
67. $5x^2y + x^3 + y^3 + 5xy^2$ by $x^2 + y^2 + 4xy$.
68. $m^4 - 6m^3 + 7m^2 - 4m - 12$ by $m^2 - 2m + 3$.
69. $x^5 + 2x^4 + 7x^3 + 6x^2 - 46x - 120$ by $x^2 + 4x + 5$.
70. $x^5 + x^4 - 4x^3 - x^2 + x$ by $x^2 - x - 1$.
71. $a^6 - a^4 + a^3 - a^2 + 2a - 1$ by $a^2 + a - 1$.
72. $x^5 - 5x^4y + 10x^3y^2 - 10x^2y^3 + 5xy^4 - y^5$ by $x^2 + y^2 - 2xy$.
73. $6x^3 - 16x^2y + 14xy^2 - 4y^3 - 2x^2 + 4xy - 2y^2$ by $3x - 2y - 1$.
74. $x^7 - 6x^6y + 21x^5y^2 - 31x^4y^3 + 23x^3y^4 + 3x^2y^5 - 8xy^6 + 6y^7$ by $x^2 - 2xy + 2y^2$.
75. $a^2c - ab^2 + acd - ad^2 - abc + b^3 - bcd + bd^2 - ac^2 + cb^2 - c^2d + cd^2$ by $ac - b^2 + cd - d^2$.

Remove parentheses and simplify by collecting like terms:

76. $(a^2 - \overline{2ax - x^2}) - \{a^2 - (2ax - \overline{x^2 + 7})\}$.

77. $a^3 - [(-a^3 - a^2b + ab^2 - b^3) - \{a^2b - (ab^2 - \overline{b^3 - a^2b})\}]$.

78. $4a^2 - (a^2 - b^2) - [-\{b^2 - (a^2 - \overline{b^2 - a^2})\}]$.

79. $1 - [4 - \{7 - (2 - \{3 - \overline{4-a}\})\}]$.

80. $14 - \{-7 + (3 - \overline{4+x}) - x\} - [-\{-(-\overline{x-4})\}]$.

81. $a - [2b - (3c + 2b) - a]$.

82. $a - \{2a - [3b - (4c - 2a)]\}$.

83. $7x - [3x - (4x - \overline{5x - 2x})]$.

84. $-[5a - (11b - 3a)] - [5b - (3a - 6b)]$.

85. $x - \{2x + 3x - [4x - 5x - (6x - 7x)]\}$.

86. $8x - \{16y - [3x - (12y - x) - 8y] + x\}$.

87. $a - 2b + \{3c - [3a - (a + b + 2a - b)]\}$.

88. $2x - \{2x - [2x - (2x - \overline{2x - x})]\}$.

89. $a - \{2b - [3c - 2a - (a + b + 2a)]\}$.

90. $2x - \{3y + (2y - z) - 4z + [2x - (3y - \overline{z - 2y})]\}$.

Collect in parentheses the coëfficients of x, y, and z in the following:

91. $ax - by + bz - ay - cx - cz$.

92. $2ax - 3bz + 4cy + 3az - 5x - y$.

93. $cy + 4ax - 3dz + 3az - 5bx + by$.

94. $3ax - 6bx + 2ay - 4y + 5az - 3bz$.

95. $4by + 2cz - 3bx - 4cy + az - 2cx - ay$.

96. $ax - 2by + 3cz - 4bx - 3cy + az - 2cx - ay$.

97. $6ax + 5ay + 2by - 6bz - 5cx + 6cy + 3cz$.

98. $3ax - 2by - 4cz - 2bx - 2bx + 8cz - 2cx - 2cy$.

REVIEW.

Expand the following:

99. $(x^3 + y^3)^2$.
100. $(x^5 - y^5)^2$.
101. $(2x + 1)^2$.
102. $(3x^2 + 4y^3)^2$.
103. $(5a^n - 2c^2)^2$.
104. $(2ad + 3cd)^2$.
105. $(6x^2 - 3y^3)^2$.
106. $(3a^2y^n + 2c^3x^n)^2$.
107. $(2x^2 + 3y^2)(2x^2 - 3y^2)$.
108. $(2a^3 + b^2)(2a^3 - b^2)$.
109. $(3x + 2a)(3x - 2a)$.
110. $(5x^2 + 3)(5x^2 - 3)$.
111. $(7bc + d^2)(7bc - d^2)$.
112. $(2x + 6)(2x + 1)$.
113. $(2x - 5)(2x - 3)$.
114. $(ax + 3b)(ax - 2b)$.
115. $(x + y + z)^2$.
116. $(x + 2y - 3z)^2$.
117. $(x - 3y + 2z)^2$.
118. $(m - n + p - q)^2$.

Find the value of x in the following equations:

119. $7x - 36 - 10x + 12 = 80 - 33x + 46$.
120. $6x - 2(9 - 4x) + 3(5x - 7) = 10x - 4 - 16x$.
121. $3x - 6(x - 5) + 2x = 2(x + 5) + 5(x - 4)$.
122. $x - 7(4x - 11) + 30 = 14(x - 5) - 19(8 - x) - 31$.
123. $17x - (8x - 9) - [4 - 3x - (2x - 3)] = 30$.
124. $(x + 12)(x - 8) = (x + 1)(x - 6)$.
125. $4(x - 3) - 3(x - 2) + 2(x - 1) + 2 = 0$.
126. $(x - 1)(2x + 1) + 14 = (2x + 3)(x + 3)$.
127. $(x + 2)(x + 1)(x + 6) - 9x^2 = x^3 + 4(7x - 1)$.
128. $a(x - a) - 2ab = -b(x - b)$.
129. $(a^2 + x)(b^2 + x) = (ab + x)^2 + x^2$.
130. $ax + b^2 = a^2 + bx$.
131. $bx + 3b^2 = 7bc + 3cx$.
132. $a(x + a) + b(b - x) = 2ab$.
133. $2ax + 12ab - 4a^2 = 9b^2 + 3bx$.
134. $3x - 9 - 3c = 12a - 2ax + 4a^2 + 2ac$.
135. $2ax + 9c^2 + 3cd = 4a^2 + 3cx + 2ad$.

FACTORING.

97. 1. What is the product of $4 \times 5a$? What relation do 4 and $5a$ bear to their product?

2. What are the integral factors of 5? x? 6? $4a$?

3. Since 5 and x cannot be separated into any integral factors except themselves and 1, what kind of quantities are they called?

4. Since 6 and $4a$ can be separated into other factors besides themselves and 1, what kind of quantities are they called?

5. Since 3 and 2, the factors of 6, are prime numbers, what kind of factors are they called?

98. The **Factors** of a quantity are the quantities which multiplied together will produce the quantity.

Thus, a, b, and $(x+y)$ are the factors of $ab(x+y)$.
Factors of a quantity are exact divisors of it.

99. A **Prime Quantity** is a quantity that has no integral factors except itself and 1.

100. A **Composite Quantity** is a quantity that has integral factors besides itself and 1.

101. A **Prime Factor** is a factor that is a prime quantity.

102. Factoring is the process of separating a quantity into its factors.

103. To separate a monomial into its factors.

1. What are the prime factors of $24\,x^2y^3z$?

SOLUTION. $\quad 24\,x^2y^3z = 2 \cdot 2 \cdot 2 \cdot 3 \cdot x \cdot x \cdot y \cdot y \cdot y \cdot z$

FACTORING.

RULE. — *Separate the numerical coëfficient into its prime factors.*

Separate the literal quantities into their prime factors by writing each quantity as a factor as many times as there are units in its exponent.

Find the prime factors of the following:

2. $8\,a^2b$.   4. $15\,a^3y^2z$.   6. $42\,axy^3$.   8. $28\,a^2c^2x$.

3. $10\,x^2y^3$.   5. $20\,ax^3y$.   7. $36\,xy^2z^3$.   9. $35\,x^2z^2c^3$.

104. To separate a polynomial into monomial and polynomial factors.

1. What are the factors of $5\,a^2bc + 10\,a^2c - 20\,a^2bc$?

$5\,a^2c\,)\,\overline{5\,a^2bc + 10\,a^2c - 20\,a^2bc}$
$\,b + 2 - 4b$

$5\,a^2c\,(b + 2 - 4b)$

EXPLANATION. — By examining the terms of the polynomial it is found that $5\,a^2c$ is a factor of every term. Dividing by this common factor the other is found. Hence, the factors are $5\,a^2c$ and $(b + 2 - 4b)$.

RULE. — *Divide the polynomial by the highest factor common to all the terms. The divisor and the quotient will be the factors sought.*

Find the factors of the following polynomials:

2. $5\,a^2b + 6\,a^2c$.
3. $8\,x^2y^2 + 12\,x^2z^2$.
4. $6\,xyz + 12\,x^2y^2z$.
5. $9\,x^3y^2z + 18\,xy^2z^3$.
6. $a^2x^2y^2z + a^2xyz^2$.
7. $a^2c + b^2c + c^2d^2$.
8. $4\,x^2y + cxy^2 + 3\,xy^3$.
9. $4\,abx + 6\,a^2x^2 + 8\,ax$.
10. $3\,a^3y - 6\,a^2y^2 + 9\,ay^3$.
11. $2\,a^2c - 2\,a^2c^2 + 3\,ac$.
12. $5\,acd - 2\,c^2d^2 + bcd$.
13. $4\,b^2c^2 - 12\,abc - 9\,c^2$.
14. $3\,a^2b + abc - abd$.
15. $5\,a^3x^2 - 5\,a^2x^3 + 10\,a^2x^2z$.
16. $6\,x^2y^3z - 3\,xy^2z^2 + xy$.
17. $12\,a^2cx + 15\,ac^2x - 12\,acx$.
18. $6\,a^4b^5 + 21\,a^3b^3 - 24\,a^2b^4$.
19. $20\,c^3x^2 - 15\,c^4x + 5\,c^2x^2$.
20. $56\,x^3y^6 + 112\,x^4y^8 - 216\,x^6y^9$.
21. $65\,x^5y^9 - 85\,x^7y^6 + 255\,x^4y^8$.
22. $75\,a^3y^4 + 150\,a^2y^6 - 225\,ay^7$.
23. $48\,b^2c^6 - 144\,b^6c^8 - 192\,b^8c^9$

HIGH SCHOOL ALGEBRA.

105. To separate a trinomial which is a perfect square into two equal factors.

$$(a+b)(a+b) = a^2 + 2ab + b^2$$
$$(a-b)(a-b) = a^2 - 2ab + b^2$$

1. What is the product of $(a+b)(a+b)$? What, then, are the factors of $a^2 + 2ab + b^2$?

2. What is the product of $(a-b)(a-b)$? What, then, are the factors of $a^2 - 2ab + b^2$? What term of the trinomial determines the sign which connects the binomial factors?

One of the two equal factors of a quantity is called its **Square Root**.

RULE. — *Arrange the terms of the trinomial according to the ascending or descending powers of one of the quantities. The square roots of the first and third terms connected by the sign of the second term will be one of the equal factors.*

Find the equal factors of the following trinomials:

1. $a^2 + 2ab + b^2$.
2. $x^2 + 2xy + y^2$.
3. $b^2 - 2bc + c^2$.
4. $r^2 + 2rs + s^2$.
5. $x^2 + 2x + 1$.
6. $x^2 + 4x + 4$.
7. $y^2 - 2y + 1$.
8. $4y^2 - 4y + 1$.
9. $9x^2 + 6x + 1$.
10. $16x^2 + 16x + 4$.
11. $9y^2 - 18y + 9$.
12. $z^6 + 16z^3 + 64$.
13. $16a^2b^4 - 8ab^3c^2 + b^2c^4$.
14. $9m^2 + 18mn + 9n^2$.
15. $9 + 6x + x^2$.
16. $1 - 2x^2 + x^4$.
17. $16n^2 - 8n + 1$.
18. $16 + 16a + 4a^2$.
19. $36 + 12a^2 + a^4$.
20. $49 - 14x^3 + x^6$.
21. $81x^2 - 18ax + a^2$.
22. $4a^{2n} + 12a^n b^n + 9b^{2n}$.
23. $x^2 - 30x + 225$.
24. $4x^4 - 64x^2 + 256$.
25. $49x^2 - 112xy + 64y^2$.
26. $196a^2b^2c^2 + 112ab^2c^2d + 16b^2c^2d^2$.

FACTORING. 75

Perfect squares containing terms in parenthesis may be factored in a similar manner.

27. Factor $x^2 + 2x(x-y) + (x-y)^2$.

SOLUTION.
$$x^2 + 2x(x-y) + (x-y)^2$$
$$= [x + (x-y)][x + (x-y)]$$
$$= (x + x - y)(x + x - y)$$
$$= (2x - y)(2x - y)$$

28. Factor $(a+b)^2 - 2(a+b)(b-c) + (b-c)^2$.

SOLUTION.
$$(a+b)^2 - 2(a+b)(b-c) + (b-c)^2$$
$$= [(a+b) - (b-c)][(a+b) - (b-c)]$$
$$= (a+b-b+c)(a+b-b+c)$$
$$= (a+c)(a+c)$$

Find the equal factors of the following:

29. $x^2 + 4x(x-y) + 4(x-y)^2$.

30. $4 + 4(x-y) + (x-y)^2$.

31. $(a+b)^2 + 6(a+b) + 9$.

32. $16 + 8(a-b) + (a-b)^2$.

33. $4x(x-y) + 4x^2 + (x-y)^2$.

34. $9x^2 - 6x(x+y) + (x+y)^2$.

35. $4x^2 + 9(x-y)^2 + 12x(x-y)$.

36. $12y(x-y) + 4(x-y)^2 + 9y^2$.

37. $(a+2b)^2 - 4(a+2b)(b-c) + 4(b-c)^2$.

38. $4(a+b)^2 - 4(a+b)(2b-c) + (2b-c)^2$.

39. $(x+y)^2 - 2(x+y)(y-z) + (y-z)^2$.

40. $9(x+y)^2 - 6(x+y)(3y-z) + (3y-z)^2$.

41. $(a+5b)^2 - 10(a+5b)(b-c) + 25(b-c)^2$.

42. $4(a+3b)^2 - 24(a+3b)(b-c) + 36(b-c)^2$.

43. $16(a+x)^2 - 32(a+x)(x-y) + 16(x-y)^2$.

44. $25(x+y)^2 - 50(x+y)(y-z) + 25(y-z)^2$.

HIGH SCHOOL ALGEBRA.

106. To resolve a binomial which is the difference of **two squares** into two binomial factors.

$$(a+b)(a-b) = a^2 - b^2$$
$$(x^2+y^2)(x^2-y^2) = x^4 - y^4$$

1. What is the product of $(a+b)(a-b)$? What, then, are the factors of $a^2 - b^2$? How do these two factors differ?

2. What is the product of $(x^2+y^2)(x^2-y^2)$? What, then, are the factors of $x^4 - y^4$? What is the difference between these two factors?

RULE. — *Find the square root of each term of the binomial, and make the sum of these square roots one factor, and their difference the other.*

Sometimes the factors of a quantity may themselves be factored.

Thus,
$$x^4 - y^4 = (x^2+y^2)(x^2-y^2)$$
$$= (x^2+y^2)(x+y)(x-y)$$

Resolve into their factors:

1. $a^2 - b^2$.
2. $c^2 - d^2$.
3. $m^2 - n^2$.
4. $4x^2 - 4y^2$.
5. $9x^2 - y^2$.
6. $x^2 - 9y^2$.
7. $16x^2 - 16y^2$.
8. $9c^2 - 16d^2$.
9. $25a^2 - 9b^2$.
10. $9y^2 - 1$.
11. $25x^4 - 16y^6$.
12. $36y^2 - 49z^6$.
13. $4x^2 - 9y^2$.
14. $x^2y^2 - 4y^2z^2$.
15. $m^4 - n^4$.
16. $a^8 - b^8$.
17. $m^{2n} - n^{2m}$.
18. $9a^{2n} - 4b^{4n}$.
19. $a^{16} - b^8$.
20. $9a^2b^2 - 4c^4$.
21. $a^4x^4 - 9c^2d^4$.
22. $x^{12} - y^{10}$.
23. $4b^2c^4 - 25d^6$.
24. $81b^4 - 16$.
25. $121a^2 - 36b^4$.
26. $169c^4 - 49d^6$.
27. $100a^2d^4 - 64$.
28. $64 - 25x^4y^4$.
29. $49 - 36x^2y^4$.
30. $16m^{2n} - n^{4m}$.
31. $289x^6 - y^{6n}$.
32. $256a^4 - 36z^8$.
33. $196x^{4n} - 9$.
34. $225x^2y^{2n} - 121z^4$.
35. $324x^2y^4z^6 - 81$.
36. $225x^2y^8 - 144$.

FACTORING.

37. Factor $9a^2 - (2a+b)^2$.

SOLUTION.—One factor is $3a + (2a+b)$, and $3a - (2a+b)$ is the other.

$$3a + (2a+b) = 5a+b$$
and $$3a - (2a+b) = (a-b)$$
$$\therefore 9a^2 - (2a+b)^2 = (5a+b)(a-b)$$

38. Factor $(5x+3y)^2 - (3x-2y)^2$.

SOLUTION.—One factor is $(5x+3y) + (3x-2y)$, and $(5x+3y) - (3x-2y)$ is the other factor.

$$(5x+3y) + (3x-2y) = 8x+y$$
and $$(5x+3y) - (3x-2y) = 2x+5y$$
$$\therefore (5x+3y)^2 - (3x-2y)^2 = (8x+y)(2x+5y)$$

Factor:

39. $c^2 - (a+b)^2$.
40. $b^2 - (x-y)^2$.
41. $4c^2 - (x+y)^2$.
42. $9c^2 - (x-y)^2$.
43. $x^2 - (a+2b)^2$.
44. $c^2 - (3a+b)^2$.
45. $b^2 - (2a+3c)^2$.
46. $4c^2 - (2a+3b)^2$.
47. $9c^2 - (2a-3b)^2$.

48. $(a+b)^2 - (a-b)^2$.
49. $(x+y)^2 - (2x+3)^2$.
50. $(x+y)^2 - (2x-4)^2$.
51. $(2x+3)^2 - (3x-4)^2$.
52. $(2x+3y)^2 - (3x-4y)^2$.
53. $(5x+2y)^2 - (4x-3y)^2$.
54. $(7x-3y)^2 - (2x+y)^2$.
55. $(8x+5y)^2 - (3x-2y)^2$.
56. $(9x+3y)^2 - (2x-5y)^2$.

Polynomials may sometimes be expressed in the form of the difference of two perfect squares, and factored like the preceding examples.

57. Factor $a^2 + 2ab + b^2 - c^2$.

SOLUTION. $\quad a^2 + 2ab + b^2 - c^2 = (a+b)^2 - c^2$
$$(a+b)^2 - c^2 = [(a+b) + c][(a+b) - c]$$
$$\text{or } (a+b+c)(a+b-c)$$
$$\therefore a^2 + 2ab + b^2 - c^2 = (a+b+c)(a+b-c)$$

HIGH SCHOOL ALGEBRA.

58. Factor $2cd + b^2 - c^2 - 2ab - d^2 + a^2$.

SOLUTION.
$$2cd + b^2 - c^2 - 2ab - d^2 + a^2$$
Arranging terms,
$$= a^2 - 2ab + b^2 - c^2 + 2cd - d^2$$
$$= a^2 - 2ab + b^2 - (c^2 - 2cd + d^2)$$
$$= (a-b)^2 - (c-d)^2$$
$$= [(a-b) + (c-d)][(a-b) - (c-d)]$$
$$= (a - b + c - d)(a - b - c + d)$$

Factor:

59. $b^2 + 2bc + c^2 - d^2$.

60. $a^2 + 2ax + x^2 - 4y^2$.

61. $4a^2 - x^2 + 2xy - y^2$.

62. $a^2 - 6ax + 9x^2 - 4c^2$.

63. $4x^2 + 4xy + y^2 - 4z^2$.

64. $x^2 - b^2 - 2xy + y^2$.

65. $x^2 + 4xy - 4z^2 + 4y^2$.

66. $x^2 - 9y^2 + 6x + 9$.

67. $4x^2 + 12xy - 9c^2 + 9y^2$.

68. $4a^2 + 9b^2 - 16c^2 - 12ab$.

69. $9x^2 - 25z^2 + 16y^2 + 24xy$.

70. $a^2 + 2bd + c^2 - b^2 + 2ac - d^2$.

71. $4b^2 - x^2 + 4xy + a^2 + 4ab - 4y^2$.

72. $d^2 - a^2 + 4ax - 4cd + 4c^2 - 4x^2$.

107. To factor a quadratic trinomial.

$$(x+3)(x+4) = x^2 + 7x + 12$$
$$(x-3)(x+4) = x^2 + x - 12$$
$$(x+3)(x-4) = x^2 - x - 12$$
$$(x-3)(x-4) = x^2 - 7x + 12$$
$$(x-2)(x+6) = x^2 + 4x - 12$$

1. In what respects do the factors differ in the first four examples?

2. What terms in the products are alike in each instance?

3. How may the first term of each factor be found from the product?

4. How is the last term of the product found from the factors?

5. How is the *coëfficient* of the second term of the product found from the last terms of the factors?

FACTORING.

108. A trinomial of the form of $x^2 \pm ax \pm b$, in which b is the product of two quantities and a their algebraic sum, is called a **Quadratic Trinomial**.

1. Resolve $x^2 - 9x - 36$ into two factors.

EXPLANATION. — The first term is evidently x. Since 36 is the product of the two other quantities, 6 and 6, or 4 and 9, or 3 and 12, or 36 and 1, or 18 and 2 are the other quantities.

Since their sum is -9, the quantities must be 3 and -12, for the other sets of factors of 36 cannot be combined so as to produce this result.

Therefore, $(x + 3)$ and $(x - 12)$ are the factors.

RULE. — *Arrange the terms of the trinomial according to the ascending or descending powers of one of the quantities.*

For the first term of each factor take the square root of the first term of the trinomial, and for the second term such quantities that their product will equal the third term of the trinomial, and their algebraic sum multiplied by the first term of the factor shall equal the second term of the trinomial.

Separate into their simplest factors:

2. $x^2 + 3x + 2$.
3. $x^2 + 7x + 12$.
4. $x^2 - 4x - 21$.
5. $x^2 - 7x - 18$.
6. $x^2 + 6x + 8$.
7. $x^2 + 12x + 32$.
8. $b^2 - 8b + 15$.
9. $b^2 + b - 12$.
10. $b^2 - b - 12$.
11. $b^2 + 2b - 35$.
12. $y^2 + y - 56$.
13. $y^2 + 3y - 40$.
14. $b^2 + 19bc + 48c^2$.
15. $c^8d^4 + 7c^4d^2 + 12$.
16. $x^2y^2 + 7xyz + 10z^2$.
17. $x^2 + 9x - 36$.
18. $x^2 - 57x + 56$.
19. $x^2 - 10x - 39$.
20. $x^2 - 12x - 64$.
21. $4x^2 - 10x + 6$.
22. $9x^2 - 27x + 18$.
23. $4x^2 + 16ax + 12a^2$.
24. $9a^2 + 30ab + 24b^2$.
25. $4a^2 + 4a - 3$.
26. $4b^2 + 10b - 14$.
27. $9a^2 - 15a - 14$.
28. $16c^2 - 48c + 35$.
29. $9x^2 + 6xy - 15y^2$.

30. $16y^2 + 24yz - 7z^2$.   32. $25y^2 - 25yz + 6z^2$.

31. $4x^2 - 14xy + 10y^2$.   33. $49x^6 + 14x^3y - 15y^2$.

109. To separate other trinomials into binomial factors.

First, *when the first term is not a perfect square.*

1. Factor $8x^2 + x - 7$.

SOLUTION.

$$8x^2 + x - 7 = (8x^2 + x - 7) \times \frac{8}{8}$$

$$= \frac{64x^2 + 8x - 56}{8} = \frac{(8x+8)(8x-7)}{8}$$

$$= \frac{8(x+1)(8x-7)}{8} = (x+1)(8x-7).$$

EXPLANATION. — Since the first term is not a perfect square, it may be made so by multiplying by 8, and the value of the quantity will not be changed by multiplying by $\frac{8}{8}$.

Multiplying by $\frac{8}{8}$ and factoring in accordance with Art. 108 and simplifying the result, the factors are $(x+1)(8x-7)$.

Second, *when the second term does not exactly contain the square root of the first.*

1. Factor $4x^2 - 7x + 3$.

SOLUTION.

$$4x^2 - 7x + 3 = (4x^2 - 7x + 3) \times \frac{4}{4}$$

$$= \frac{16x^2 - 28x + 12}{4} = \frac{(4x-4)(4x-3)}{4}$$

$$= \frac{4(x-1)(4x-3)}{4} = (x-1)(4x-3).$$

EXPLANATION. — Although the first term is a perfect square, its square root is not an integral factor of the second term. The second term may be made to contain the square root of the first term by multiplying by 4, and the value of the expression will not be changed by multiplying by $\frac{4}{4}$.

Multiplying by $\frac{4}{4}$ and factoring in accordance with Art. 108 and simplifying, the factors are $(x-1)(4x-3)$.

FACTORING.

Find the binomial factors of:

2. $2x^2 + 3x - 2$.
3. $3x^2 + 6x + 3$.
4. $3x^2 - 17x + 10$.
5. $5x^2 - 38x + 21$.
6. $3x^2 - x - 2$.
7. $3x^2 + 7x - 6$.
8. $3x^2 + 11x - 20$.
9. $2x^2 + 11x + 12$.
10. $4x^2 + xy - 3y^2$.
11. $4x^2 - 5xy + y^2$.
12. $5x^2 - 29xy + 36y^2$.
13. $7x^2 + 123x - 54$.
14. $10x^2 + 3xy - y^2$.
15. $3x^2 + 7xy - 6y^2$.
16. $4x^2 + 7x - 15$.
17. $6x^2 - xy - 2y^2$.

110. To factor by grouping terms.

1. Factor $xy + ay + bx + ab$.

SOLUTION. $\quad xy + ay + bx + ab = y(x+a) + b(x+a)$
$\qquad\qquad\qquad\qquad\qquad = (y+b)(x+a)$

2. Factor $ax - bx - ay + by$.

SOLUTION. $\quad ax - bx - ay + by = ax - bx - (ay - by)$
$\qquad\qquad\qquad\qquad\qquad = x(a-b) - y(a-b)$
$\qquad\qquad\qquad\qquad\qquad = (x-y)(a-b)$

Factor the following:

3. $x^2 + bx + cx + bc$.
4. $ax - by + bx - ay$.
5. $b^2 - bc + ab - ac$.
6. $ay - by + ab - y^2$.
7. $by - bx + 3ax - 3ay$.
8. $abc + acd + bd + a^2c^2$.
9. $2a + bx^2 + 2b + ax^2$.
10. $x^2 - xy + 3y - 3x$.
11. $bx - cx + bc - x^2$.
12. $dxy - cxy + cdx^2 - y^2$.
13. $abx^2 - axy + bxy - y^2$.
14. $bd^2x - b^2dy + abcy - acdx$.
15. $3bc - 4ad + 6ac - 2bd$.
16. $15ax - 20ay + 9bx - 12by$.
17. $45ab - 18b^2 - 18a^2$.
18. $12ax - 9ay - 8bx + 6by$.

19. $acx + bcx + adx + bdx + acy + bcy + ady + bdy$.

20. $2axz - 2avx + 3bvx - 3bxz - 4avy + 4ayz - 6byz + 6bvy$.

ALGEBRA. — 6.

111. To factor a trinomial of the form of $a^4 + a^2b^2 + b^4$.

1. Factor $x^4 + x^2y^2 + y^4$.

EXPLANATION. — $x^4 + x^2y^2 + y^4$ lacks x^2y^2 of being a perfect square. If x^2y^2 is added to the expression and then subtracted from it, the value of the expression will not be changed. The quantity added and then subtracted must be such as will render the trinomial a perfect square. Adding and subtracting x^2y^2, the expression becomes

$$\begin{aligned} x^4 + x^2y^2 + y^4 &= x^4 + 2x^2y^2 + y^4 - x^2y^2 \\ &= (x^2 + y^2)^2 - x^2y^2 \\ &= [(x^2 + y^2) + xy][(x^2 + y^2) - xy] \\ &= (x^2 + y^2 + xy)(x^2 + y^2 - xy) \\ &= (x^2 + xy + y^2)(x^2 - xy + y^2) \end{aligned}$$

2. Factor $a^4 + a^2 + 1$.

SOLUTION.
$$\begin{aligned} a^4 + a^2 + 1 &= a^4 + 2a^2 + 1 - a^2 \\ &= (a^2 + 1)^2 - a^2 \\ &= [(a^2 + 1) + a][(a^2 + 1) - a] \\ &= (a^2 + 1 + a)(a^2 + 1 - a) \\ &= (a^2 + a + 1)(a^2 - a + 1) \end{aligned}$$

Factor the following:

3. $b^4 + b^2c^2 + c^4$.
4. $x^4 + x^2 + 1$.
5. $9b^4 + 21b^2c^2 + 25c^4$.
6. $9b^4 + 3b^2x^2 + 4x^4$.
7. $25x^4 - 9x^2y^2 + 16y^4$.
8. $16a^4 - 17a^2b^2 + b^4$.
9. $36x^4 + 23x^2y^2 + 16y^4$.
10. $49b^4 - 11b^2c^2 + 25c^4$.
11. $64a^4 + 128a^2b^2 + 81b^4$
12. $49x^4 + 34x^2y^2 + 25y^4$.
13. $9c^4 + 38c^2d^2 + 49d^4$.
14. $36a^4 - 16a^2b^2 + 49b^4$.
15. $25b^4 + 31b^2c^2 + 64c^4$.
16. $25b^4 - b^2c^2 + 64c^4$.
17. $81x^4 + 20x^2y^2 + 4y^4$.
18. $81x^4 - 64x^2y^2 + 4y^4$.

112. To factor the difference of the same powers of two quantities.

$$(a^2 - b^2) \div (a - b) = a + b$$
$$(a^3 - b^3) \div (a - b) = a^2 + ab + b^2$$
$$(a^4 - b^4) \div (\ - b) = a^3 + a^2b + ab^2 + b^3$$

1. How does the first term of the quotient compare with the first term of the dividend? What quantities does the second term of the quotient contain? The third term? The fourth term?

FACTORING.

2. What is the sign of each term?

3. What are the exponents of a and b in each term when the difference of the same powers of two quantities is divided by the difference of the quantities?

113. PRINCIPLE. — *The difference of the same powers of two quantities is always divisible by the difference of the quantities.*

Write out the quotient of: Factor:

1. $(x^8 - y^8) \div (x - y)$.
2. $(x^9 - y^9) \div (x - y)$.
3. $(x^4 - 1) \div (x - 1)$.
4. $(x^4 - 16) \div (x - 2)$.
5. $(x^8 - y^8) \div (x^2 - y^2)$.

6. $(a^3 - 27)$.
7. $(x^6 - 64)$.
8. $(8a^3 - b^3)$.
9. $(27x^3 - y^3)$.
10. $(343 - x^3)$.

114. A course of reasoning which discloses the truth or falsity of a statement is a **Demonstration**.

115. The following is a general demonstration of the principle given in Art. 113:

Let x and y represent any two quantities, and n the exponent of any power. Then, $x^n - y^n$ will be the difference of the same powers of two quantities, and $x - y$ the difference of the two quantities.

PROCESS.

$$\begin{array}{r|l} x^n - y^n & x - y \\ x^n - x^{n-1}y & \overline{x^{n-1} + x^{n-2}y} \end{array}$$

1st Rem., $x^{n-1}y - y^n$
           $x^{n-1}y - x^{n-2}y^2$

2d Rem., $x^{n-2}y^2 - y^n$
nth Rem., $x^{n-n}y^n - y^n$
                $x^0 y^n - y^n$
                $y^n - y^n = 0$

DEMONSTRATION. — By dividing $x^n - y^n$ until several remainders are obtained, it is found that the first term of the first remainder is $x^{n-1}y$; of the second, $x^{n-2}y^2$; of the third, $x^{n-3}y^3$; of the fourth, $x^{n-4}y^4$, and, consequently, of the nth, $x^{n-n}y^n$. But x^{n-n} is x^0, which equals 1 (Art. 94, 1). Therefore, the first term of the nth remainder reduces to y^n.

Since the second term in the nth remainder is $-y^n$, the entire nth remainder is $y^n - y^n$, or 0; that is, there is no remainder, and the division is exact. Therefore, $x^n - y^n$ is divisible by $x - y$ when x and y represent any two quantities, and n the exponent of any power; or,

The difference of the same powers of two quantities is always divisible by the difference of the quantities.

116. It has been shown that the *difference* of the *same powers* of two quantities is *always* divisible by the *difference* of the quantities; it is yet to be shown *when* it is divisible by the *sum* of the quantities.

$$(x^2 - y^2) \div (x + y) = x - y.$$
$$(x^3 - y^3) \div (x + y) = x^2 - xy + y^2. \quad \text{Rem. } -2y^3.$$
$$(x^4 - y^4) \div (x + y) = x^3 - x^2y + xy^2 - y^3.$$
$$(x^5 - y^5) \div (x + y) = x^4 - x^3y + x^2y^2 - xy^3 + y^4.$$
Rem. $-2y^5$.

1. What are the signs of the terms of the quotient when the difference of the same powers is divided by the sum of the quantities?

2. What is the law of the exponents in the quotient?

3. What are the exponents of x and y when the difference of the same powers of two quantities is exactly divisible by the sum of the quantities?

117. PRINCIPLE. — *The difference of the same powers of two quantities is divisible by the sum of the quantities only when the exponents are even.*

Write out the quotient of:

1. $(x^8 - y^8) \div (x + y)$.
2. $(x^{10} - y^{10}) \div (x + y)$.
3. $(x^4 - 1) \div (x + 1)$.
4. $(x^4 - 16) \div (x + 2)$.
5. $(x^8 - y^8) \div (x^2 + y^2)$.

Factor:

6. $x^4 - 81$.
7. $x^6 - 64$.
8. $16a^4 - 16b^4$.
9. $a^6x^6 - y^6$.
10. $a^8b^8 - c^8d^8$.

FACTORING.

11. Demonstrate the truth of the above principle.

PROCESS.

$$
\begin{array}{r|l}
x^n - y^n & x + y \\
x^n + x^{n-1}y & x^{n-1} - x^{n-2}y + x^{n-3}y^2 - x^{n-4}y^3 \\
\end{array}
$$

1st Rem., $\quad -x^{n-1}y - y^n$
$\quad\quad\quad\;\; -x^{n-1}y - x^{n-2}y^2$

2d Rem., $\quad\quad\quad x^{n-2}y^2 - y^n$
$\quad\quad\quad\quad\quad x^{n-2}y^2 + x^{n-3}y^3$

3d Rem., $\quad\quad\quad\quad -x^{n-3}y^3 - y^n$
$\quad\quad\quad\quad\quad\quad -x^{n-3}y^3 - x^{n-4}y^4$

4th Rem., $\quad\quad\quad\quad\quad\quad x^{n-4}y^4 - y^n$

SUGGESTION.—Since the second term of each remainder is a *negative* quantity, the remainders can reduce to 0 only when the first term is a *positive* quantity. Show what remainders reduce to 0.

118. To factor the sum of the same powers of two quantities.

$(x^2 + y^2) \div (x + y) = x - y$. Rem. $2y^2$.

$(x^3 + y^3) \div (x + y) = x^2 - xy + y^2$.

$(x^4 + y^4) \div (x + y) = x^3 - x^2y + xy^2 - y^3$. Rem. $2y^4$.

$(x^5 + y^5) \div (x + y) = x^4 - x^3y + x^2y^2 - xy^3 + y^4$.

1. What are the signs of the terms of the quotient?

2. What is the law of the exponents?

3. What are the exponents of x and y when the sum of the same powers of two quantities is exactly divisible by the sum of the quantities?

119. PRINCIPLE. — *The sum of the same powers of two quantities is divisible by the sum of the quantities only when the exponents are odd.*

Write out the quotient of:

1. $(x^7 + y^7) \div (x + y)$.

2. $(x^9 + y^9) \div (x + y)$.

3. $(x^5 + 1) \div (x + 1)$.

4. $(x^3y^3 + a^3b^3) \div (xy + ab)$.

5. $(x^{10} + y^5z^5) \div (x^2 + yz)$.

6. $(x^{10} + 1) \div (x^2 + 1)$.

HIGH SCHOOL ALGEBRA.

Factor:

7. $a^5 + b^{10}c^{10}$.
8. $x^7 + y^7 z^7$.
9. $m^5 n^5 + r^5 s^5$.
10. $a^5 b^{10} + c^5 d^{10}$.
11. $a^5 x^{10} + b^{10} y^{10}$.
12. $x^{10} y^5 + z^5 y^{10}$.

13. Demonstrate the truth of the above principle.

120. REVIEW IN FACTORING.

Factor the following:

1. $a^2 - 2a - 8$.
2. $9a^2 - b^2$.
3. $(a+b)^2 - (a-b)^2$.
4. $y^2 + 2ym + m^2$.
5. $b^2 - 12b - 45$.
6. $x^4 - y^4$.
7. $n^2 + 2n - 80$.
8. $m^4 - n^4$.
9. $ac + bc + ad + bd$.
10. $q^2 - 4qr + 4r^2$.
11. $9x^2 - 24xc + 16c^2$.
12. $x^5 + y^5$.
13. $a^2 - 7a - 30$.
14. $a^2 - 9a + 14$.
15. $x^{18} - y^9$.
16. $ax - ay - bx + by$.
17. $az - z + 2a - 2$.
18. $p^2 - 4p - 5$.
19. $a^{12} - b^{12}$.
20. $6a^2 - 31a + 35$.
21. $(x-y)^2 - z^2$.
22. $x^2 + 13xy + 36y^2$.
23. $a^{2m} - b^{2n}$.
24. $4a^2 c^2 - 12ac^3 + 9c^4$.
25. $c^2 + 5c - 14$.
26. $y^2 + 6y + 8$.
27. $x^6 - 1$.
28. $xy - y^2 + xz - yz$.
29. $9a^2 + 12az^2 + 4z^4$.
30. $4a^6 - 4b^6$.
31. $x^{2m} + 2x^m + 1$.
32. $4x^{6m} + 4x^{3m} y^m + y^{2m}$.
33. $4p^2 - 8pr + 4r^2$.
34. $16x^{4n} - 81y^{8n}$.
35. $m^2 - m - 90$.
36. $x^2 + 2x - 35$.
37. $ac - ad^2 + bc - bd^2$.
38. $m^2 - 3m - 28$.
39. $x^2 + 14x + 40$.
40. $85 + 12y - y^2$.

FACTORING.

41. $4x^2 + 8x + 3$.

42. $x^4 + x^2 + 1$.

43. $a^2 + b^2 - c^2 - 2ab$.

44. $4c^4 - 25c^2$.

45. $c^4 - 8c^2 + 16$.

46. $c^4 - c^2 - 12$.

47. $y^2 - 3y + 2$.

48. $z^4 - 10z^2 - 24$.

49. $p^2 + 7p + 10$.

50. $x^8 - y^8$.

51. $6b^2 - 7b - 3$.

52. $100x^2 - (10x + y)^2$.

53. $6a^2 - 3am - 2an + mn$.

54. $4c^2 + 40cd + 100d^2$

55. $4x^2 - 9y^4$.

56. $2a^2 + 4a + ab + 2b$.

57. $x^2 - 2x - 360$.

58. $25p^m - 36x^{2p}$.

59. $12c^2 + 7c - 12$.

60. $z^2 + (x-y)^2 - 2z(x-y)$.

61. $(x^2 - xy)^2 - 2(x^2 - xy)(xy - 1) + (xy - 1)^2$.

62. $(3x - 4y)^2 + (2x - 3y)^2 - 2(3x - 4y)(2x - 3y)$.

63. $(a + b + c + d)^2 - (a - b + c - d)^2$.

64. $(x^2 + y^2 + z^2 - 2xy + 2xz - 2yz) - (y + z)^2$.

65. $x^2y - x^2z - xy^2 + xyz$.

66. $x^2z^2 + ax^2 - a^2z^2 - a^3$.

67. $abxy + b^2y^2 + acx - c^2$.

68. $y^3 - 2ay^2 - by^2 + 2aby + a^2y - a^2b$.

69. $2mn - m^2 - n^2 + a^2 + b^2 - 2ab$.

70. $a^2 - 2cd + b^2 - c^2 - d^2 - 2ab$.

71. $(a^2 - b^2)^2 - m^2 - 2mn - n^2$.

72. $3a^2 - 6ab + 3b^2 + 6ac - 6bc$.

73. $(x^2 - 1)^2 - 2(x^2 - 1)(y^2 - 1) + (y^2 - 1)^2$.

74. $(a^2 + b^2 + 2ab)^2 - (a^2 + b^2 - 2ab)^2$.

75. $2ax + 2ay - 2az + 2bx + 2by - 2bz$.

COMMON DIVISORS.

121. 1. What quantity will exactly divide both $3a^2$ and $2ax$?

2. Give all the exact divisors of $12a^2xy$ and $18ax^2y$. Which is the highest of these common divisors?

3. What is the highest common divisor of $24a^2b^2c$ and $48a^2bc^2$?

4. What prime factors, or divisors, are common to $24a^2b^2c$ and $48a^2bc^2$?

5. How may the highest common divisor of $24a^2b^2c$ and $48a^2bc^2$ be obtained from their factors?

122. A **Common Divisor** of two or more quantities is an exact divisor of each of them.

Thus, $6a$ is a common divisor of $12a$, $24a^2c$, and $30a^2y$.

123. A divisor of a quantity is a factor of it.

124. The **Highest*** **Common Divisor** of two or more quantities is the quantity of the highest degree that is an exact divisor of each of them.

Thus, $4a^2x$ is the highest common divisor of $12a^3xy$ and $8a^2x^2y$.

125. When quantities have no common divisor except themselves and **1**, they are **prime to each other.**

Thus, $5x$, $3y$, and $8z$ are prime to each other.

126. PRINCIPLE. — *The highest common divisor of two or more quantities is the product of all their common prime factors.*

---

* The letters H. C. D. are used to abbreviate the expression *Highest Common Divisor.*

COMMON DIVISORS.

127. To find the highest common divisor of quantities that can be factored readily.

1. What is the highest common divisor of $8\,a^2b^2c^3$ and $12\,ab^2c^2$?

PROCESS.

$$8\,a^2b^2c^3 = 2 \times 2 \times 2 \times aa \times bb \times ccc$$
$$12\,ab^2c^2 = 3 \times 2 \times 2 \times a \times bb \times cc$$

H. C. D. $= 2 \times 2 \times a \times bb \times cc = 4\,ab^2c^2$

EXPLANATION. — Since the highest common divisor is the product of all the common prime factors (Prin.), the quantities are separated into their prime factors. The only prime factors common to the given quantities are 2, 2, a, b, b, c, c; and their product, $4\,ab^2c^2$, is therefore the highest common divisor.

2. What is the highest common divisor of $a(x^2 - y^2)$ and $a(x^2 + 2xy + y^2)$?

PROCESS.

$$a(x^2 - y^2) \quad\quad = a(x+y)(x-y)$$
$$a(x^2 + 2xy + y^2) = a(x+y)(x+y)$$

H. C. D. $\quad\quad = a \times (x+y) = a(x+y)$

EXPLANATION. — Reasoning as in the preceding example, the quantities are separated into their prime factors, and the product of the common factors will be the highest common divisor.

The common factors are a and $(x+y)$; therefore, $a(x+y)$ is the highest common divisor.

RULE. — *Separate the quantities into their prime factors, and find the product of all the common factors.*

Find the highest common divisor of the following:

    3. $12\,m^2n^3x^2$ and $18\,m^2nx^2$.

    4. $16\,r^4s^3x^2$ and $20\,r^2s^2x^2$.

    5. $21\,x^4y^2z^3$ and $14\,x^2y^3z^3$.

    6. $15\,x^5y^2z^5$ and $20\,x^5yz^2$.

    7. $11\,a^2xy$, $8\,ax^2y$, and $9\,axy$.

    8. $15\,a^3x^2y^2$, $9\,a^2x^2y^3$, and $8\,a^2xy^2$.

9. $18b^2c^2d^3$, $8b^2c^2d^2$, and $12ab^2c$.

10. $10c^3x^3y^3$, $8a^2x^2y^3$, and $12a^2xy^2$.

11. $18r^2s^2t^2$, $10r^2s^3t$, and $16r^2s^2t^2$.

12. $20a^3x^3y^3$, $15a^2x^2y^3$, and $10a^2xy^2$.

13. $12x^3y^2z^2$, $18x^4y^3z^3$, and $15x^2y^4z^2$.

14. $a^2 - b^2$ and $a^2 - 2ab + b^2$.

15. $x^2 - 2x$ and $2xy^2 - 4y^2$.

16. $16x^2 - y^2$ and $16x^2 - 8xy + y^2$.

17. $x^2 - 2x - 15$ and $x^2 + 9x + 18$.

18. $x^2 + 9x + 20$ and $x^2 + 2x - 15$.

19. $x^2 + x - 30$ and $x^2 + 12x + 36$.

20. $x^2 - x - 12$ and $x^2 - 4x - 21$.

21. $x^2 + 9x + 14$ and $x^2 + 2x - 35$.

22. $x^2 + x - 30$ and $x^2 + 9x + 18$.

23. $a(x^4 - y^4)$ and $x^4 + 2x^3y + x^2y^2$.

24. $4a^2 - 9x^2$ and $4a^2 - 12ax + 9x^2$.

25. $b^4 - 27a^3b$ and $(b - 3a)^2$.

26. $4ab^3 - 4abc^2$ and $12a^2b^2 - 12a^2c^2$.

27. $yz - z$ and $y^4z - yz$.

28. $5a^2 - 2a - 3$ and $5a^2 - 11a + 6$.

29. $9y^2 - 4$ and $9y^2 - 15y - 14$.

30. $a^2 - b^2$, $a^2 - ab$, and $a^2 - 2ab + b^2$.

31. $8(x - y)^4$, $10(x^2 - y^2)^2$, and $12(x^4 - y^4)$.

32. $a^2 - b^2$, $a^2 - 2ab + b^2$, and $a^2b - ab^2$.

33. $z^2 + z - 6$, $z^2 + 7z + 12$, and $z^2 - 2z - 15$.

34. $x^2 + 5x + 6$, $x^2 + 7x + 10$, and $x^2 + 12x + 20$.

35. $1 - a^2$, $1 + a^3$, and $a^2 + 5a + 4$.

COMMON DIVISORS.

128. To find the highest common divisor of quantities that cannot be factored readily by inspection.

1. What are the exact divisors of ax^2? What are these also of 2 times ax^2 or $2ax^2$? Of c times ax^2 or cax^2?

2. If a quantity is an exact divisor of some quantity, what will it also be of any number of times that quantity?

3. Since a is a divisor of $2ab$ and $3ab$, what will it be also of their sum? What of their difference?

4. What is the H. C. D. of $2a^2$ and $6a^3$? Of $2a^2$ and b times $6a^3$, or $6a^3b$? Of $2a^2$ and $5bc$ times $6a^3$, or $30a^3bc$? Of $2a^2$ and $\frac{1}{3}$ of $6a^3$, or $2a^3$? Of 12 times $2a^2$ or $24a^2$ and $6a^3$?

5. When, then, is the H. C. D. of two or more quantities not affected by multiplying or dividing any one of them by other quantities?

129. PRINCIPLES. — 1. *A divisor of any quantity is a divisor of any number of times that quantity.*

2. *A divisor of two or more quantities is a divisor of their sum and of the difference between any two of them.*

3. *The highest common divisor of two or more quantities is not affected by multiplying or dividing any of them by quantities which are not factors of the others.*

130. 1. Find the H. C. D. of $2x^2-5x+2$ and $4x^3+12x^2-x-3$.

PROCESS.

$$2x^2-5x+2\overline{)4x^3+12x^2-x-3}(2x+11$$
$$\underline{4x^3-10x^2+4x}$$
$$22x^2-5x-3$$
$$\underline{22x^2-55x+22}$$
$$25\overline{)50x-25}$$
$$2x-1\overline{)2x^2-5x+2}(x-2$$
$$\underline{2x^2-x}$$
$$-4x+2$$
$$\underline{-4x+2}$$

$\therefore 2x-1$ is the H. C. D.

92   HIGH SCHOOL ALGEBRA.

EXPLANATION. — It is evident that the highest common divisor cannot be higher than $2x^2 - 5x + 2$. The highest common divisor will be $2x^2 - 5x + 2$ if it is contained without a remainder in $4x^3 + 12x^2 - x - 3$. By trial it is found that it is not an exact divisor of $4x^3 + 12x^2 - x - 3$, inasmuch as there is a remainder of $50x - 25$. Therefore, $2x^2 - 5x + 2$ is not the highest common divisor.

Since $2x^2 - 5x + 2$ contains the highest common divisor, $(2x + 11)$ times $(2x^2 - 5x + 2)$ will also contain the highest common divisor (Prin. 1); and since that product and $4x^3 + 12x^2 - x - 3$ both contain the highest common divisor, their difference, or $50x - 25$, will contain their highest common divisor (Prin. 2). Hence, the highest common divisor cannot be higher than $50x - 25$.

Since 25 is a factor of $50x - 25$, but not of the quantities whose highest common divisor is sought, it may be divided by 25 without changing the highest common divisor (Prin. 3). Therefore, the highest common divisor cannot be higher than $2x - 1$.

$2x - 1$ will be the highest common divisor if it is an exact divisor of $2x^2 - 5x + 2$, since, if it is contained in $2x^2 - 5x + 2$, it will be contained in any number of times $2x^2 - 5x + 2$ (Prin. 1) and in the sum of $(2x + 11)$ times $2x^2 - 5x + 2$ and $50x - 25$ (Prin. 2). By trial it is found to be an exact divisor of $2x^2 - 5x + 2$. Therefore, $2x - 1$ is the highest common divisor of the quantities.

2. Find the H. C. D. of $3x^2 + 11x + 6$ and $2x^2 + 11x + 15$.

SOLUTION.

$$2x^2 + 11x + 15 \overline{)3x^2 + 11x + 6}$$
$$2$$
$$\overline{6x^2 + 22x + 12}(3$$
$$6x^2 + 33x + 45$$
$$\overline{-11) -11x - 33}$$
$$x + 3)2x^2 + 11x + 15(2x + 5$$
$$2x^2 + 6x$$
$$\overline{5x + 15}$$
$$5x + 15$$

SUGGESTIONS. — Since the divisor is not contained in the dividend an entire or integral number of times, the terms of the quantity which contains the highest coëfficient of x^2 are multiplied by 2 so that the quotient may be integral or entire. The same result might have been obtained by multiplying the quantity $2x^2 + 11x + 15$ by 3 (Prin. 3).

The remainder $-11x - 33$ may be divided by -11 since -11 is not a factor of $2x^2 + 11x + 15$.

COMMON DIVISORS.

3. Find the H. C. D. of $2x^4 - 12x^3 + 17x^2 + 6x - 9$ and $4x^3 - 18x^2 + 19x - 3$.

SOLUTION.

$$4x^3-18x^2+19x-3)2x^4-12x^3+17x^2+6x-9$$
$$\underline{\qquad\qquad\qquad\qquad\qquad\qquad 2}$$
$$4x^4-24x^3+34x^2+12x-18(x$$
$$\underline{4x^4-18x^3+19x^2-3x}$$
$$-3)-6x^3+15x^2+15x-18$$
$$\overline{2x^3-5x^2-5x+6)4x^3-18x^2+19x-3(2}$$
$$\underline{4x^3-10x^2-10x+12}$$
$$-8x^2+29x-15$$

$$2x^3-5x^2-5x+6$$
$$\underline{\qquad\qquad\qquad\qquad\; 4}$$
$$-8x^2+29x-15)8x^3-20x^2-20x+24(-x-1$$
$$\underline{8x^3-29x^2+15x}$$
$$9x^2-35x+24$$
$$\underline{8x^2-29x+15}$$
$$x^2-6x+9)-8x^2+29x-15(-8$$
$$\underline{-8x^2+48x-72}$$
$$-19)-19x+57$$
$$\overline{x-3)x^2-6x+9(x-3}$$
$$\underline{x^2-3x}$$
$$-3x+9$$
$$\underline{-3x+9}$$

∴ The H. C. D. is $x - 3$.

SUGGESTIONS. — The highest quantity is multiplied by 2 so that the other quantity may be contained in it an entire or integral number of times. Then the first remainder is divided by -3 because -3 is not a factor of the preceding divisor and consequently not of the H. C. D. Another dividend is multiplied by 4, and another divisor divided by -19.

RULE. — *Divide the higher quantity by the lower, and if there be a remainder, divide the lower quantity by it, then the preceding divisor by the last remainder, and so on until nothing remains. The last divisor will be the highest common divisor.*

If more than two quantities are given, find the highest common divisor of any two, then of this divisor and another, and so on. The last divisor will be the highest common divisor.

1. If any quantity contains a factor not found in the other, the factor may be omitted before beginning the process.

2. A common factor of the quantities may be removed before beginning the division, but it must appear as a factor of the highest common divisor.

3. When necessary, the dividend or divisor may be multiplied or divided by any quantity not a factor of the other quantity.

4. The signs of all the terms of either dividend or divisor, or both, may be changed without changing the highest common divisor.

Find the H. C. D. of the following:

4. $2x^2 - 16x + 14$ and $x^2 - 5x - 14$.

5. $3x^2 + 14x + 8$ and $4x^2 + 19x + 12$.

6. $6x^2 - 23x + 15$ and $2x^2 - 12x + 18$.

7. $4x^2 + 21x - 18$ and $2x^2 + 15x + 18$.

8. $21x^2 - 26x + 8$ and $6x^2 - x - 2$.

9. $x^2 - 6xy + 8y^2$ and $x^2 - 8xy + 16y^2$.

10. $x^3 - y^3$ and $x^2 - 2xy + y^2$.

11. $x^4 - 2x^2 + 1$ and $x^4 - 4x^3 + 6x^2 - 4x + 1$.

12. $2x^3 + 6x^2 + 6x + 2$ and $6x^3 + 6x^2 - 6x - 6$.

13. $3x^3 + 3x^2 - 15x + 9$ and $3x^4 + 3x^3 - 21x^2 - 9x$.

14. $20x^4 + x^2 - 1$ and $25x^4 + 5x^3 - x - 1$.

15. $x^2 - 9$, $x^2 - 3x - 18$, and $x^2 + 11x + 24$.

16. $x^2 - 3x - 28$, $x^2 - 11x + 28$, and $x^2 - 15x + 56$.

17. $x^2 + 6x + 9$, $x^3 - x^2 - 12x$, and $x^2 - 4x - 21$.

18. $a^4 - b^4$, $a^3 + a^2b - ab^2 - b^3$, and $a^4 - 2a^2b^2 + b^4$.

19. $x^4 + 5x^3 + 6x^2$, $x^3 + 3x^2 + 3x + 2$, and $3x^3 + 8x^2 + 5x + 2$

20. $a^3 + 3a^2b + 3ab^2 + b^3$, $4a^2b^2 + 12ab^3 + 8b^4$, and $a^2 - b^2$.

21. $9x^4 + 12x^3 + 10x^2 + 4x + 1$ and $3x^4 + 8x^3 + 14x^2 + 8x + 3$

22. $x^4 + 3x^3 + 9x^2 + 12x + 20$ and $x^5 + 6x^3 + 6x^2 + 8x + 24$.

23. $3a^2x^2 + a^2x + 2a^2 + 12x^2 + 4x + 8$ and $a^2x^2 + 3a^2x + 4a^2 + 4x^2 + 12x + 16$.

24. $15x^3 + 9x^2 - 3x + 3$ and $40x^3 - 21x^2 + 10x - 1$.

25. $16ax^3 - 20ax^2 + 10ax - 6a$ and $3ax^2 - 15ax + 12a$.

26. $2a^3x - 2a^2bx - 2ab^2x + 2b^3x$ and $4a^3 + 4a^2b - 4ab^2 - 4b^3$.

27. $2x^2 - 14x + 20$ and $4x^3 - 25x^2 + 20x + 25$.

28. $3x^2 + 21x - 132$ and $6x^3y + 54x^2y - 138xy - 66y$.

29. $x^5 - x^4 - 3x^3 + x^2 + 3x + 1$ and $x^4 - x^3 - 3x^2 + 2x + 2$.

30. $a^3 - 5a^2 - 99a + 40$ and $a^3 - 6a^2 - 86a + 35$.

31. $14x^4 + x^3 + 8x^2 - x + 2$ and $6x^4 + 7x^3 + 7x^2 + 3x + 1$.

32. $3x^3 - x^2 - 2x - 16$ and $2x^3 - 2x^2 - 3x - 2$.

COMMON MULTIPLES.

131. 1. What quantity exactly contains 2, 3, a, and b? What relation does each of them bear to their product?

2. What is the lowest quantity that exactly contains $3a$ and $4ab$?

3. What factor of $3a$ is not found in $4ab$? What is the product of 3 multiplied by $4ab$?

4. To what, then, is the lowest common multiple of several quantities equal?

132. A **Multiple** of a quantity is a quantity that will exactly contain it.

Thus, a^2x is a multiple of a, a^2, and x

133. A **Common Multiple** of two or more quantities is a quantity that will exactly contain each of them.

Thus, $4b^2c$ is a common multiple of $2b$ and c.

134. The **Lowest* Common Multiple** of two or more quantities is the lowest quantity that will exactly contain each of them.

Thus, $2bc$ is the lowest common multiple of $2b$ and c.

135. PRINCIPLE. — *The lowest common multiple of two or more quantities is the product of all their different prime factors, using each factor the greatest number of times it occurs in any of the quantities.*

136. 1. What is the lowest common multiple of $3x^2y^2zv$ and $5x^2y^3z^2$.

PROCESS.

$$3x^2y^2zv = 3 \times x^2 \times y^2 \times z \times v$$
$$5x^2y^3z^2 = 5 \times x^2 \times y^3 \times z^2$$

$$\text{L. C. M.} = 5 \times 3 \times x^2 \times y^3 \times z^2 \times v = 15\,x^2y^3z^2v$$

EXPLANATION. — Since the lowest common multiple is a product made up of the prime factors of the quantities, the quantities are separated into their prime factors. Since each prime factor is used as a factor of the lowest common multiple the highest number of times it occurs as a factor in any of the quantities (Prin.), the factors of the lowest common multiple are seen to be 5, 3, x^2, y^3, z^2, v, and their product $15\,x^2y^3z^2v$ is their L. C. M.

2. What is the lowest common multiple of $a^2 - a - 12$ and $a^2 - 4a - 21$?

PROCESS.

$$\frac{(a^2 - a - 12)(a^2 - 4a - 21)}{a + 3} = (a - 4)(a^2 - 4a - 21)$$
$$= a^3 - 8a^2 - 5a + 84$$

EXPLANATION. — Since the product of any two quantities is their *common multiple*, it follows that if their common factors are omitted from the product, the result will be the lowest common multiple. Since their common factors or divisors will be the highest common divisor of the quantities, the product of the two quantities divided by their highest common divisor will be their lowest common multiple.

---

* The abbreviations L. C. M. are used for *Lowest Common Multiple*.

COMMON MULTIPLES.

Their highest common divisor is $a + 3$; omitting this factor from dividend and divisor, the result is $(a-4)(a^2-4a-21)$, which is equal to $a^3 - 8a^2 - 5a + 84$, their lowest common multiple.

RULE. — *Find the product of all the different prime factors, using each factor the greatest number of times it occurs in any of the given quantities.*

The product of *two* quantities divided by their H. C. D. is their L. C. M.

Find the L. C. M. of the following:

3. $8a^2b^2c^3$ and $10a^2bc$.

4. $10x^2y^2z$, $20x^3y^2z$, and $25x^2y^3z^3$.

5. $14a^2b^2c^2$, $7b^2x^2y$, and $35abcx$.

6. $12m^2n^2y^2$, $18mny^3$, and $24m^2n^3y$.

7. $18r^2s^2z^3$, $9r^3sz^2$, and $36rs^3z^4$.

8. $x^2 - y^2$ and $x^2 - 2xy + y^2$.

9. $x^2 - y^2$ and $x^2 + 2xy + y^2$.

10. $x^2 - y^2$, $x^2 - 2xy + y^2$, and $x^2 + 2xy + y^2$.

11. $x^2 - y^2$ and $x^3 - y^3$.

12. $a^2(x-z)$ and $y^2(x^2-z^2)$.

13. $x^2 - 1$, $x^2 + 1$, and $x^4 - 1$.

14. $2x(x-y)$, $4xy(x^2-y^2)$, and $6xy^2(x+y)$.

15. $x^2 - x$, $x^3 - 1$, and $x^3 + 1$.

16. $x^2 - 1$, $x^2 - x$, and $x^3 - 1$.

17. $4(1+x)$, $4(1-x)$, and $8(1-x^2)$.

18. $x^2 + 5x + 6$ and $x^2 + 6x + 8$.

19. $a^2 - a - 20$ and $a^2 + a - 12$.

20. $x^2 - 9x - 22$ and $x^2 - 13x + 22$.

21. $x^2 - 8x + 15$ and $x^2 + 2x - 5$.

22. $x^3 + x^2y + xy^2 + y^3$ and $x^3 - x^2y + xy^2 - y^3$.

ALGEBRA. — 7

98 HIGH SCHOOL ALGEBRA.

23. $x^3 - x^2y + xy^2 - y^3$ and $x^3 + x^2y - xy^2 - y^3$.

24. $a^3 - 2a^2 + 4a - 8$ and $a^3 + 2a^2 - 4a - 8$.

25. $x^2 + y^2$, $x^3 - xy^2$, and $x^3 + xy^2 + x^2y + y^3$.

26. $x^2 - 4$, $x^2 - x - 6$, and $x^3 - 3x^2 - 4x + 12$.

27. $x - 5$, $x^2 - 2ax + a^2$, $x^2 - 10x + 25$, and $x^2 + 5a - bx - ax$.

28. $x^4 - 16$, $x^2 + 4x + 4$, and $x^2 - 4$.

29. $m^2 - 3m + 2$ and $m^2 - 4m + 4$.

30. $p^2 - 7p + 10$ and $p^2 - p - 20$.

31. $5c - c^2 - 4$ and $3 - 4c + c^2$.

32. $1 - p + p^2$, $1 + p - p^2$, and $1 - p^2$.

33. $4ac - 4c - 4c^2$ and $2ab - 2bc - 2b$.

34. $3x^2 - 19x + 6$ and $x^2 - 10x + 24$.

35. $m^2 - m - 90$, $m^2 - 13m + 30$, and $m^2 - 100$.

36. $y^2 - 2y - 35$, $y^2 - 8y + 7$, and $y^2 + 4y - 5$.

37. $2x^3 - 12x^2 + 19x - 12$ and $2x^3 - 6x^2 + 7x - 3$.

38. $a^4 + 2a^3 - 2a^2 - 2a + 1$ and $a^4 - 1$.

39. $c^2 - 5c + 4$, $c^2 - 6c + 8$, and $c^2 - 8c + 16$.

40. $a^2 + 3a - 4$, $a^2 - 6a + 5$, and $a^2 - a - 20$.

41. $x^2 - (y + z)^2$, $y^2 - (x + z)^2$, and $z^2 - (x + y)^2$.

42. $a^3 - 8a^2 + 19a - 12$, $a^3 - 9a^2 + 26a - 24$, and $a^3 - 6a^2 + 11a - 6$.

43. $4c^2 - 9$, $6c^2 - 5c - 6$, and $6c^2 + 13c + 6$.

44. $1 - x + x^2$, $1 + x + x^2$, and $1 + x^2 + x^4$.

45. $x^3 + 3x^2y + 3xy^2 + y^3$ and $x^3 - xy^2 + x^2y - y^3$.

46. $(a+b)(a+c)$, $(a+b)(b+c)$, and $(a+c)(b+c)$.

47. $a^2 + ab + b^2$, $a^3 - b^3$, and $a - b$.

48. $4a^3(a+x)$, $4a^3(a-x)$, and $2a^2(a^2+x^2)$.

FRACTIONS.

137. 1. When anything is divided into two equal parts, what is one part called? How is it expressed?

2. What does $\frac{a}{2}$ represent? $\frac{2a}{5}$? $\frac{2a}{7}$? $\frac{3a}{9}$?

3. How may one fifth of x be expressed? Two thirds of b? Three sevenths of y? Eight elevenths of z?

138. A **Fraction** is one or more of the equal parts of a unit.

139. Since a fraction is one or more of the equal parts of anything, to express a fraction, two numbers, or quantities, are necessary, one to express the number of equal parts into which the unit has been divided; the other to express how many parts form the fraction. These numbers, or quantities, are written one above the other, with a horizontal line between them.

140. The **Denominator** is the number, or quantity, which shows into how many equal parts the unit is divided.

It is written below the line.

Thus, in the fraction $\frac{a}{b}$, b is the denominator. It shows that the unit has been divided into b equal parts.

141. The **Numerator** is the number, or quantity, which shows how many parts form the fraction.

It is written above the line.

Thus, in the fraction $\frac{a}{b}$, a is the numerator. It shows how many parts form the fraction.

142. The numerator and denominator are called the **Terms of a Fraction.**

143. An indicated process in division may be written in the *form of a fraction*, the numerator being the dividend and the denominator the divisor.

144. An **Entire Quantity** is a quantity, no part of which is in the form of a fraction.

Thus, $2a$, $3c$, $2x+y$, etc., are entire quantities.

145. A **Mixed Quantity** is a quantity composed of an entire quantity and a fraction.

Thus, $2a + \dfrac{3b}{7}$, $2x + 2y - \dfrac{3z+2}{x+7}$, are mixed quantities.

146. The **Sign of a Fraction** is the sign written before the dividing line. This sign belongs to the fraction as a *whole*, and not to either the numerator or denominator.

Thus, in $-\dfrac{x+y}{2z}$ the sign of the fraction is $-$, while the signs of the quantities x, y, and $2z$ are $+$. The sign before the dividing line shows whether the fraction is to be added or subtracted.

REDUCTION OF FRACTIONS.

147. To reduce fractions to higher or lower terms.

1. How many fourths are there in one half?

2. How many fourths are there in $\dfrac{b}{2}$? $\dfrac{x}{2}$? $\dfrac{y}{2}$?

3. How many sixths are there in $\dfrac{b}{3}$? How many ninths? How are the terms of the fraction $\dfrac{2b}{6}$ obtained from $\dfrac{b}{3}$?

FRACTIONS. 101

4. What, then, may be done to the terms of a fraction without changing the value of the fraction?

5. How many thirds are there in $\frac{2b}{6}$? In $\frac{3b}{9}$? In $\frac{4b}{12}$? How are the terms of the equivalent fraction $\frac{b}{3}$ obtained from these fractions?

6. What else may be done to the terms of a fraction besides multiplying them by the same quantity, that will not change the value of the fraction?

148. A fraction is expressed in its **Lowest Terms** when its numerator and denominator have no *common divisor*.

149. PRINCIPLE. — *Multiplying or dividing both terms of a fraction by the same quantity does not change the value of the fraction.*

EXAMPLES.

1. Change $\frac{3}{2b}$ to a fraction whose denominator is $6b^2$.

PROCESS.

$$\frac{3}{2b}$$

$$6b^2 \div 2b = 3b$$

$$\frac{3 \times 3b}{2b \times 3b} = \frac{9b}{6b^2}$$

EXPLANATION. — Since the fraction is to be changed to an equivalent fraction expressed in higher terms, the terms of the fraction must be multiplied by the same quantity, so that the value of the fraction may not be changed (Prin.). In order to produce the required denominator, the given denominator must be multiplied by $3b$; consequently, the numerator must be multiplied by $3b$ also.

2. Reduce $\frac{15x^4y^2}{25x^5y}$ to its lowest terms.

PROCESS.

$$\frac{15x^4y^2}{25x^5y} = \frac{3y}{5x}$$

EXPLANATION. — Since the fraction is to be changed to an equivalent fraction expressed in its lowest terms, the terms of the fraction may be divided by any quantity that will exactly divide them (Prin.). Dividing by the quantity $5x^4y$, the expression is reduced to its lowest terms, for the terms are then *prime to each other;* or,

The terms may be divided by their highest common divisor.

150. To express a fraction in higher terms.

RULE. — *Multiply the terms of the fraction by such a quantity as will change the given term to the required term.*

151. To express a fraction in its lowest terms.

RULE. — *Divide the terms of the fraction by any common divisor, and continue to divide thus until they have no common divisor;* or,

Divide the terms of the fraction by their highest common divisor.

3. Change $\dfrac{3a}{7}$ to a fraction whose denominator is 28.

4. Change $\dfrac{5x^2}{6}$ to a fraction whose denominator is 36.

5. Change $\dfrac{2a+4b}{3}$ to a fraction whose denominator is 15.

6. Change $\dfrac{3x+7}{6}$ to a fraction whose denominator is 30.

7. Change $\dfrac{2x}{6x+3}$ to a fraction whose numerator is $6x$.

8. Change $\dfrac{3x}{6x-8}$ to a fraction whose numerator is $9x$.

9. Change $\dfrac{2ax}{3+2y}$ to a fraction whose numerator is $4ax^2$.

10. Change $\dfrac{2a+x}{a+b}$ to a fraction whose denominator is a^2-b^2.

11. Change $\dfrac{3x-y}{a+b}$ to a fraction whose denominator is $a^2+2ab+b^2$.

FRACTIONS. 103

Reduce the following to their lowest terms:

12. $\dfrac{15 x^2 y^2 z}{75 xy^2 z}$.

13. $\dfrac{21 x^2 y^2 z^2}{28 x^2 y^3 z^2}$.

14. $\dfrac{10 abx^2 y}{25 abx^3 y^2}$.

15. $\dfrac{16 xyz^3 m}{24 x^2 y^2 zm^2}$.

16. $\dfrac{21 m^2 n^2 z^3}{12 m^2 z^4}$.

17. $\dfrac{24 x^3 y^2 z^5}{12 x^3 y^3 z^4}$.

18. $\dfrac{35 x^5 y^4 z^7}{49 x^4 y^5 z^8}$.

19. $\dfrac{22 a^5 x^2 yz^4}{33 a^4 x^2 y^2 z^2}$.

20. $\dfrac{a^2 - b^2}{a^2 - 2ab + b^2}$.

21. $\dfrac{a^2 - b^2}{a^2 + 2ab + b^2}$.

22. $\dfrac{2x^2}{6x - 4x^2}$.

23. $\dfrac{2a^2 - 2}{a^2 - 2a + 1}$.

24. $\dfrac{3a^2 - 4ab + b^2}{a^2 - ab}$.

25. $\dfrac{x^2 - z^2}{x^3 - z^3}$.

26. $\dfrac{a^2 - 2ab}{a^2 - 4ab + 4b^2}$.

27. $\dfrac{2 x^4 y^4 + 2}{3 x^8 y^8 - 3}$.

28. $\dfrac{18 a^2 c - 6 bc}{42 a^2 d - 14 bd}$.

29. $\dfrac{(x+y)^2}{x^2 - y^2}$.

30. $\dfrac{m^3 + n^3}{(m+n)^3}$.

31. $\dfrac{(1 - x^2)^2}{(1 + x)^3}$.

32. $\dfrac{a^2 - 4a - 12}{a^2 + 5a + 6}$.

33. $\dfrac{x^2 - 5x + 6}{x^2 + 4x - 21}$.

34. $\dfrac{a^2 + a - 90}{a^2 + 8a - 20}$.

35. $\dfrac{2x^2 - 5x + 3}{x^2 + x - 2}$.

36. $\dfrac{m^2 - 12m + 32}{m^2 + 2m - 24}$.

37. $\dfrac{3a^2 - 6a}{2ax - 4x}$.

38. $\dfrac{7 a^2 b + 7 abx}{a^2 - x^2}$.

39. $\dfrac{x^2 - 1}{x^2 + 2x + 1}$.

40. $\dfrac{x^2 - 1}{2xy + 2y}$.

41. $\dfrac{x^3 - a^2 x}{x^2 - 2ax + a^2}$.

42. $\dfrac{x^5 - x^3 y^2}{x^4 - y^4}$.

43. $\dfrac{x^2 + 6x + 9}{x^3 - x^2 - 12x}$.

44. $\dfrac{x^2 - 3x - 28}{x^2 - 11x + 28}$.

45. $\dfrac{m^2 - n^2}{m^3 - n^3}$.

46. $\dfrac{a^6 + b^6}{a^4 - b^4}$.

47. $\dfrac{m^2 - n^2}{m^3 + n^3}$.

48. $\dfrac{m^8 - n^8}{m^6 - n^6}$.

49. $\dfrac{9a^2 b + 9ab^2}{3a^2 + 6ab + 3b^2}$.

50. $\dfrac{x^3 - y^3}{x^2 - 2xy + y^2}$.

51. $\dfrac{y^3 - b^2 y}{y^2 + 2by + b^2}$.

52. $\dfrac{a^2 - 4b^2}{(a - 2b)^2}$.

53. $\dfrac{a^2 + 2a - 3}{a^2 + 5a + 6}$.

54. $\dfrac{a^2 + 5a + 6}{a^2 + 7a + 12}$.

55. $\dfrac{6c^2 - 5c - 6}{8c^2 - 2c - 15}$.

56. $\dfrac{a^2 - (b + c)^2}{a^2 + ab + ac}$.

57. $\dfrac{15x^2 + xy - 2y^2}{9x^2 + 3xy - 2y^2}$.

58. $\dfrac{a^2 - 12a + 35}{a^2 - 15a + 56}$.

59. $\dfrac{x^3 + 5x^2 + 5x + 1}{x^3 + 1}$.

152. To reduce an entire or mixed quantity to a fraction.

1. How many fifths are there in 3? In 4? In 10? In a?

2. How many sevenths are there in 2? In 4? In b^2?

3. How many fourths are there in $2\tfrac{1}{4}$? In $3\tfrac{3}{4}$? In $a + \dfrac{a}{4}$?

EXAMPLES.

1. Reduce $a + \dfrac{b}{c}$ to a fractional form.

PROCESS.

$a = \dfrac{ac}{c}$

$a + \dfrac{b}{c} = \dfrac{ac}{c} + \dfrac{b}{c} = \dfrac{ac + b}{c}$

EXPLANATION.—Since 1 is equal to $\dfrac{c}{c}$, a is equal to $\dfrac{ac}{c}$; consequently, $a + \dfrac{b}{c} = \dfrac{ac}{c} + \dfrac{b}{c}$, or $\dfrac{ac + b}{c}$.

FRACTIONS.

Rule. — *Multiply the entire part by the denominator of the fraction; to this product add the numerator when the sign of the fraction is plus, and subtract it when it is minus, and write the result over the denominator.*

If the *sign of the fraction* is —, the signs of all the terms in the numerator must be changed when it is subtracted.

Reduce the following to fractional forms:

2. $2x + \dfrac{4y}{5}$.

3. $5x - \dfrac{3y}{4}$.

4. $4x - \dfrac{6z}{2}$.

5. $x + \dfrac{4y+3}{4}$.

6. $2a + \dfrac{3x+4}{4}$.

7. $2x + \dfrac{3y-4}{8}$.

8. $3x - \dfrac{2y+3}{6}$.

9. $5a - \dfrac{3x+4}{2}$.

10. $6a - \dfrac{3y+7}{4}$.

11. $3c + \dfrac{4a+b}{d}$.

12. $4a + \dfrac{3c-d}{cd}$.

13. $3x + \dfrac{6a-x}{ax}$.

14. $x + 4 + \dfrac{2c-d}{5}$.

15. $a - \dfrac{2ac-c^2}{a}$.

16. $2x - 5 - \dfrac{x^2+4}{x-2}$.

17. $a + x + \dfrac{a^2+x^2}{a-x}$.

18. $a + c + \dfrac{2ac-c^2}{a-c}$.

19. $x - y + \dfrac{x^2-y^2}{x+y}$.

20. $x + 4 - \dfrac{x^2-2}{x-4}$.

21. $a + x - \dfrac{4ax-5x^2}{a-x}$.

22. $a - b - \dfrac{a^2+b^2}{a-b}$.

23. $m + n - \dfrac{2mn+n^2}{m-n}$.

153. To reduce a fraction to an entire or a mixed quantity.

1. How many units are there in $\frac{8}{3}$? In $\frac{25}{9}$? In $\frac{31}{7}$?

2. How many units in $\dfrac{6a+6b}{6}$? In $\dfrac{8x-4}{4}$? In $\dfrac{5m+n}{5}$?

HIGH SCHOOL ALGEBRA.

EXAMPLES.

1. Reduce $\dfrac{bx+d}{b}$ to a mixed quantity.

PROCESS.
$$\dfrac{bx+d}{b} = x + \dfrac{d}{b}$$

EXPLANATION.—Since a fraction may be regarded as an expression of unexecuted division, by performing the division indicated, the fraction is changed into the form of a mixed quantity.

Reduce the following to entire or mixed quantities:

2. $\dfrac{a^2+c^2}{a}$.

3. $\dfrac{bx+cd}{b}$.

4. $\dfrac{2ab+b^2}{a+b}$.

5. $\dfrac{a^2-x^2}{a-x}$.

6. $\dfrac{a^2-x^2}{a+x}$.

7. $\dfrac{x^3+1}{x+1}$.

8. $\dfrac{x^3+1}{x-1}$.

9. $\dfrac{1}{1-x}$.

10. $\dfrac{x^3}{x+y}$.

11. $\dfrac{7a^2}{a-b}$.

12. $\dfrac{x-y}{x+y}$.

13. $\dfrac{2x^2+7}{x-4}$.

14. $\dfrac{b^4-a^2}{b^2+a}$.

15. $\dfrac{x^8-y^8}{x^3-y^3}$.

16. $\dfrac{x^2+2ax+a^2}{a+x}$.

17. $\dfrac{a^3+b^3}{a-b}$.

18. $\dfrac{5ay+ax+x}{ax}$.

19. $\dfrac{2a^2-2b^2}{a+b}$.

20. $\dfrac{x^2+2xy+2y^2+x}{x+y}$.

21. $\dfrac{a^3-b^3}{a-b}$.

22. $\dfrac{x^2+xy+y^2}{x+y}$.

23. $\dfrac{(m+n)^3}{m^2-n^2}$.

24. $\dfrac{x^3-y^3-z^3}{x-z}$.

25. $\dfrac{4x^3+4x^2y^2-6y^4}{2x-y^2}$.

26. $\dfrac{x^4+2x^3-x^2-4x-6}{x^2-2}$.

27. $\dfrac{5a^3-a^2+5}{5a^2+4a-1}$.

28. $\dfrac{a^2+b^2+ab+a}{a+b+1}$.

29. $\dfrac{a^3-3a^2b+3ab^2+b^3}{a^2-2ab+b^2}$.

FRACTIONS.

154. To reduce dissimilar to similar fractions.

1. Into what fractions having the same fractional unit may $\frac{1}{3}$, $\frac{1}{4}$, and $\frac{1}{6}$ be changed?

2. Into what fractions having the same fractional unit may $\frac{1}{2a}$ and $\frac{1}{5a}$ be changed?

3. Express $\frac{1}{2a}$, $\frac{1}{5a}$, and $\frac{1}{10a}$ in equivalent fractions having their lowest common denominator.

155. Similar Fractions are those which have the same fractional unit.

156. Dissimilar Fractions are those which have not the same fractional unit.

Similar fractions have, therefore, a common denominator.

157. When similar fractions are expressed in their lowest terms, they have their **Lowest Common Denominator**.

158. Principles. — 1. *A common denominator of two or more fractions is a common multiple of their denominators.*

2. *The lowest common denominator of two or more fractions is the lowest common multiple of their denominators.*

1. Reduce $\dfrac{d}{2ac}$ and $\dfrac{2c}{3a^2d}$ to similar fractions having their lowest common denominator.

PROCESS.

$$\frac{d}{2ac} = \frac{d \times 3ad}{2ac \times 3ad} = \frac{3ad^2}{6a^2cd}$$

$$\frac{2c}{3a^2d} = \frac{2c \times 2c}{3a^2d \times 2c} = \frac{4c^2}{6a^2cd}$$

Explanation. — Since the lowest common denominator of several fractions is the lowest common multiple of their denominators (Prin. 2), the lowest common multiple of the denominators $2ac$ and $3a^2d$ must be found, which is $6a^2cd$. The fractions are then reduced to fractions having the denominator $6a^2cd$, by multiplying the numerator and denominator of each fraction by the quotient of $6a^2cd$, divided by its denominator (Art. 149). $6a^2cd \div 2ac = 3ad$, the multiplier of the terms of the first fraction. $6a^2cd \div 3a^2d = 2c$, the multiplier of the terms of the second fraction.

HIGH SCHOOL ALGEBRA.

RULE. — *Find the lowest common multiple of the denominators of the fractions for a lowest common denominator.*

Divide this denominator by the denominator of each fraction, and multiply the terms of the fraction by the quotient.

1. Any multiple of the lowest common denominator will be a *common* multiple of the denominators.
2. All mixed quantities should be changed to the fractional form, and all fractions to their lowest terms before finding their lowest common denominator.

Reduce the following to similar fractions having their lowest common denominator:

2. $\dfrac{3x}{4}$ and $\dfrac{5x}{6}$.

3. $\dfrac{7a}{8}$ and $\dfrac{5a}{6}$.

4. $\dfrac{3x^2y}{4}$ and $\dfrac{2xy^2}{16}$.

5. $\dfrac{2x}{3a}$ and $\dfrac{4y}{6a}$.

6. $\dfrac{2b}{3y}$ and $\dfrac{2c}{6y^2}$.

7. $\dfrac{3ac}{2x^2y}$ and $\dfrac{2bd}{3x^2z}$.

8. $\dfrac{2x-4y}{5x^2}$ and $\dfrac{3x-8y}{10x}$.

9. $\dfrac{4a+5b}{3a^2}$ and $\dfrac{3a+4b}{4a}$.

10. $\dfrac{3x-2y}{5ac}$ and $\dfrac{4x-3y}{10a^2c}$.

11. $\dfrac{3}{2xy}$, $\dfrac{4a}{4x^2y}$, $\dfrac{5c}{3yx^2z}$.

12. $\dfrac{c}{4x}$, $\dfrac{d}{4xy}$, $\dfrac{d}{8x^2y^2}$.

13. $\dfrac{d}{a^2c}$, $\dfrac{c}{3ac^2}$, 4.

14. $\dfrac{x+y}{4}$, $\dfrac{x-y}{2c}$, $\dfrac{x^2+y^2}{2a}$.

15. $\dfrac{x+2}{x-1}$, $\dfrac{x-2}{x+1}$, $\dfrac{x+3}{x^2-1}$.

16. $\dfrac{x^2y}{a+b}$, $\dfrac{xy}{a-b}$, $\dfrac{xy^2}{a^2-b^2}$.

17. $\dfrac{x+y}{x-y}$, $\dfrac{x-y}{x+y}$, $\dfrac{x^2+y^2}{x^2-y^2}$.

18. $\dfrac{x^2-1}{x^2+1}$, $\dfrac{x^2+1}{x^2-1}$, and $\dfrac{x^4+1}{x^4-1}$.

19. $\dfrac{1}{(a-b)(b-c)}$ and $\dfrac{1}{(a-b)(a-c)}$.

FRACTIONS. 109

20. $\dfrac{a}{a+1},\ \dfrac{1}{a+3},\ \dfrac{2}{a^2-9},\ \dfrac{a-1}{a-3}.$

21. $\dfrac{3x-1}{x+2},\ \dfrac{2x+5}{2x+4},\ \dfrac{4x-1}{6x+12}.$

22. $\dfrac{8}{x-5},\ \dfrac{25}{4x^2-25},\ \dfrac{3x}{2x+5}.$

23. $\dfrac{1}{m-n},\ \dfrac{2m}{m^2-n^2},\ \dfrac{m+n}{m^3-n^3}.$

24. $\dfrac{x+3}{x^2+x-2},\ \dfrac{x+1}{x^2-x-6},\ \dfrac{x-2}{x^2-4x+3},\ \dfrac{1}{x^2+2x+4}.$

CLEARING EQUATIONS OF FRACTIONS.

159. 1. Ten is one half of what number?

2. If one third of a number is 12, what is the number?

3. If $\tfrac{1}{2}x$ equals 4, what is the value of x?

4. If $\tfrac{1}{4}x = 8$, what is the value of x?

5. If both members of an equation are multiplied by the same quantity, how is the equality of the members affected?

6. When $\dfrac{x}{3} = 6$, what is the resulting equation when each member is multiplied by 3? By 6? By 9? By 12? By 15?

7. How may an equation containing fractions be changed into an equation without fractions?

160. Clearing an equation of Fractions is changing it into another equation without the fractions.

161. PRINCIPLE. — *An equation may be cleared of fractions by multiplying each member by some multiple of the denominators of the fractions.* (Art. 57, Ax. 4.)

HIGH SCHOOL ALGEBRA.

1. Find the value of x in the following $x + \dfrac{x}{5} = 12$.

PROCESS.

$$x + \dfrac{x}{5} = 12$$

Clearing of fractions, $5x + x = 60$
Uniting terms, $\qquad 6x = 60$
Therefore, $\qquad\qquad x = 10$

giving as a resulting equation $5x + x = 60$.
therefore, $x = 10$.

EXPLANATION. — Since the equation contains a fraction, it may be cleared of fractions by multiplying each member by the denominator of the fraction (Prin.). The denominator is 5; therefore each member is multiplied by 5. Uniting similar terms, $6x = 60$.

2. Given $x + \dfrac{x}{3} + \dfrac{x}{5} + \dfrac{x}{6} = \dfrac{153}{10}$, to find the value of x.

PROCESS.

$$x + \dfrac{x}{3} + \dfrac{x}{5} + \dfrac{x}{6} = \dfrac{153}{10}$$

Clearing of fractions, $30x + 10x + 6x + 5x = 459$
Therefore, $\qquad\qquad\qquad\qquad\quad 51x = 459$
And, $\qquad\qquad\qquad\qquad\qquad\qquad x = 9$

EXPLANATION. — Since the equation may be cleared of fractions by multiplying by some multiple of the denominators (Prin.), this equation may be cleared of fractions by multiplying both members by 3, 5, 6, and 10 successively, or by their product, or by any multiple of 3, 5, 6, and 10.

Since the multiplier will be the smallest when we multiply by the lowest common multiple of the denominators, for convenience we multiply both members by 30, the lowest common multiple of 3, 5, 6, and 10. Uniting terms, and dividing, the result is $x = 10$.

RULE. — *Multiply both members of the equation by the least, or lowest, common multiple of the denominators.*

1. An equation may also be cleared of fractions by multiplying each member by all the denominators successively.
2. If a fraction has the minus sign before it, the signs of all the terms of the numerator must be changed when the denominator is removed.
3. Multiplying a fraction by its denominator removes the denominator

Find the value of x, and verify the result in the following:

3. $x + \dfrac{x}{5} = 24$.

4. $\dfrac{x}{6} + x = 21$.

5. $2x + \dfrac{x}{3} = 28$.

6. $4x + \dfrac{x}{5} = 42$.

FRACTIONS. 111

7. $3x - \dfrac{x}{7} = 40.$

8. $x - \dfrac{x}{6} = 25.$

9. $\dfrac{4x}{6} - x = -24.$

10. $\dfrac{3x}{5} + 7x = 38.$

11. $\dfrac{x}{2} + \dfrac{x}{3} + \dfrac{x}{4} = 26.$

12. $\dfrac{x}{3} + \dfrac{x}{4} + \dfrac{x}{6} = 18.$

13. $x + \dfrac{2x}{3} + \dfrac{3x}{4} = 29.$

14. $2x + \dfrac{x}{3} - \dfrac{x}{4} = 50.$

15. $3x - \dfrac{2x}{3} - \dfrac{5x}{6} = 18.$

16. $4x + \dfrac{x}{3} - \dfrac{x}{9} = 74.$

17. $3x - \dfrac{x}{6} + \dfrac{x}{12} = 70.$

18. $\dfrac{x}{4} + \dfrac{x}{6} + \dfrac{x}{8} = 26.$

19. $\dfrac{x+9}{4} + \dfrac{2x}{7} = \dfrac{3x-6}{5} + 3.$

20. $\dfrac{3x+4}{7} + \dfrac{4x-51}{47} = 0.$

21. $\dfrac{3x}{4} + \dfrac{180-5x}{6} = 29.$

22. $\dfrac{1}{4}x + \dfrac{1}{10}x + \dfrac{1}{8}x = 19.$

23. $\dfrac{x+3}{2} + \dfrac{3x}{5} = 2 + \dfrac{4x-5}{3}.$

24. $\dfrac{x+2}{5} + \dfrac{x-1}{7} = \dfrac{x-2}{2}.$

25. $\dfrac{7x+2}{10} - 12 = \dfrac{3x+3}{5} - \dfrac{x}{2}.$

26. $\dfrac{3x}{4} + a - \dfrac{x}{5} = a + \dfrac{x}{5} + 5\tfrac{1}{2}.$

27. $\dfrac{2x+4}{3} - 3\tfrac{1}{3} = \dfrac{x-3}{4} + \dfrac{x+2}{3}.$

28. $\dfrac{3x-4}{2} = \dfrac{6x-5}{8} + \dfrac{3x-1}{16}.$

29. $\dfrac{2x-5}{6} + \dfrac{6x+3}{4} = 5x - 17\tfrac{1}{2}.$

30. $\dfrac{5-3x}{4} + \dfrac{3-5x}{3} = \dfrac{3}{2} - \dfrac{5x}{3}.$

31. $\dfrac{x+3}{2} + \dfrac{x+4}{3} + \dfrac{x+5}{4} = 16.$

32. $\dfrac{2x-1}{5} + \dfrac{6x-4}{7} = \dfrac{7x+12}{11}.$

33. $\dfrac{x+1}{3} - \dfrac{3x-1}{5} = x - 2.$*

34. $\dfrac{x-3}{4} - \dfrac{x-1}{9} = \dfrac{x-5}{6}.$

35. $\dfrac{x}{2} + 3 = \dfrac{x}{4} - \dfrac{x-2}{5}.$

36. $\dfrac{x-1}{2} + \dfrac{x-3}{4} - \dfrac{x-2}{3} = \dfrac{2}{3}.$

37. $\dfrac{1-2x}{3} - \dfrac{4-5x}{4} = -\dfrac{13}{42}.$

38. $\dfrac{x+3}{2} - \dfrac{x-2}{3} - \dfrac{1}{4} = \dfrac{3x-5}{12}.$

39. $\dfrac{4x-2}{11} + 4 - \dfrac{3x-5}{13} = 5.$

40. $\dfrac{3x-3}{4} - \dfrac{3x-3}{3} = \dfrac{15}{3} - \dfrac{27+4x}{9}.$

41. A spent $\tfrac{1}{4}$ of his money, and then received $2. He then spent $\tfrac{1}{2}$ of what he had, and had $7 remaining. How much had he at first?

---

* In clearing this and the following equations of fractions, the signs should be changed as indicated in Note 2, under the Rule.

FRACTIONS.

Solution.

Let $x=$ the money he had at first.

Then $\dfrac{x}{4}=$ what he spent at first.

$\dfrac{3x}{4}+2=$ what he had after he received \$2.

$\dfrac{1}{2}\left(\dfrac{3x}{4}+2\right)=\dfrac{3x}{8}+1=$ what he spent the second time.

Therefore, $\quad \dfrac{x}{4}+\dfrac{3x}{8}+1+7=x+2$

Clearing of fractions, $\quad 2x+3x+8+56=8x+16$

Transposing, $\quad 2x+3x-8x=-48$

$$-3x=-48$$
$$x=16$$

42. What number is there to which, if $\frac{1}{4}$ of it be added, the sum will be 15?

43. Find a number such that the sum of $\frac{1}{2}$ of it and $\frac{1}{3}$ of it is 15.

44. One third of A's age plus two fifths of A's age equals 22 years. How old is he?

45. Three sons were left a legacy, of which the eldest received $\frac{2}{3}$, the second $\frac{1}{5}$, and the third the rest, which was \$200. How much did each receive?

46. A's capital was $\frac{3}{4}$ of B's. If A's had been \$500 less, it would have been but $\frac{1}{2}$ of B's. What was the capital of each?

47. A horse and carriage cost \$420. If the horse cost $\frac{3}{4}$ as much as the carriage, what was the cost of each?

48. A had twice as much money as B, C $1\frac{1}{3}$ times as much as A, D $\frac{1}{4}$ as much as A, and they all had \$50. How much had each?

49. There is a number such that $\frac{1}{5}$ of it is 3 greater than $\frac{1}{6}$ of it. What is the number?

50. A clerk spent $\frac{1}{3}$ of his salary for board, $\frac{1}{3}$ of the rest for other expenses, and saved annually \$280. **What was his salary?**

51. There is a number such that, if $\frac{1}{5}$ of it is subtracted from 50, and the remainder multiplied by 4, the result will be 70 less than the number. What is the number?

52. Divide 100 into two parts such that, if $\frac{1}{3}$ of one part be subtracted from $\frac{1}{4}$ of the other, the remainder will be 11.

53. The second of two numbers is 1 greater than the first, and $\frac{1}{2}$ of the first plus $\frac{1}{5}$ of the first is equal to the sum of $\frac{1}{3}$ of the second and $\frac{1}{4}$ of the second. What are the numbers?

54. Five years ago A's age was $2\frac{1}{3}$ times B's. One year hence it will be $1\frac{4}{9}$ times B's. How old is each now?

55. The difference between two numbers is 20, and $\frac{1}{7}$ of one is equal to $\frac{1}{3}$ of the other. What are the numbers?

56. When the sum of the fourth, fifth, and tenth parts of a certain number is taken from 33 the remainder is nothing. What is the number?

57. The difference between two numbers is 8, and the quotient arising from dividing the greater by the less is 3. What are the numbers?

ADDITION AND SUBTRACTION OF FRACTIONS.

162. 1. Find the value of $\frac{1}{3}+\frac{1}{6}$; $\frac{1}{4}-\frac{1}{8}$; $\frac{2a}{3}+\frac{a}{6}$.

2. What kind of fractions can be added or subtracted without changing their form?

3. What must be done to dissimilar fractions before they can be added or subtracted? How are dissimilar fractions made similar?

163. PRINCIPLES. — 1. *Only similar fractions can be added or subtracted.*

2. Dissimilar fractions must be reduced to similar fractions before they can be united by addition or subtraction into one term.

FRACTIONS.

EXAMPLES.

1. What is the sum of $\dfrac{5a}{6}$, $\dfrac{3a}{4}$, and $\dfrac{2b}{9}$?

PROCESS.

$$\frac{5a}{6}+\frac{3a}{4}+\frac{2b}{9}=\frac{30a}{36}+\frac{27a}{36}+\frac{8b}{36}=\frac{57a+8b}{36}; \text{ or, } a+\frac{21a+8b}{36}$$

EXPLANATION. — Since the fractions to be added are dissimilar, they must be made similar before they are added.

The lowest common denominator of the given fraction is 36. $\dfrac{5a}{6}=\dfrac{30a}{36}$, $\dfrac{3a}{4}=\dfrac{27a}{36}$; $\dfrac{2b}{9}=\dfrac{8b}{36}$. Therefore their sum is $\dfrac{57a+8b}{36}$, which, expressed as a mixed quantity, is $a+\dfrac{21a+8b}{36}$.

2. Subtract $\dfrac{2a}{7b}$ from $\dfrac{6b}{11a}$.

PROCESS.

$$\frac{6b}{11a}-\frac{2a}{7b}=\frac{42b^2}{77ab}-\frac{22a^2}{77ab}=\frac{42b^2-22a^2}{77ab}$$

EXPLANATION. — Since the fractions are not similar, before subtracting they must be changed to similar fractions. The lowest common denominator of the fractions is $77ab$. Therefore, $\dfrac{6b}{11a}=\dfrac{42b^2}{77ab}$, and $\dfrac{2a}{7b}=\dfrac{22a^2}{77ab}$. Subtracting the numerator of the subtrahend from the numerator of the minuend, the remainder is $\dfrac{42b^2-22a^2}{77ab}$.

3. Find the sum of $a+\dfrac{2ax}{7}$ and $3a+\dfrac{3x}{z}$.

PROCESS.

$$a+3a=4a$$

$$\frac{2ax}{7}+\frac{3x}{z}=\frac{2axz}{7z}+\frac{21x}{7z}; \text{ or, } \frac{2axz+21x}{7z}$$

Entire sum $=4a+\dfrac{2axz+21x}{7z}$.

RULE FOR ADDITION. — *Reduce the given fractions to similar fractions.*

Add their numerators, and write the sum over the common denominator.

When there are entire or mixed quantities, add the entire and fractional parts separately, and then add their results.

RULE FOR SUBTRACTION. — *Reduce the given fractions to similar fractions. Subtract the numerator of the subtrahend from the numerator of the minuend, and place the result over the common denominator.*

When there are entire or mixed quantities, subtract the entire and fractional parts separately, and unite the results.

Add:

4. $\dfrac{a}{a+x}$ and $\dfrac{z}{a-z}$.

5. $\dfrac{1+x}{1-x}$ and $\dfrac{1-x}{1+x}$.

6. $\dfrac{1+x^2}{1-x^2}$ and $\dfrac{1-x^2}{1+x^2}$.

7. $\dfrac{4x^2}{1-x^4}$ and $\dfrac{1-x^2}{1+x^2}$.

8. $\dfrac{x}{x^2-y^2}$ and $\dfrac{y}{x+y}$.

9. $\dfrac{2}{1+a}$ and $\dfrac{a^2+1}{a+a^2}$.

Subtract:

10. $\dfrac{2ab}{3xy}$ from $\dfrac{5ad}{2xy}$.

11. $\dfrac{3mn}{4y^2}$ from $\dfrac{2mn}{4y}$.

12. $\dfrac{a+b}{3}$ from $\dfrac{a-b}{2}$.

13. $\dfrac{3}{a+b}$ from $\dfrac{2}{a-b}$.

14. $\dfrac{4}{x-1}$ from $\dfrac{5}{x+1}$.

15. $\dfrac{x+1}{x-1}$ from $\dfrac{x-1}{x+1}$.

Simplify:

16. $\dfrac{x^2}{x^2-1} + \dfrac{x}{x-1} - \dfrac{x}{x+1}$.

17. $a + \dfrac{1}{a^2-b^2} + 2a - b$.

18. $\dfrac{a}{a-b} + \dfrac{b}{a+b} - \dfrac{a-b}{a+b}$.

19. $\dfrac{y^2-2xy-x^2}{x^2-xy} + \dfrac{x}{x-y}$.

20. $\dfrac{1}{2(x-1)} - \dfrac{1}{2(x+1)} + \dfrac{1}{x^2}$.

21. $\dfrac{1+x}{1+x+x^2} + \dfrac{1-x}{1-x+x^2}$.

FRACTIONS. 117

22. $\dfrac{(a+b)^2}{a-b} - \dfrac{(a-b)^2}{a+b}.$

23. $\left(6x + \dfrac{2a-3b}{5a}\right) - \left(3x - \dfrac{3a+2b}{6a}\right).$

24. $\left(7x + \dfrac{2x}{y}\right) - \left(3x - \dfrac{x-3z}{y}\right).$

25. $\left(3b + \dfrac{a}{ax}\right) - \left(2b - \dfrac{b}{ax}\right).$

26. $\left(5x + \dfrac{x+2}{3}\right) - \left(2x - \dfrac{2x-3}{4}\right).$

27. $\left(a + \dfrac{a+x}{a-x}\right) - \left(b + \dfrac{a-x}{a+x}\right).$

28. $\dfrac{2x+5y}{x^2y} + \dfrac{4xy-3y^2}{xy^2} - \dfrac{5xy-2y^2}{x^2y^2}.$

29. $\dfrac{3ab-4}{a^2b^2} - \dfrac{6a^2-1}{a^3b} - \dfrac{5b^2+7}{ab^3}.$

30. $\dfrac{x}{1-x} - \dfrac{x^2}{1-x^2} + \dfrac{x}{1+x^2}.$

31. $\dfrac{1}{x-a} + \dfrac{4a}{(x-a)^2} - \dfrac{5a^2}{(x-a)^3}.$

32. $\dfrac{3}{x} - \dfrac{5}{2x-1} - \dfrac{2x-7}{4x^2-1}.$

33. $\dfrac{x^2+y^2}{xy} - \dfrac{x^2}{xy+y^2} - \dfrac{y^2}{x^2+xy}.$

34. $\dfrac{1}{x+1} + \dfrac{2}{x+2} + \dfrac{3}{x-3}.$

35. $\dfrac{a-b}{ab} + \dfrac{b-c}{bc} + \dfrac{c-a}{ac}.$

36. $\dfrac{3}{2x^2-2x} + \dfrac{6}{3x^2+6x}.$

37. $\dfrac{x-3}{x^2-1}+\dfrac{4}{x+2}+\dfrac{5}{x-1}$.

38. $\dfrac{a+1}{a^2+a+1}+\dfrac{1-a}{a^2-a+1}$.

39. $\dfrac{1}{x^2-y^2}-\dfrac{1}{x^2+y^2}+\dfrac{2y^2}{x^4-y^4}$.

40. $\dfrac{a-b}{a^2+ab+b^2}+\dfrac{2}{a-b}+\dfrac{a^2-3b^2}{a^3+b^3}$.

41. $\dfrac{x^3+2x^2y}{x^4-y^4}+\dfrac{x-y}{x^2+y^2}-\dfrac{1}{x+y}$.

42. $\left(2-\dfrac{4}{x+2}\right)-\left(x+4-\dfrac{3x}{x-1}\right)$.

43. $\dfrac{x+2}{x^2-9x+18}-\dfrac{x-1}{x^2+x-12}$.

44. $\left(m-\dfrac{4n-p}{3}\right)-\left(4m-\dfrac{2n+3p}{2}\right)$.

45. $\dfrac{x+1}{x^2+x+1}-\dfrac{x-1}{x^2-x+1}$.

46. $a+x-\dfrac{3a^2-2x^2}{2a+x}$.

47. $\dfrac{a+2}{a^2-2a-3}-\dfrac{a+1}{a^2-a-6}$.

48. $\dfrac{3m}{m^2-5m-14}-\dfrac{2m}{m^2-3m-10}$.

49. $\dfrac{1}{x+2}-\dfrac{9x-13-x^2}{x^2+5x+6}$.

50. $\dfrac{x^2-y}{x^2-y^2}-\dfrac{x^2+y}{x^2+y^2}$.

51. $\dfrac{(x^2+y^2)^2}{(x-y)^2xy}-\left(\dfrac{y}{x}+2+\dfrac{x}{y}\right)$.

52. $\dfrac{x^2+x-1}{2x^2-x-3}-\dfrac{x^2-x-1}{3x^2-x-4}$.

FRACTIONS.   119

53. $\dfrac{x-2}{x^2+4x-21} + \dfrac{2-x}{x^2+2x-15} + \dfrac{2}{x^2+12x+35}.$

54. $\dfrac{x-2}{2x^3-6x^2-4x+12} + \dfrac{x+6}{4x^3+2x^2-8x-4}.$

55. $\dfrac{1}{x+y} + \dfrac{1}{x-y} + \dfrac{x-y}{(x+y)^2} + \dfrac{x}{x^2-y^2} + \dfrac{x+y}{(x+y)^2}.$

MULTIPLICATION OF FRACTIONS.

164. 1. How much is 2 times $\dfrac{3}{5}$? $3 \times \dfrac{4}{7}$? $5 \times \dfrac{2a}{3}$?

2. Express $2 \times \dfrac{3}{6}$ in its lowest terms, $3 \times \dfrac{4}{9}$, $4 \times \dfrac{5a}{8}$.

3. How may these products be obtained from the given fractions?

4. In what two ways, then, may a fraction be multiplied by an integer?

5. How much is $\dfrac{1}{2}$ of $\dfrac{4}{5}$, or $\dfrac{4}{5} \div 2$? $\dfrac{1}{3}$ of $\dfrac{3a}{7}$, or $\dfrac{3a}{7} \div 3$?

6. How much is $\dfrac{1}{2}$ of $\dfrac{1}{2}$, or $\dfrac{1}{2} \div 2$? $\dfrac{1}{3}$ of $\dfrac{a}{2}$, or $\dfrac{a}{2} \div 3$?

7. In what two ways, then, may a fraction be divided by an integer?

165. PRINCIPLE.—1. *Multiplying the numerator, or dividing the denominator of a fraction by any quantity, multiplies the fraction by that quantity.*

2. *Dividing the numerator, or multiplying the denominator of a fraction by any quantity, divides the fraction by that quantity.*

HIGH SCHOOL ALGEBRA.

EXAMPLES.

1. What is the product of $\dfrac{a}{b}$ multiplied by $\dfrac{c}{d}$?

PROCESS. EXPLANATION. — To multiply $\dfrac{a}{b}$ by $\dfrac{c}{d}$ is to find c times
$\dfrac{a}{b} \times \dfrac{c}{d} = \dfrac{ac}{bd}$ $\dfrac{1}{d}$ part of $\dfrac{a}{b}$. $\dfrac{1}{d}$ part of $\dfrac{a}{b} = \dfrac{a}{bd}$ (Prin. 2) and c times $\dfrac{a}{bd} = \dfrac{ac}{bd}$
(Prin. 1). That is, the product of the numerators is the numerator of the product, and the product of the denominators is the denominator of the product.

RULE. — *Multiply the numerators together for the numerator of the product, and the denominators together for its denominator.*

1. Reduce all entire and mixed quantities to the fractional form before multiplying.

2. Entire quantities may be expressed in the form of a fraction by writing 1 as a denominator. Thus, a may be written $\dfrac{a}{1}$.

3. Cancel equal factors from both numerator and denominator.

2. Find the product of $\dfrac{a^2 - 2a}{a^2 - 2a - 3} \times \dfrac{a^2 - 9}{a^2 - a} \times \dfrac{a^2 + a}{a^2 + a - 6}$.

SOLUTION. $\dfrac{a^2 - 2a}{a^2 - 2a - 3} \times \dfrac{a^2 - 9}{a^2 - a} \times \dfrac{a^2 + a}{a^2 + a - 6}$

$= \dfrac{a(a-2)}{(a+1)(a-3)} \times \dfrac{(a+3)(a-3)}{a(a-1)} \times \dfrac{a(a+1)}{(a+3)(a-2)}$

Cancelling equal factors from dividend and divisor, the result is $\dfrac{a}{a-1}$

Multiply:

3. $\dfrac{3ac}{4b}$ by $\dfrac{4x}{2ay}$.

4. $\dfrac{5x^2y^2}{a^2x^2}$ by $\dfrac{3ax^2}{2ay^2}$.

5. $\dfrac{a^4b^4}{2a^2y^n}$ by $\dfrac{a^2x}{xy^n}$.

6. $\dfrac{x-y}{a^2}$ by $\dfrac{a^3y}{2x}$.

7. $\dfrac{x+y}{10}$ by $\dfrac{ax}{3(x+y)}$.

8. $\dfrac{2a+3b}{2x}$ by $\dfrac{2x}{4b}$.

9. $\dfrac{x^2-a^2}{xy}$ by $\dfrac{xy}{x+a}$.

10. $\dfrac{a}{x-y}$ by $\dfrac{b}{x+y}$.

FRACTIONS. 121

11. $\dfrac{x^2 - xy}{a+c}$ by $\dfrac{a+c}{x-y}$.

12. $\dfrac{x+y}{x-y}$ by $\dfrac{x^2-y^2}{(x+y)^2}$.

13. $\dfrac{x^4-y^4}{a^2x^2}$ by $\dfrac{ax^3}{x^2+y^2}$.

14. $\dfrac{c}{x^2-y^2}$ by $\dfrac{d}{x^2+y^2}$.

15. $\dfrac{3a^2}{5x-15}$ by $\dfrac{15x-45}{2a}$.

16. $\dfrac{4ax}{2+3x}$ by $\dfrac{12+18x}{8x^2}$.

Simplify the following:

17. $\dfrac{(a-x)^2}{2c} \times \dfrac{3ax}{a-x} \times \dfrac{2}{a(a-x)}$.

18. $\dfrac{x+y}{(x-y)^2} \times \dfrac{x-y}{(x+y)^2} \times \dfrac{x-y}{x+y}$.

19. $\dfrac{x^2+4x}{x^2-3x} \times \dfrac{6x^2-18x}{4x^2+16x}$.

20. $\dfrac{x^2-11x+30}{x^2-6x+9} \times \dfrac{x^2-3x}{x^2-5x}$.

21. $\dfrac{x^2+x-2}{x^2-7x} \times \dfrac{x^2-13x+42}{x^2+2x}$.

22. $\dfrac{x^2+3x+2}{x^2-5x+6} \times \dfrac{x^2-7x+12}{x^2+x}$.

23. $\left(x + \dfrac{xy}{x-y}\right) \times \left(y - \dfrac{xy}{x+y}\right)$.

24. $\left(4 + \dfrac{2x}{3c}\right) \times \left(2 - \dfrac{2x}{6c}\right)$.

25. $\dfrac{y^2-2y+1}{y^2-4} \times \dfrac{y-2}{y^2+9y+10}$.

26. $(a^2-a+1)\left(\dfrac{1}{a^2} + \dfrac{1}{a} + 1\right)$.

27. $\dfrac{x^2-6x-16}{x^2+4x-21} \times \dfrac{x^2-8x+15}{x^2+9x+14}$.

28. $\left(\dfrac{x}{y}+1\right) \times \left(\dfrac{x}{y^2} - \dfrac{1}{y} + \dfrac{1}{x}\right)$.

29. $\dfrac{x^2+x-2}{x^2-3x} \times \dfrac{x^2+9x-36}{x^3+2x^2}$.

DIVISION OF FRACTIONS.

166. 1. How many times is $\frac{1}{8}$ contained in 1? $\frac{1}{7}$ in 1? $\frac{1}{10}$ in 1? $\frac{1}{20}$ in 1?

2. How does the number of times a fraction, having **1** for its numerator, is contained in 1 compare with its denominator?

3. How many times is $\frac{1}{d}$ contained in 1? $\frac{1}{4a}$ in 1? $\frac{1}{x+y}$ in 1?

4. How many times is $\frac{2}{3}$ contained in 1? $\frac{2}{d}$ in 1? $\frac{3}{d}$ in 1?

EXAMPLES.

1. What is the value of $\frac{a}{c} \div \frac{b}{d}$?

PROCESS.
$$\frac{a}{c} \div \frac{b}{d} = \frac{a}{c} \times \frac{d}{b} = \frac{ad}{bc}$$

EXPLANATION. — The fraction $\frac{1}{d}$ is contained in 1, d times; and $\frac{b}{d}$ is contained in 1, $\frac{1}{b}$ part of d times, or $\frac{d}{b}$ times.

And, since $\frac{b}{d}$ is contained in 1, $\frac{d}{b}$ times, it will be contained in $\frac{a}{c}$, $\frac{a}{c}$ times $\frac{d}{b}$, or $\frac{ad}{bc}$ times. That is, the quotient of one fraction divided by another is equal to the product of the dividend by the divisor inverted.

RULE. — *Multiply the dividend by the divisor inverted.*

1. Change entire and mixed quantities to the fractional form.
2. An entire quantity may be expressed as a fraction by writing 1 for its denominator.
3. When possible, use cancellation.

2. Divide $\frac{x^2-1}{x^2-4x-5}$ by $\frac{x^2+2x-3}{x^2-25}$.

SOLUTION.

$$\frac{x^2-1}{x^2-4x-5} \div \frac{x^2+2x-3}{x^2-25} = \frac{(x+1)(x-1)}{(x-5)(x+1)} \times \frac{(x-5)(x+5)}{(x+3)(x-1)}$$

Cancelling, the quotient is $\frac{x+5}{x+3}$.

FRACTIONS. 123

Divide:

3. $\dfrac{4a^3x}{6dy^2}$ by $\dfrac{2a^2x^2}{8a^2y}$.

4. $\dfrac{7x^2y}{3ad}$ by $\dfrac{2xy^2}{3a^2d}$.

5. $\dfrac{5x^2y^3z}{6a^2b^2c}$ by $\dfrac{10xy^3z^2}{8ab^2c^2}$.

6. $\dfrac{4ayz}{8bcd}$ by $\dfrac{6a^2y^2z^2}{16bcd}$.

7. $\dfrac{axyz}{cmn}$ by $\dfrac{8ax^2y}{dmn}$.

8. $\dfrac{mny^2}{abc}$ by $\dfrac{m^3n^2y^2}{a^2b^2c^2}$.

9. $\dfrac{5xy}{a-x}$ by $\dfrac{10xy}{a^2-x^2}$.

10. $\dfrac{2ax+x^2}{a^3-x^3}$ by $\dfrac{x}{a-x}$.

11. $\dfrac{m^2-n^2}{6}$ by $\dfrac{3m+n}{12}$.

12. $\dfrac{a^2-c^2}{1+x}$ by $a+c$.

13. $\dfrac{a^2+4x+4}{d+c}$ by $a+z$.

14. $\dfrac{5(x+y)^2}{x-y}$ by $x+y$.

15. $\dfrac{ab+cd}{a+c}$ by $a-c$.

16. $\dfrac{3an+cm}{x^2-y^2}$ by x^2+y^2.

17. $\dfrac{4ax+4by}{c^2+d^2}$ by $a+b$.

18. $\dfrac{5xy+5xz}{a+n}$ by $5(x+z)$.

19. $\dfrac{1}{x^2-17x+30}$ by $\dfrac{1}{x-15}$.

20. $\dfrac{a}{x^2-5x-6}$ by $\dfrac{b}{x^2+x}$.

21. $\dfrac{b^2p+pxy}{a^2m}$ by $\dfrac{b^2c+cxy}{am^2}$.

22. 16 by $\dfrac{(a-c)^2}{a}-a+2c$.

23. $\dfrac{a^3-b^3}{a^2+ab+b^2}$ by $\dfrac{a-b}{a+b}$.

24. $\dfrac{a^2+b^2}{a^2+2ab+b^2}$ by $\dfrac{a-b}{a+b}$.

25. $\dfrac{a^2+ab+b^2}{a+b}$ by $\dfrac{a^2-ab+b^2}{a-b}$.

26. $\dfrac{x^2+6x+8}{x^2-9}$ by $\dfrac{x^2-2x-8}{x-3}$.

27. $\dfrac{a^2-3a+2}{a^2-7a+10}$ by $\dfrac{a^2-1}{a^2-4a+4}$.

28. $\dfrac{x^2-4x-12}{x^2+3x-28}$ by $\dfrac{x^2-3x-18}{x^2+6x-40}$.

29. $\dfrac{(x+y)^2 - z^2}{z^2 - (x-y)^2}$ by $\dfrac{z+y+x}{z-y+x}$.

30. $\dfrac{x^6 - y^6}{x^4 - y^4}$ by $\dfrac{(x^3 - y^3)(x+y)}{x^2 - y^2}$.

31. $\dfrac{a^4 - 1}{a^3 + 1}$ by $\dfrac{a^2 + 1}{a + 1}$ and then by $\dfrac{a-1}{a^2 - a + 1}$.

32. $\dfrac{m^2 - 7m - 18}{n^2 - 11n + 18}$ by $\dfrac{m^2 - 5m - 14}{n^2 - 8n - 9}$ by $\dfrac{an + a}{6n - 12}$.

167. Complex Fractions are expressions which have a fraction in either the numerator or denominator, or in both.

They are simply expressions of unexecuted division.

1. Find the value of the expression $\dfrac{\dfrac{a}{b}}{\dfrac{c}{d}}$.

PROCESS.

$$\dfrac{\dfrac{a}{b}}{\dfrac{c}{d}} = \dfrac{a}{b} \div \dfrac{c}{d} = \dfrac{a}{b} \times \dfrac{d}{c} = \dfrac{ad}{bc}$$

EXPLANATION. — Complex fractions or fractional forms are simply expressions of division; and, therefore, the given fractional form is the same as though it were written $\dfrac{a}{b} \div \dfrac{c}{d}$. Performing the division according to the principles already given, the quotient is $\dfrac{ad}{bc}$.

Find the value of the following:

2. $\dfrac{x + \dfrac{a}{c}}{x + \dfrac{b}{d}}$.

3. $\dfrac{a^2 + \dfrac{x}{3}}{4 + \dfrac{x}{5}}$.

4. $\dfrac{3a^2 - 3y^2}{\dfrac{a+y}{3}}$.

5. $\dfrac{\dfrac{4x - 4y}{5ab}}{\dfrac{5x - 3y}{5xy}}$.

6. $\dfrac{\dfrac{x+y}{4ax}}{\dfrac{x^2 - y^2}{8ax^2}}$.

7. $\dfrac{4a^2 - 4x^2}{\dfrac{a+x}{a-x}}$.

8. $\dfrac{x + \dfrac{2d}{3ac}}{x + \dfrac{3d}{2ac}}$.

9. $\dfrac{x^2 - \dfrac{y^2}{2}}{\dfrac{x - 3y}{2}}$.

10. $\dfrac{xy - \dfrac{3x}{ac}}{\dfrac{ac}{x} + 2c}$.

FRACTIONS.

11. $\dfrac{\dfrac{a}{b}+\dfrac{x}{y}}{\dfrac{a}{z}-\dfrac{x}{c}}.$

12. $\dfrac{1+\dfrac{x}{y}}{1-\dfrac{x^2}{y^2}}.$

13. $\dfrac{x+\dfrac{2x}{x-3}}{x-\dfrac{2x}{x-3}}.$

14. $\dfrac{x+\dfrac{y-x}{1+xy}}{1-\dfrac{xy-x^2}{1+xy}}.$

15. $\dfrac{a+5+\dfrac{1}{a-5}}{a-5+\dfrac{1}{a+5}}.$

16. $\dfrac{4}{x+1}-\dfrac{3x-1}{x+\dfrac{x}{2}}.$

17. $1-\dfrac{1}{1+\dfrac{1}{a}}.$

18. $\dfrac{1}{a+\dfrac{1}{a+\dfrac{1}{a}}}.$

19. $\dfrac{\dfrac{x+y}{y}+\dfrac{y}{x+y}}{\dfrac{1}{x}+\dfrac{1}{y}}.$

20. $\dfrac{\left(\dfrac{a}{b}+\dfrac{b}{a}\right)\left(\dfrac{a}{b}-\dfrac{b}{a}\right)}{1-\dfrac{b-a}{b+a}}.$

21. $\dfrac{3x-2+\dfrac{1}{x}}{\dfrac{3x-1}{x}}.$

22. $1+\dfrac{c}{1+c+\dfrac{2c^2}{1-c}}.$

REVIEW OF FRACTIONS.

168. Reduce to their lowest terms:

1. $\dfrac{x^3-6x^2+11x-6}{x^3-2x^2-x+2}.$

2. $\dfrac{m^3+m^2+m-3}{m^3+3m^2+5m+3}.$

3. $\dfrac{x^3-2x^2+4x-3}{x^3-5x^2+13x-9}.$

4. $\dfrac{a^3-a^2-a-2}{a^3+3a^2+3a+2}.$

5. $\dfrac{c^2+8c+15}{c^3-3c^2-10c+24}.$

6. $\dfrac{x^4-x^3-4x^2-x+1}{4x^3-3x^2-8x-1}.$

7. $\dfrac{a^3-7a^2+16a-12}{3a^3-14a^2+16a}.$

8. $\dfrac{x^3+3x^2+3x+1}{3x^3-x^2-11x-7}.$

9. $\dfrac{2x^3-3x^2+5x-2}{2x^4-x^3+2x^2+x-1}.$

10. $\dfrac{2a^3-4a^2-13a-7}{6a^3-11a^2-37a-20}.$

Find the value of the following:

11. $\dfrac{x}{x+1}-\dfrac{x}{1-x}+\dfrac{x^2}{x^2-1}.$

12. $\dfrac{3+2x}{2-x} - \dfrac{2-3x}{2+x} + \dfrac{16x-x^2}{x^2-4}$.

13. $\dfrac{1}{(x-y)(y-z)} + \dfrac{1}{(y-x)(x-z)} + \dfrac{1}{(x-z)(x-y)}$.

Suggestion.—By dividing the terms of the second fraction by -1 it becomes $+\dfrac{-1}{(x-y)(x-z)}$ or $-\dfrac{1}{(x-y)(x-z)}$.

14. $\dfrac{1}{a(a-b)(a-c)} + \dfrac{1}{b(b-a)(b-c)} + \dfrac{1}{c(c-a)(c-b)}$.

15. $\left(1+\dfrac{x+y}{x-y}\right) \times \left(1-\dfrac{x-y}{x+y}\right)$.

16. $\left(x-y+\dfrac{x^2+y^2}{x+y}\right) \times \left(x-y+\dfrac{x^2+y^2}{x+y}\right)$.

17. $(a^2+1+a) \times \left(1-\dfrac{1}{a}+\dfrac{1}{a^2}\right)$.

18. $\left(\dfrac{a+1}{a-1} - \dfrac{a-1}{a+1}\right) \div \dfrac{x}{a-1}$.

19. $\left(\dfrac{2x}{x-1}\right) \times \left(1+\dfrac{x-1}{x+1}\right) \div \left\{\left(1-\dfrac{x-1}{x+1}\right) \times \left(1+\dfrac{x+1}{x-1}\right)\right\}$.

20. $\dfrac{1+\dfrac{a-x}{a+x}}{1-\dfrac{a-x}{a+x}} \div \dfrac{1+\dfrac{a^2-x^2}{a^2+x^2}}{1-\dfrac{a^2-x^2}{a^2+x^2}}$.

21. $\dfrac{\dfrac{a-1}{6} - \dfrac{2a-7}{2}}{\dfrac{3a}{4}-3}$.

22. $1 - \dfrac{1}{1-\dfrac{1}{1-\dfrac{1}{x}}}$.

23. $\dfrac{4}{1+\dfrac{4}{1+\dfrac{4}{1+\dfrac{4}{x}}}}$.

24. $\dfrac{\dfrac{x^2+y^2}{x^2-y^2} - \dfrac{x^2-y^2}{x^2+y^2}}{\dfrac{x+y}{x-y} - \dfrac{x-y}{x+y}}$.

FRACTIONS.

25. $\dfrac{x^2 + xy + y^2 + \dfrac{x^3 + y^3}{x - y}}{x - \dfrac{(x^2 + 2y)^2}{x^2 - y^2} - 4}.$

26. $\dfrac{\dfrac{x^2 + 2x}{x^2 - 8x + 12} - \dfrac{3x^2 - 1}{3x^2 - 20x + 12}}{\dfrac{3}{3x^2 - 8x + 4} - \dfrac{1}{x^2 + 3x - 10}}.$

27. $\dfrac{1 + \dfrac{a - c}{a + c}}{1 - \dfrac{a - c}{a + c}} \div \dfrac{1 + \dfrac{a^2 - c^2}{a^2 + c^2}}{1 - \dfrac{a^2 - c^2}{a^2 + c^2}}.$

28. $\dfrac{x - 5 - \dfrac{3x - 15}{x + 3}}{x - 1} \times \dfrac{x + 4 - \dfrac{x^2 - 13}{x - 1}}{\dfrac{1}{5} - \dfrac{1}{x}}.$

29. $\left\{ 1 - \dfrac{\dfrac{x^2}{y^2} - \dfrac{y^2}{x^2}}{\dfrac{x^2}{y^2} + \dfrac{y^2}{x^2}} \right\} \times \left(\dfrac{x^3}{y} + \dfrac{y^3}{x} \right).$

30. $\left(\dfrac{x^2 y^2}{(x - y)^2 + 2xy} - y^2 \right) \left(\dfrac{y^4 - x^4}{y^6} - \dfrac{2xy - 2y^3}{y^2} \right).$

31. $\dfrac{(a^2 - b^2)(2a^2 - 2ab)}{4(a - b)^2 \div \dfrac{ab}{a + b}}.$

32. $\dfrac{\dfrac{a - 1}{a} - \dfrac{b + 1}{b} - \dfrac{1}{c}}{\dfrac{a}{bc} - \dfrac{b}{ac}} + \dfrac{b}{ac + cb} + \dfrac{b}{a + b} \times \dfrac{b^2}{a - b} \times \dfrac{1}{bc}.$

33. $\dfrac{x^2 - 14x + 24}{y^2 + 9y - 36} \div \left(\dfrac{x^2 + 4x - 12}{y^2 + 2y - 15} \times \dfrac{xy + 5x}{xy + 6y} \right).$

34. $\dfrac{a^4 - a^3b + a^2b^2 - ab^3 + b^4}{a-b} \div \left(\dfrac{a-b}{a+b} \times \dfrac{a^5 + b^5}{a^2 + b^2}\right).$

35. $1 \div \left\{\left(\dfrac{1+a}{1+a+a^2} + \dfrac{1-a}{1-a+a^2}\right) \times \dfrac{2}{a}\right\}.$

36. $\left(1 + \dfrac{x}{x-2}\right) \times \dfrac{4-x^2}{1-x^2} \times \dfrac{x^2 - 3x - 4}{x^2 + x - 2} - \dfrac{2x}{x-4}.$

37. $\left(\dfrac{x-2}{x^2-4} + \dfrac{x-3}{x-2} - \dfrac{x}{x+2}\right) \times \left(\dfrac{x^2}{2} - 2\right) \div \dfrac{x^2 - 36}{6x}.$

38. $\dfrac{a+x}{x} - \dfrac{2a}{a+x} + \dfrac{a^2 x - a^3}{x(a^2 - x^2)}.$

39. $\dfrac{x+y}{xy} \times \dfrac{x^2 - y^2}{3(x^2 + y^2)} \times \dfrac{3x^3 y + 3xy^3}{(x-y)^2 (x+y)^2}.$

40. $\dfrac{2x^2 y}{x^3 + y^3} - \dfrac{2xy^2}{x^3 - y^3} + \dfrac{x^2 + y^2}{x^2 - y^2}.$

41. $\dfrac{ax}{ax+c} - \dfrac{c}{ax-c} + \dfrac{ax(3c - ax)}{a^2 x^2 - c^2}.$

42. $\dfrac{10ax + 3a^2 + 3x^2}{10ax - 3a^2 - 3x^2} \div \left(\dfrac{3a+x}{x-3a} \cdot \dfrac{x}{a}\right).$

43. $\dfrac{a^2 - b^2}{a^3 + b^3} \times \left(\dfrac{a^2 + b^2}{b} - a\right) \div \left(\dfrac{1}{b} - \dfrac{1}{a}\right).$

44. $\left(\dfrac{1-a^2}{1-a^3} + \dfrac{1-a}{1-a+a^2}\right) \div \left(\dfrac{1+a}{1+a+a^2} - \dfrac{1-a^2}{1+a^3}\right).$

45. $\dfrac{a - x^2 - y^2}{ax - x^2 - xy} \times \left(\dfrac{(a+x)^2 - y^2}{a^2 + ax - ay} \div \dfrac{(a+y)^2 - x^2}{a}\right)$

46. $\dfrac{a^2 - b^2}{ab^2 x} \times \dfrac{b(a-b)}{a^2 + 2ab + b^2} \times \dfrac{b(a+b)}{a^2 - 2ab + b^2}.$

47. $\left(\dfrac{a-x}{a+x} - \dfrac{a^3 - x^3}{a^3 + x^3}\right) \div \left(\dfrac{a+x}{a-x} + \dfrac{a^2 + x^2}{a^2 - x^2}\right).$

SIMPLE EQUATIONS.

169. REVIEW. — 1. Definition of an Equation.
2. Definition of Members of an Equation.
3. Definition of First Member; Second Member.
4. Definition of Clearing of Fractions. Rule.
5. Definition of Transposing. Rule.
6. Definition of an **Axiom**. State Axioms.
7. Definition of a Statement of a Problem.
8. Definition of a Solution of a Problem.

170. The **Degree** of an equation which has no unknown quantity in any denominator or under the radical sign, is determined from the highest number of factors of unknown quantities contained in any term.

Thus, $x + b = c$, $3ax + y = n$, $4b^2x + 3a^3x = a$, are equations of the *first* degree.

$x^2 + a = c$, $bx^2 + 3y = d$, $x + xy = 7$, $axy + 3y^2 = n$, are equations of the *second* degree.

$x^3 = a$, $x^2y = a$, $xy^2 = a$, $x + x^2 + x^3 = a$, are equations of the *third* degree.

171. A **Simple Equation** is an equation of the *first* degree.

172. A **Quadratic Equation** is an equation of the *second* degree.

173. A **Cubic Equation** is an equation of the *third* degree.

174. A **Numerical Equation** is an equation in which all the known quantities are expressed by figures.

175. A **Literal Equation** is an equation in which some or all of the known quantities are expressed by letters.

EXAMPLES.

Solve the following:

1. $4x - \dfrac{x+2}{2} = 3x + 3.$

2. $x - \dfrac{3x+4}{3} = \dfrac{x}{9} + \dfrac{x-12}{6}.$

3. $\dfrac{6x-8}{2} + 2 = x - \dfrac{5-2x}{4}.$

4. $x - 3 - \dfrac{x+2}{8} = \dfrac{x}{3}.$

5. $\dfrac{15x}{4} = 2\tfrac{1}{4} - \dfrac{3-x}{2}.$

6. $\dfrac{x}{3} - x = \dfrac{x-1}{11} - 9.$

7. $\dfrac{9x}{7} - \dfrac{x+3}{5} = 2x - 21.$

8. $\dfrac{ax-b}{c} + a = \dfrac{x+ac}{c}.$

9. $ax - \dfrac{3a-bx}{2} = \dfrac{1}{4}.$

10. $\dfrac{x}{a} - b = \dfrac{c}{d} - x.$

11. $\dfrac{3x-5}{2} - 12 = \dfrac{4-2x}{3} - x$

12. $\dfrac{x}{a-1} - \dfrac{x}{a+1} = b.$

13. $\dfrac{x^2 + 2ax + a^2}{x+a} = \dfrac{4ab}{16b}.$

14. $2 - 2x = \dfrac{x+8}{4} - \dfrac{x+6}{3}.$

15. $\dfrac{4x}{5} + \dfrac{3b}{2} = \dfrac{a}{6} + \dfrac{12b}{2}.$

16. $\dfrac{x}{a} - a = \dfrac{a}{c} - \dfrac{x}{c-a}.$

17. $\dfrac{2x-8}{4} + \dfrac{x}{3} = 30 - \dfrac{x+32}{2}$

18. $10 - \dfrac{3x+4}{3} = 2x - 3\tfrac{1}{3}.$

19. $4 + 10x + 5 - 6x\left(\dfrac{1}{x} - \dfrac{1}{3}\right) = 27.$

20. $x - \dfrac{x-2}{3} + \dfrac{x-4}{5} = 7 + \dfrac{x-5}{6}.$

21. $\dfrac{x}{3} - \dfrac{x^2-5x}{3x-7} = \dfrac{2}{3}.$

22. $\dfrac{3}{x+1} - \dfrac{x+1}{x-1} = \dfrac{x^2}{1-x^2}.$

23. $\dfrac{2x^4 + 2x^3 - 9x^2 + 12}{x^2 + 3x - 4} = 2x^2 - 4x - 3.$

SIMPLE EQUATIONS.

24. $\dfrac{2}{3} + \dfrac{3x-3}{4} - \dfrac{3x-4}{3} = 6 - \dfrac{27+4x}{9}.$

25. $\dfrac{9x+20}{36} - \dfrac{x}{4} = \dfrac{4x-12}{5x-4}.$

Suggestion. — The equation may be expressed as follows:

$$\dfrac{9x}{36} + \dfrac{20}{36} - \dfrac{x}{4} = \dfrac{4x-12}{5x-4}.$$

Simplifying, $\qquad \dfrac{5}{9} = \dfrac{4x-12}{5x-4}.$

26. $\dfrac{2x+8}{5} + \dfrac{x}{2} - 8 = \dfrac{x - \dfrac{4x-9}{3}}{6} - 8\tfrac{5}{6}.$

27. $\dfrac{a+b}{x-c} = \dfrac{a}{x-a} + \dfrac{b}{x-b}.$

28. $\dfrac{6x+13}{15} - \dfrac{9x+15}{5x-25} + 3 = \dfrac{2x+15}{5}.$

See suggestion, Example 25.

29. $\dfrac{x-a}{a-b} - \dfrac{x+a}{a+b} = \dfrac{2ax}{a^2-b^2}.$

30. $x + 3 + \dfrac{3(x+3)}{7} = \dfrac{3(x+3)}{2} - \dfrac{1}{2}.$

Suggestion. — Combine similar terms before clearing of fractions.

31. $x + 6 - \dfrac{3(x+6)}{4} = \dfrac{1}{3}(x+6) - 6.$

32. $x - 7 + \dfrac{x-7}{2} + \dfrac{3(x-7)}{4} = 2\tfrac{1}{4}.$

33. $\dfrac{3(x+4)}{2} + \dfrac{x+4}{4} - \dfrac{3(x+4)}{5} = 11\tfrac{1}{2}.$

34. $\dfrac{2(x-3)}{3} + 2(x-3) = 5 - \dfrac{x-3}{4}.$

132 HIGH SCHOOL ALGEBRA.

35. $\dfrac{21-3x}{3} - \dfrac{2(2x+3)}{9} = 6 - \dfrac{5x+1}{4}.$

36. $x - 4 - \dfrac{3x-5}{2} = 8 - \dfrac{2x-4}{3}.$

37. $\dfrac{4x+3}{9} + \dfrac{7x-29}{5x-12} = \dfrac{8x+19}{18}.$

See suggestion, Example 25.

38. $\dfrac{9x+5}{14} + \dfrac{8x-7}{6x+2} = \dfrac{36x+15}{56} + \dfrac{10\frac{1}{4}}{14}.$

39. $\dfrac{6x+7}{15} - \dfrac{2x-2}{7x-6} = \dfrac{2x+1}{5}.$

40. $\dfrac{6x+1}{15} - \dfrac{2x-4}{7x-16} = \dfrac{2x-1}{5}.$

41. $\dfrac{6x+13}{15} - \dfrac{3x+5}{5x-25} = \dfrac{2x}{5}.$

42. $\dfrac{x-12}{x-7} + \dfrac{x-4}{x-12} = 2 + \dfrac{7}{x-7}.$

43. $\dfrac{2x-10}{4} - \dfrac{x-8}{5} - \dfrac{x-5}{2} = x - 14.$

44. $\dfrac{2x-5}{3} - \dfrac{x-3}{4} = x - 2 - \dfrac{x-1}{2}.$

45. $\dfrac{x}{5} - \dfrac{2x-14}{3} = 4\frac{1}{3} - \dfrac{x}{2}.$

46. $\dfrac{3x}{4} - \dfrac{3x-11}{2} = 6x - \dfrac{20x+13}{4}.$

47. $\dfrac{7x+8}{3x-1} - 8 = \dfrac{27x-36}{3x-1} + 4.$

48. $\dfrac{3x-3}{4} - \dfrac{3x-4}{3} = \dfrac{16}{3} - \dfrac{27+4x}{9}.$

SIMPLE EQUATIONS.

49. $\dfrac{6x+18}{13} - \dfrac{11-3x}{36} = 5x - 43\tfrac{1}{6} - \dfrac{13-x}{12} - \dfrac{21-2x}{18}.$

50. $\dfrac{4x+3}{9} = \dfrac{8x+19}{18} - \dfrac{7x-29}{5x-12}.$

51. $\dfrac{ab+x}{b^2} - \dfrac{b^2-x}{a^2 b} = \dfrac{x-b}{a^2} - \dfrac{ab-x}{b^2}.$

52. $(a+x)(b+x) - a(b+c) = \dfrac{a^2 c}{b} + x^2.$

53. $\dfrac{4x-8}{8} + \dfrac{3x-15}{12} - \dfrac{2x-16}{16} = 0.$

54. $\dfrac{4x-16}{24} - \dfrac{2x+6}{60} + \dfrac{9x}{12} = x - \dfrac{x}{4} + \dfrac{5}{6}.$

55. $\dfrac{13-4x}{15} - \dfrac{3+4x}{3} + 6x - 8 = \dfrac{12x}{5} + \dfrac{58}{15}.$

56. $\dfrac{y+1}{y-1} - \dfrac{y-1}{y+1} = \dfrac{3}{y-1}.$

57. $\dfrac{3}{x-4} + \dfrac{7}{x+6} = \dfrac{80}{x^2+2x-24}.$

58. $\dfrac{3x+2}{3} - \dfrac{6x-3}{x+4} - \dfrac{3x}{5} = \dfrac{2x}{5} - \dfrac{101}{24}.$

59. $\dfrac{2x}{x+3} + \dfrac{x}{x-5} - 3 = \dfrac{1}{2x-10}.$

60. $\dfrac{x-1}{a} + \dfrac{x-1}{b} + \dfrac{x-1}{c} = 0.$

61. $a^2 y - 2ab - a^2 = b^2 y + b^2.$

62. $(x-a)^2 - (x-b)^2 = (a-b)^2.$

63. $\dfrac{a(b^2 - z^2)}{cz} + \dfrac{2z}{c} - cz = \dfrac{ab^2}{cz} - \dfrac{8a}{c}.$

64. $\dfrac{1}{m+n} - \dfrac{2mn}{(m+n)^3} - \dfrac{m}{(m+n)^2} = \dfrac{x-m}{(m+n)^2}.$

65. $\dfrac{3ax-2b}{3b} - \dfrac{ax-a}{2b} = \dfrac{ax}{b} - \dfrac{2}{3}.$

66. $\dfrac{x}{a-b} - \dfrac{2+x}{a+b} = \dfrac{c}{a^2-b^2} + \dfrac{d}{a-b}.$

67. $\dfrac{x-ab}{3} - \dfrac{5b^2}{6} + \dfrac{x-b^2}{4} = \dfrac{7a^2}{12} + \dfrac{11ab}{6} - \dfrac{x-a^2}{2}.$

PROBLEMS.

176. Directions for Solving. — *Represent one of the unknown quantities by x, and from the conditions of the problem find an expression for each of the other quantities given.*

Find from the problem two expressions that are equal, and express them as an equation.

Solve the equation.

68. When the half of a certain number is added to the number, the sum is as much more than 60 as the number is less than 65. What is the number?

69. The difference between two numbers is 8, and the quotient arising from dividing the greater by the less is 3. What are the numbers?

70. A man left one half of his property to his wife, one sixth to his children, a twelfth to his brother, and the rest, which was $600, to charitable purposes. How much property had he?

71. Find two numbers whose sum is 70, such that the first, divided by the second, gives a quotient of 2 and a remainder of 1.

72. Out of a cask of wine, one fifth part had leaked away. Afterward, 10 gallons were drawn out, when the cask was found to be two thirds full. How much did it hold?

73. A can do a piece of work in 5 days, and B can do the same work in 6 days. How long will it take both working together to do it?

Let $x =$ the number of days it will take both to do it.

Then, $\frac{1}{x} =$ the part both can do in a day.

$\frac{1}{5} =$ the part of the work which A can do in a day.

$\frac{1}{6} =$ the part of the work which B can do in a day.

Therefore, $\frac{1}{5} + \frac{1}{6} = \frac{1}{x}$

And $x = 2\frac{8}{11}$

74. A can do a piece of work in 9 days, and B can do the same in 10 days. How long will it take both to do it?

75. A can do a piece of work in 5 days, B in 7 days, and C in 9 days. In how many days can they all together do it?

76. Two pipes empty into a cistern. One can fill it in 8 hours, and the other in 9 hours. How soon will it be filled, if both empty into it at the same time?

77. A cistern can be filled by a pipe in 3 hours, and emptied by another pipe in 4 hours. How much time will be required to fill the cistern if both are running?

78. A fish was caught whose tail weighed 9 pounds. His head weighed as much as his tail and half his body, and his body weighed as much as his head and tail. How much did the fish weigh?

79. Of a detachment of soldiers, $\frac{2}{3}$ are on duty, $\frac{1}{8}$ of them sick, $\frac{1}{5}$ of the remainder absent on leave, and the rest, 380, have deserted. How many were there in the detachment?

80. A person spends one fourth of his annual income for his board, one third for clothes, one twelfth for other expenses, and saves $500. What is his income?

Fractions may be avoided in this and similar examples, by letting x, with a coëfficient which is a multiple of the denominators, represent the number sought. Thus, in the above example, let $12x$ represent the annual income.

Let $\qquad 12\,x =$ his annual income.

Then, $\qquad 3\,x =$ what he paid for board.

$\qquad 4\,x =$ what he paid for clothes.

$\qquad x =$ what he paid for his other expenses.

Therefore, $3\,x + 4\,x + x + 500 = 12\,x$

$$4\,x = 500$$
$$x = 125$$
$$12\,x = 1500, \text{ his income.}$$

81. A farm of 392 acres was divided among four heirs, so that A had four fifths as much as B, C as much as A and B, and D one half as much as A and C. What was the share of each?

82. A farmer wishes to mix 300 bushels of provender, containing rye, corn, and oats, so that the mixture may contain $\frac{2}{3}$ as much oats as corn, and $\frac{1}{2}$ as much rye as oats. How many bushels of each should he use?

83. Into what two parts can the number 204 be divided, such that $\frac{2}{5}$ of the greater being taken from the less, the remainder will be equal to $\frac{3}{7}$ of the less subtracted from the greater?

84. A man spent $14 more than $\frac{3}{7}$ of his money, and had $6 more than $\frac{1}{3}$ of it left. How much had he at first?

85. A merchant lost $\frac{1}{5}$ of his capital during the first year. The second year he gained $\frac{3}{8}$ as much as he had left at the end of the first. The third year he gained $\frac{3}{11}$ of what he had at the close of the second, making his capital $7000. What was his original capital?

86. An officer wished to arrange his men in a solid square. He found by his first arrangement that he had 39 men over. He then increased the number on a side by 1 man, and found he needed 50 men to complete the square. How many men had he?

Let $\quad x =$ the number of men in each side in the first arrangement.

Then, $\quad x^2 =$ the number of men in the first square.

Then, $x + 1 =$ the number of men in each side in the second arrangement.

PROBLEMS.

And $(x+1)^2 =$ the number of men in the second square.

$x^2 + 39 =$ the entire number of men.

$(x+1)^2 - 50 =$ the entire number of men.

Therefore, $(x+1)^2 - 50 = x^2 + 39$

$x^2 + 2x + 1 - 50 = x^2 + 39$

$x = 44$

$x^2 + 39 = 1975$

87. A regiment of troops was drawn up in a solid square with a certain number on a side, when it was found that there were 295 men left. Upon arranging them so that each side contained 5 men more, it was found that there were none left. How many men were there in the regiment?

88. A colonel, upon attempting to draw up his troops in the form of a solid square, found that he had 31 men over. If he had increased the side of the square by 1 man there would have been a deficiency of 24 men. How many men were there in the regiment?

89. A person in purchasing sugar found that if he bought sugar at 11 cents he would lack 30 cents of having money enough to pay for it; so he bought sugar at $10\frac{1}{2}$ cents, and had 15 cents left. How many pounds did he buy?

90. Into what two parts may the number 56 be divided, so that one may be to the other as 3 to 4?

Since one number is to the other as 3 to 4, one is $\frac{3}{4}$ of the other. Therefore, to avoid fractions,

Let $\qquad 4x =$ one part.

Then, $\qquad 3x =$ the other part.

Therefore, $\qquad 4x + 3x = 56$

$x = 8$

$4x = 32$, one part.

$3x = 24$, the other part.

91. Find two numbers which are to each other as 5 to 7, and whose sum is 72.

92. A's age is to B's as 3 to 8, and the sum of their ages is 44 years. How old is each?

93. An estate of $15,000 was divided between two sons, so that the elder's share was to the younger's as 8 to 7. What was the share of each?

94. A sum of money was divided between A and B, so that the share of A was to that of B as 5 to 3. The share of A also exceeded $\frac{5}{9}$ of the whole sum by $50. What was the share of each?

95. A and B started out together with equal sums of money. B paid A a debt of $20, but afterwards A made a purchase of B which cost him half of all he then had, when he found that he had just half as much as B. How much had each at first?

96. A lady distributed $252 among some poor people, giving to the men $12 each, the women $6 each, and the children $3 each. The number of women was 2 less than twice the number of men, and the number of children was 4 less than 3 times the number of women. To how many persons did she give the money?

97. I bought a number of apples at the rate of 5 for 2 cents. I sold half of them at 2 for a cent, and the remainder at 3 for a cent, gaining 1 cent. How many did I buy?

98. A merchant engaged in business with a certain capital. His gain the first year lacked $1000 of being as much as his original capital. His gain the second year lacked $1000 of being as much as he had at the end of the first year, and the third year his gain lacked $1000 of being as much as he had at the end of the second year. He found that at the end of the third year his capital was 3 times his original capital. What was his original capital?

99. A and B began business with equal capital. The first year A gained a sum equal to $\frac{1}{3}$ of his capital, and B lost $\frac{1}{4}$ of his. The second year A lost $72 and B gained $36, when it was found that B's capital was $\frac{3}{4}$ of A's. What was the original capital of each?

PROBLEMS.

100. A cistern, which held 648 gallons of water, was filled in 18 minutes by two pipes, one of which conveyed 6 gallons more per minute than the other. How much did each convey per minute?

101. A farmer had 90 sheep in four fields. If the number in the first be increased by 2, the number in the second diminished by 2, the number in the third multiplied by 2, and the number in the fourth divided by 2, the results will be equal. How many were there in each flock?

102. A gentleman who had $10,000, used a portion of it in building a house, and put the rest out at interest for one year: $\frac{1}{3}$ of it at 6% and $\frac{2}{3}$ of it at 5%. The income from both investments was $320. What was the cost of the house?

103. Paving a square court with stone at 40 cents a square yard will cost as much as inclosing it with a fence at a dollar per yard. What is the length of a side of the court?

104. Two soldiers start together for a fort. One, who travels 12 miles per day, after traveling 9 days, turns back as far as the other had traveled during those 9 days. He then turns and pursues his way toward the fort, where both arrive together 18 days from the time they set out. At what rate did the other travel?

105. A boy bought a certain number of apples at the rate of 4 for 5 cents, and sold them at the rate of 3 for 4 cents. He gained 60 cents. How many did he buy?

106. A gentleman left $315 to be divided among four servants, as follows: B was to receive as much as A and $\frac{1}{2}$ as much more; C was to receive as much as A and B and $\frac{1}{3}$ as much more; D was to receive as much as the other three and $\frac{1}{4}$ as much more. What was the share of each?

107. Two numbers are to each other as 2 to 3; but if 50 be subtracted from each, one will be $\frac{1}{2}$ the other. What are the numbers?

108. A woman sold eggs and apples. The eggs were worth 5 cents a dozen more than the apples; and 8 dozen eggs were worth as much as $13\frac{5}{7}$ dozen apples. What was the price of each per dozen?

109. Three men, A, B, and C, build 318 rods of wall. A builds 7 rods per day, B 6 rods, and C 5 rods. B works twice as many days as A, and C works $\frac{1}{2}$ as many days as both A and B. How many days does each work?

110. A gentleman has two horses, and a carriage worth $150. The value of the poorer horse and carriage is twice the value of the better horse; and the value of the better horse and carriage is three times the value of the poorer horse. What is the value of each horse?

111. A man bought two pieces of cloth, one of which lacked 12 yards of being 4 times as long as the other. The longer cost $5 per yard, and the shorter $4 per yard. Twenty-three yards being cut off from the longer, and 5 from the shorter, and each remainder being sold for a dollar a yard more than it cost, he received $142. How many yards of each were there?

112. When, after 2 o'clock, will the hour and minute hands of a clock be together?

Let $x =$ the number of minute-spaces that the minute hand travels before they come together.

Then, $\dfrac{x}{12} =$ the number of minute-spaces that the hour hand travels.

Then, since they were 10 minute-spaces apart at 2 o'clock,

$$x - \frac{x}{12} = 10$$

$$\frac{11x}{12} = 10$$

$$11x = 120$$

$x = 10\frac{10}{11}$, the number of minutes after 2.

113. When, after 5 o'clock, will the hour and minute hands of a clock be together?

PROBLEMS.

114. When, after 8 o'clock, will the hour and minute hands of a clock be together?

115. When, after 4 o'clock, will the hour and minute hands of a clock make a straight line?

116. When, after 5 o'clock, will the hour and minute hands of a clock make a straight line?

117. When, first, after 6 o'clock, will the hour and minute hands of a clock be 15 minute-spaces apart?

118. When, after half-past 8 o'clock, will the hour and minute hands of a clock be 15 minute-spaces apart?

119. A man has two horses worth together $460, and a carriage worth $210. When he drives the better horse the outfit is worth three times the other horse, increased by $\frac{1}{10}$ of the value of the first horse. Find the value of each horse.

120. A manufacturer hired two skilled mechanics and a common laborer. The first mechanic earns twice and the second three times as much per day as the laborer. The laborer worked ten days, the first mechanic seven days, and the second mechanic four days, and they all earned $72. Find the daily wages of each.

121. A boy started from home on his bicycle at 7 a.m., going at the rate of 8 miles an hour. After riding a certain distance, the machine broke down, and he was compelled to return home afoot. When he reached home he found it was 6.30 p.m. How far did he go, if he walked back at the rate of $3\frac{1}{2}$ mi. an hour?

122. The sum of two numbers is 80, and their difference is 6. Find the numbers.

123. Said James to Isaac, "We had just the same amount of money when we left home, but you gave me $60, and I gave you $10, and now I have three times as much as you." How much had each at first?

124. The sum of three numbers is 230. The second is $\frac{2}{3}$ of the remainder left after decreasing the first by 20, and the third is 50 less than twice the first. What are the numbers?

125. A carpenter received $3.50 a day for his labor, and paid $1.00 a day for his board. At the end of 24 days he had saved $39. How many days did he work?

126. A certain lot is twice as long as it is wide. If its length were increased 1 rod, and its width decreased 1 rod, the area would be decreased 6 sq. rds. Find the dimensions of the lot.

127. A man going from a certain place traveled at the rate of 5 miles an hour. After he had been gone 6 hours, a horseman, going at the rate of 8 miles an hour, was sent after him. How far did the latter travel before overtaking the former?

128. A man has four times as many dollar pieces as he has dimes, and the worth of all is $28.70. How many pieces of each has he?

129. A young man spends one fifth of the money he has in bank each year, and adds to it an annuity of $4000. How much had he in bank at first, if at the beginning of the fourth year he has $13,600 to his credit?

130. A messenger going at the rate of $3\frac{1}{2}$ miles an hour was sent with secret messages from a king to his army 70 miles away. After he had been gone 6 hours, a second person going at the rate of 5 miles an hour was sent to countermand the orders; 7 hours later a fleet horseman going at the rate of 10 miles an hour was sent with new and special orders. When did the horseman overtake each of the footmen?

131. A boy spent $\frac{1}{2}$ of his money and $\frac{1}{2}$ a cent more, then $\frac{1}{2}$ of the remainder and $\frac{1}{2}$ a cent more, and finally $\frac{1}{2}$ of what he had left and $\frac{1}{2}$ a cent more, when he found he had two cents remaining. How much had he at first?

132. After paying out $\dfrac{1}{m}$ and $\dfrac{1}{n}$ of my money, I had b dollars left. How much had I at first?

PROBLEMS.

Let $x =$ the amount I had at first.

Then, $\dfrac{x}{m} + \dfrac{x}{n} =$ the amount I spent.

Therefore, $\dfrac{x}{m} + \dfrac{x}{n} + b = x$

$$mx + nx + mnb = mnx$$
$$mnx - mx - nx = mnb$$
$$(mn - m - n)x = mnb$$
$$x = \dfrac{mnb}{mn - m - n}$$

177. A problem in which *literal notation* is used, is called a General Problem.

Such problems give an infinite number of numerical results, by assigning different numerical values to the literal quantities.

Thus, in problem 132, when $m = 4$, $n = 5$, and $b = 66$, the value of x is 120; when $m = 5$, $n = 8$, and $b = 54$, the value of x is 80.

133. A horse and saddle are worth m dollars, and the horse is worth n times as much as the saddle. What is the value of each when $m = 200$ and $n = 9$?

134. A man gave two servants b dollars, giving A a times as much as B. How much did he give to each? How much did he give to each if $b = 75$ and $a = 4$?

135. Divide the number b into two such parts that one shall be a times the other. What will be the result when $b = 24$ and $a = 7$?

136. If A can do a piece of work in n days and B in m days, in what time can both do it working together? What will be the result when n is 5 and m is 7? What, when n is 10 and m is 8?

137. A pleasure party of a persons hired a coach. If there had been b persons more, it would have cost each d dollars less than it did. How much did each one pay? What is the result when a is 8, b is 4, and d, \$1?

138. A certain number divided by b gives a result such that the sum of the dividend, divisor, and quotient is c. What **is the number? What is the number when b is 16 and c is 84?**

SIMULTANEOUS EQUATIONS.

TWO UNKNOWN QUANTITIES.

178. 1. If the sum of two numbers is 12, what are the numbers? How many answers may be given to the question?

2. Let x and y stand for the two numbers, then in the equation $x + y = 12$, how many values has x? How many has y?

3. How many values have the unknown quantities in a single equation containing two unknown quantities? What name may be given to such equations?

4. If the equations $x + y = 6$ and $x - y = 2$, are added together (Ax. 2), what is the resulting equation? What is the value of x? Of y?

5. How many equations containing two unknown quantities are needed to determine the values of the quantities?

6. If the equation $x + y = 6$ is subtracted from $2x + 2y = 12$, what is the resulting equation? How many values has each quantity?

7. Since the equation $2x + 2y = 12$ was derived from $x + y = 6$ by multiplying the equation by 2, what may it be called?

8. Since the values of two unknown quantities cannot be determined from two equations when one of them is a *derived* equation, what kind of equations must be given to determine the values of the quantities?

SIMULTANEOUS EQUATIONS.

179. Simultaneous Equations are those in which the same unknown quantity has the same value in every equation.

Thus, $\begin{Bmatrix} x+y=12 \\ x-y=2 \end{Bmatrix}$ are simultaneous equations in which $x=7$ and $y=5$.

180. Derived Equations are those which are obtained by combining other equations or performing some operation upon them.

Thus, $2x + 2y = 8$, is an equation *derived* from $x + y = 4$, and $2x + 3y = 7$, is derived by adding $x + y = 3$ and $x + 2y = 4$.

181. Independent Equations are such as cannot be derived from one another or reduced to the same form.

Thus, $2x + y = 5$ and $x + 2y = 6$, are independent equations.

182. An Indeterminate Equation is one in which the unknown quantities may have an infinite number of values.

Thus, $x + y = 12$, is an indeterminate equation, because each of the unknown quantities may have an infinite number of values.

183. Principles. — 1. *Every single equation containing two unknown quantities is indeterminate. Consequently,*

2. *In order to solve equations containing two unknown quantities, two independent equations, involving one or both of the quantities, must be given.*

184. Elimination is the process of deducing from simultaneous equations, equations containing a less number of unknown quantities than is found in the given equations.

185. Elimination by Addition or Subtraction.

1. When $x + 2y = 10$ and $x - 2y = 6$, how may y be eliminated?

2. When $3x + 4y = 16$ and $5x - 4y = 16$, how may y be eliminated?

3. When may a quantity be eliminated by addition?

ALGEBRA. — 10.

4. When $x + 2y = 6$ and $x + y = 4$, how may x be eliminated?

5. When may a quantity be eliminated by subtraction?

186. Principle. — *Quantities may be eliminated by addition or by subtraction when they have the same coëfficients.*

EXAMPLES.

1. Find the value of x and y in the equations $2x + 3y = 13$ and $3x + 2y = 12$.

PROCESS.

$2x + 3y = 13 \quad (1)$
$3x + 2y = 12 \quad (2)$
$6x + 9y = 39 \quad (3)$
$6x + 4y = 24 \quad (4)$
$5y = 15 \quad (5)$
$y = 3 \quad (6)$
$2x + 9 = 13 \quad (7)$
$2x = 4 \quad (8)$
$x = 2 \quad (9)$

EXPLANATION. — Since the quantities in the given equations have not the same coëfficients, the first equation must be multiplied by 3 and the second by 2. Equations (3) and (4) are thus produced in which the coëfficients of x are alike. Since the coëfficients of x are alike, and they have the same sign, x may be eliminated by subtraction (Prin.). Subtracting (4) from (3), we obtain (5). Dividing equation (5) by the coëfficient of y, (6) is obtained.

Substituting the value of y in equation (1), the resulting equation is (7). Transposing and uniting, the value of $x = 2$.

Rule. — *If necessary, multiply or divide the members of one or both equations so that one unknown quantity may have the same coëfficient in each equation.*

When the signs of the equal coëfficients are the same, subtract the members of one equation from those of the other; when the signs are unlike, add the members.

Find the values of the unknown quantities in the following:

2. $\begin{cases} x + 2y = 7. \\ x + y = 5. \end{cases}$

3. $\begin{cases} 4x + 3y = 7. \\ 2x - 3y = -1. \end{cases}$

4. $\begin{cases} 4x - 5y = 3. \\ 3x + 5y = 11. \end{cases}$

5. $\begin{cases} 2x + 6y = 10. \\ 3x + 2y = 8. \end{cases}$

6. $\begin{cases} 8x + 3y = 22. \\ 4x + 5y = 18. \end{cases}$

7. $\begin{cases} 3x + 4y = 25. \\ 4x + 3y = 21. \end{cases}$

SIMULTANEOUS EQUATIONS. 147

8. $\begin{cases} 5x + 6y = 61. \\ 4x + 5y = 50. \end{cases}$

9. $\begin{cases} 4x + 3y = 32. \\ 7x - 6y = 11. \end{cases}$

10. $\begin{cases} 5x + 6y = 40. \\ 8x - 4y = 4. \end{cases}$

11. $\begin{cases} 3x + 6y = 39. \\ 5x - 3y = 13. \end{cases}$

12. $\begin{cases} \dfrac{x}{2} + \dfrac{y}{3} = 3. \\ \dfrac{x}{5} + \dfrac{y}{2} = \dfrac{23}{10}. \end{cases}$

13. $\begin{cases} \dfrac{x}{6} - \dfrac{y}{3} = -\dfrac{1}{3}. \\ \dfrac{2x}{3} - \dfrac{3y}{4} = 1. \end{cases}$

14. $\begin{cases} \dfrac{5x}{6} + \dfrac{2y}{5} = 14. \\ \dfrac{3x}{4} - \dfrac{2y}{5} = 5. \end{cases}$

15. $\begin{cases} \dfrac{3x}{4} + \dfrac{2y}{3} = 12. \\ \dfrac{3x}{5} - \dfrac{y}{2} = \dfrac{3}{10}. \end{cases}$

16. $\begin{cases} \dfrac{4x}{5} + \dfrac{2y}{3} = 6. \\ \dfrac{2x}{3} + \dfrac{3y}{4} = 5\tfrac{7}{12}. \end{cases}$

17. $\begin{cases} \dfrac{3x}{5} + \dfrac{7y}{4} = 10. \\ \dfrac{2x}{7} - \dfrac{y}{5} = \dfrac{22}{35}. \end{cases}$

18. $\begin{cases} \dfrac{2x}{5} + \dfrac{3y}{4} = 13\tfrac{1}{2}. \\ \dfrac{5x}{6} + \dfrac{3y}{5} = 13\tfrac{1}{2}. \end{cases}$

19. $\begin{cases} \dfrac{3x}{4} + \dfrac{5y}{8} = 23\tfrac{1}{4}. \\ \dfrac{7x}{12} - \dfrac{2y}{9} = 5\tfrac{1}{3}. \end{cases}$

187. Elimination by Comparison.

1. If, in the simultaneous equations $x + 2y = 8$ and $x - y = 5$, $2y$ in the first and y in the second are transposed to the second member, what will be the resulting equations?

2. Since each of the second members of these derived equations is equal to x, how do they compare with each other?

3. If these second members are formed into an equation, how many unknown quantities will it contain?

4. How may an unknown quantity be eliminated from two simultaneous equations by *comparison?*

EXAMPLES.

1. Find the value of x and of y in the equations $x + 2y = 8$ and $3x + 2y = 12$.

PROCESS.

$$x + 2y = 8 \quad (1)$$
$$3x + 2y = 12 \quad (2)$$
$$x = 8 - 2y \quad (3)$$
$$x = \frac{12 - 2y}{3} \quad (4)$$
$$\frac{12 - 2y}{3} = 8 - 2y \quad (5)$$
$$12 - 2y = 24 - 6y \quad (6)$$
$$4y = 12 \quad (7)$$
$$y = 3 \quad (8)$$
$$x + 6 = 8 \quad (9)$$
$$x = 2 \quad (10)$$

EXPLANATION. — Since, in elimination by comparison, the value of the same unknown quantity in each equation is to be found, and a new equation is to be formed from these values, $2y$ in equation (1) is transposed, giving (3). Transposing $2y$ in (2) and dividing by 3, equation (4) is obtained. Since these two values of x are equal, equation (5) is obtained (Ax. 1). Clearing of fractions, we obtain (6). Transposing and uniting, (7) is obtained. Dividing by 4, we obtain (8). Substituting this value of y in equation (1), we obtain (9). Uniting, we obtain (10).

RULE. — *Find an expression for the value of the same unknown quantity in each equation.*

Make an equation of these values and solve it.

Solve the following equations by comparison:

2. $\begin{cases} 3x + y = 9. \\ x + 2y = 8. \end{cases}$

3. $\begin{cases} 2x - y = 3. \\ x + 3y = 19. \end{cases}$

4. $\begin{cases} 4x + 2y = 26. \\ 3x + 4y = 39. \end{cases}$

5. $\begin{cases} 2x - 3y = -14. \\ 3x + 2y = 44. \end{cases}$

6. $\begin{cases} 3x + 4y = 18. \\ x + 2y = 8. \end{cases}$

7. $\begin{cases} x + 6y = 13. \\ 5x + 2y = 9. \end{cases}$

8. $\begin{cases} 4x + 2y = 26. \\ 3x - 4y = 3. \end{cases}$

9. $\begin{cases} 2x - 3y = -7. \\ 4x - 5y = -9. \end{cases}$

10. $\begin{cases} 3x + 2y = 33. \\ 9x - 4y = 9. \end{cases}$

11. $\begin{cases} 6x + y = 45. \\ 3x - 2y = 15. \end{cases}$

SIMULTANEOUS EQUATIONS. 149

12. $\begin{cases} 4x - 5y = -34. \\ 2x - 3y = -22. \end{cases}$

13. $\begin{cases} x + 4y = 11. \\ 5x - 2y = 11. \end{cases}$

14. $\begin{cases} 2x - 3y = 3. \\ 4x + 5y = 39. \end{cases}$

15. $\begin{cases} 7x - 4y = 81. \\ 5x - 3y = 57. \end{cases}$

16. $\begin{cases} \dfrac{x}{2} + \dfrac{y}{3} = 5. \\ \dfrac{x}{3} + \dfrac{y}{2} = 5. \end{cases}$

17. $\begin{cases} \dfrac{3x}{5} + \dfrac{2y}{3} = 17. \\ \dfrac{2x}{3} + \dfrac{3y}{4} = 19. \end{cases}$

18. $\begin{cases} \dfrac{2x}{7} + \dfrac{2y}{3} = 5\frac{1}{3}. \\ \dfrac{3x}{5} + \dfrac{3y}{5} = 7\frac{1}{5}. \end{cases}$

19. $\begin{cases} \dfrac{3x}{5} - \dfrac{y}{4} = 4\frac{3}{4}. \\ \dfrac{2x}{7} - \dfrac{2y}{5} = 3\frac{5}{14}. \end{cases}$

20. $\begin{cases} \dfrac{4x}{5} + \dfrac{2y}{3} = 24\frac{4}{5}. \\ \dfrac{3x}{7} - \dfrac{5y}{6} = -9. \end{cases}$

21. $\begin{cases} \dfrac{5x}{6} + \dfrac{4y}{5} = 28\frac{1}{2}. \\ \dfrac{8x}{9} + \dfrac{5y}{8} = 25\frac{5}{6}. \end{cases}$

22. $\begin{cases} \dfrac{4x}{5} - \dfrac{y}{7} = 7\frac{3}{5}. \\ \dfrac{7x}{12} + \dfrac{2y}{3} = 16\frac{1}{3}. \end{cases}$

188. Elimination by Substitution.

1. In an equation containing two unknown quantities, if the value of one quantity is known, how may the value of the other be found?

2. Express the value of x in the first of the simultaneous equations $x + y = 5$ and $x + 2y = 7$ by transposing y to the second member.

3. When this value of x is substituted in the second equation, how many unknown quantities does the resulting equation contain?

4. How may an unknown quantity be eliminated from simultaneous equations by *substitution?*

150 HIGH SCHOOL ALGEBRA.

EXAMPLES.

1. Find the value of x and of y in the equations $3x + 2y = 12$ and $2x + 3y = 13$.

PROCESS.

$3x + 2y = 12 \quad (1)$
$2x + 3y = 13 \quad (2)$
$x = \dfrac{12 - 2y}{3} \quad (3)$
$\dfrac{24 - 4y}{3} + 3y = 13 \quad (4)$
$24 - 4y + 9y = 39 \quad (5)$
$5y = 15 \quad (6)$
$y = 3 \quad (7)$
$x = \dfrac{12 - 6}{3} = 2 \quad (8)$

EXPLANATION. — Since one unknown quantity can be eliminated by finding its value in one of the given equations and substituting this value in another, we find the value of x from (1) and obtain (3). Substituting this value in (2), (4) is obtained. Clearing of fractions, the resulting equation is (5). Uniting terms, we obtain (6). Dividing, $y = 3$. Substituting this value in (3), $x = 2$.

RULE. — *Find an expression for the value of one of the unknown quantities in one of the equations.*

Substitute this value for the same unknown quantity in the other equation, and solve the equation.

Solve the following by substitution:

2. $\begin{cases} x + 2y = 10. \\ 2x - 3y = -1. \end{cases}$

3. $\begin{cases} 3x - 2y = 1. \\ x + 4y = 19. \end{cases}$

4. $\begin{cases} x - 2y = 6. \\ 2x - y = 27. \end{cases}$

5. $\begin{cases} 9x - y = 6. \\ x + y = 4. \end{cases}$

6. $\begin{cases} 3x + 5y = 2. \\ 6x + 5y = 3. \end{cases}$

7. $\begin{cases} 7x - 5y = 13. \\ 3x + 3y = 21. \end{cases}$

8. $\begin{cases} 6x + y = 60. \\ 3x + 2y = 39. \end{cases}$

9. $\begin{cases} 2x + 5y = 29. \\ 2x - 5y = -21. \end{cases}$

10. $\begin{cases} \dfrac{x}{5} + \dfrac{y}{6} = 18. \\ \dfrac{x}{2} - \dfrac{y}{4} = 21. \end{cases}$

11. $\begin{cases} x + 5y = 41. \\ 3x - 2y = 21. \end{cases}$

SIMULTANEOUS EQUATIONS.

12. $\begin{cases} \dfrac{x}{2}+\dfrac{y}{3}=7. \\ \dfrac{x}{3}+\dfrac{y}{4}=5. \end{cases}$

14. $\begin{cases} \dfrac{x+y}{3}=5. \\ \dfrac{x-y}{2}=2\tfrac{1}{2}. \end{cases}$

13. $\begin{cases} \dfrac{x}{3}+\dfrac{z}{4}=8. \\ x-z=-3. \end{cases}$

15. $\begin{cases} \dfrac{x}{7}+7y=251. \\ \dfrac{y}{7}+7x=299. \end{cases}$

Solve the following by any method:

16. $\begin{cases} \dfrac{1}{x}+\dfrac{2}{y}=10. \\ \dfrac{4}{x}+\dfrac{3}{y}=20. \end{cases}$

21. $\begin{cases} \dfrac{2x-y}{7}+3x=2y-6. \\ \dfrac{y+3}{5}+\dfrac{y-x}{6}=2x-8. \end{cases}$

17. $\begin{cases} \dfrac{7}{x}+\dfrac{5}{y}=19. \\ \dfrac{8}{x}-\dfrac{3}{y}=7. \end{cases}$

22. $\begin{cases} \dfrac{x-2}{5}-\dfrac{10-x}{3}=\dfrac{y-10}{4}. \\ \dfrac{2y+4}{3}=\dfrac{4x+y+13}{8}. \end{cases}$

18. $\begin{cases} \dfrac{5}{3x}+\dfrac{2}{5y}=7. \\ \dfrac{7}{6x}-\dfrac{1}{10y}=3. \end{cases}$

23. $\begin{cases} \dfrac{x+1}{y-1}-\dfrac{x-1}{y}=\dfrac{6}{7}. \\ x-y=1. \end{cases}$

19. $\begin{cases} \dfrac{a}{x}+\dfrac{b}{y}=m. \\ \dfrac{a}{x}-\dfrac{b}{y}=n. \end{cases}$

24. $\begin{cases} \dfrac{1-3x}{7}+\dfrac{3y-1}{5}=2. \\ \dfrac{3x+y}{11}+y=9. \end{cases}$

20. $\begin{cases} \dfrac{x}{4}+8=\dfrac{y}{2}-12. \\ \dfrac{x+y}{5}+\dfrac{y}{3}=\dfrac{2x-y}{4}+35. \end{cases}$

25. $\begin{cases} 4x+y=11. \\ \dfrac{y}{5x}=\dfrac{7x-y}{3x}-\dfrac{23}{15}. \end{cases}$

26. $\begin{cases} \dfrac{x}{a} + \dfrac{y}{b} = 2. \\ bx - ay = 0. \end{cases}$

27. $\begin{cases} \dfrac{x}{c} + \dfrac{y}{c} = 1. \\ \dfrac{ax - by}{a - b} = c. \end{cases}$

28. $\begin{cases} \dfrac{a}{b+y} = \dfrac{b}{3a+x}. \\ ax + 2by = d. \end{cases}$

29. $\begin{cases} \dfrac{a}{x} + \dfrac{b}{y} = c. \\ \dfrac{m}{x} + \dfrac{n}{y} = e. \end{cases}$

30. $\begin{cases} ax + by = r. \\ ax + cy = s. \end{cases}$

31. $\begin{cases} \dfrac{x}{m} + \dfrac{y}{n} = 2. \\ \dfrac{x}{m} - \dfrac{y}{n} = 1. \end{cases}$

32. $\begin{cases} abx + cdy = 2. \\ ax - cy = \dfrac{d-b}{bd}. \end{cases}$

33. $\begin{cases} ax - dy = c. \\ mx - ny = c. \end{cases}$

34. $\begin{cases} \dfrac{x}{a} + \dfrac{y}{b} = 2ab. \\ \dfrac{x}{ab} + \dfrac{y}{ab} = a + b. \end{cases}$

189. Fractional simultaneous equations in which the unknown quantities are found in the denominators may often be solved without clearing of fractions.

1. Solve the equations $\begin{cases} \dfrac{3}{x} + \dfrac{2}{y} = \dfrac{7}{4} & (1) \\ \dfrac{5}{x} + \dfrac{7}{y} = \dfrac{19}{4} & (2) \end{cases}$

Solution.

Multiplying (1) by 5, $\qquad \dfrac{15}{x} + \dfrac{10}{y} = \dfrac{35}{4}$ (3)

Multiplying (2) by 3, $\qquad \dfrac{15}{x} + \dfrac{21}{y} = \dfrac{57}{4}$ (4)

Subtracting (3) from (4), $\qquad \dfrac{11}{y} = \dfrac{22}{4}$ (5)

Dividing (5) by 11, $\qquad \dfrac{1}{y} = \dfrac{2}{4}$ or $\dfrac{1}{2}$

$\qquad\qquad\qquad\qquad\quad \therefore y = 2$

Substituting in (1), $\qquad x = 4$

SIMULTANEOUS EQUATIONS.

2. $\begin{cases} \dfrac{2}{x}+\dfrac{4}{y}=14. \\ \dfrac{6}{x}-\dfrac{2}{y}=14. \end{cases}$

3. $\begin{cases} \dfrac{1}{x}+\dfrac{3}{y}=11. \\ \dfrac{5}{x}+\dfrac{4}{y}=22. \end{cases}$

4. $\begin{cases} \dfrac{2}{x}+\dfrac{1}{y}=\dfrac{4}{3}. \\ \dfrac{3}{x}+\dfrac{5}{y}=\dfrac{19}{6}. \end{cases}$

5. $\begin{cases} \dfrac{1}{x}+\dfrac{3}{y}=a. \\ \dfrac{5}{x}+\dfrac{2}{y}=b. \end{cases}$

6. $\begin{cases} \dfrac{8}{x}+\dfrac{6}{y}=3. \\ \dfrac{6}{x}+\dfrac{15}{y}=4. \end{cases}$

7. $\begin{cases} \dfrac{5}{x}+\dfrac{6}{y}=7. \\ \dfrac{7}{x}+\dfrac{9}{y}=10. \end{cases}$

8. $\begin{cases} \dfrac{9}{x}+\dfrac{8}{y}=\dfrac{43}{6}. \\ \dfrac{3}{x}+\dfrac{10}{y}=\dfrac{29}{6}. \end{cases}$

9. $\begin{cases} \dfrac{a}{x}+\dfrac{b}{y}=c. \\ \dfrac{b}{x}+\dfrac{a}{y}=d. \end{cases}$

10. $\begin{cases} \dfrac{6}{x}+\dfrac{8}{y}=1. \\ \dfrac{7}{x}-\dfrac{11}{y}=-9. \end{cases}$

11. $\begin{cases} \dfrac{4}{2x}+\dfrac{6}{3y}=14. \\ \dfrac{3}{3x}+\dfrac{10}{5y}=11. \end{cases}$

12. $\begin{cases} \dfrac{3}{x+3}+\dfrac{6}{y-2}=7. \\ \dfrac{5}{x+3}+\dfrac{4}{y-2}=\dfrac{17}{3}. \end{cases}$

13. $\begin{cases} \dfrac{x}{a}+\dfrac{y}{b}=p. \\ \dfrac{x}{b}+\dfrac{y}{a}=q. \end{cases}$

PROBLEMS.

190. 1. If 7 lb. of tea and 5 lb. of coffee cost $5.50, and 6 lb. of tea and 3 lb. of coffee cost $4.20, what was the price per pound of each?

HIGH SCHOOL ALGEBRA.

Solution.

Let $x =$ the price of tea per lb.
Let $y =$ the price of coffee per lb.

$$7x + 5y = \$5.50 \qquad (1)$$
$$6x + 3y = \$4.20 \qquad (2)$$
$$x = \$ \ .50 \qquad (3)$$
$$y = \$ \ .40 \qquad (4)$$

2. There is a fraction such that if 1 be added to the numerator the value of the fraction will be 1; and if 3 be added to the denominator the value will be $\frac{1}{2}$. What is the fraction?

Solution.

Let $x =$ the numerator.
$y =$ the denominator.

Then, $\dfrac{x}{y} =$ the fraction.

$$\frac{x+1}{y} = 1 \qquad (1)$$
$$\frac{x}{y+3} = \tfrac{1}{2} \qquad (2)$$
$$x = 4 \qquad (3)$$
$$y = 5 \qquad (4)$$
$$\frac{x}{y} = \frac{4}{5} \qquad (5)$$

3. There is a number such that if it be divided by the sum of the digits which express it the quotient will be 4, and if 36 be added to it the sum will be expressed by the digits inverted. What is the number?

Solution.

Let $x =$ the digit in tens' place.
$y =$ the digit in units' place.
$10x + y =$ the number.
$10y + x =$ the number when the digits are inverted.

$$\frac{10x + y}{x + y} = 4 \qquad (1)$$
$$10x + y + 36 = 10y + x \qquad (2)$$
$$x = 4 \qquad (3)$$
$$y = 8 \qquad (4)$$
$$10x + y = 48 \qquad (5)$$

SIMULTANEOUS EQUATIONS.

4. The sum of two numbers is 24, and their difference is 8. What are the numbers?

5. The sum of two numbers is 29, and their difference is 5. What are the numbers?

6. The sum of two numbers divided by 2 gives a quotient of 24, and their difference divided by 2 gives a quotient of 17. What are the numbers?

7. A man hired 6 men and 2 boys for one day for $28, and afterward, at the same rate, 3 men and 4 boys for $20. What was paid each per day?

8. There is a fraction such that if 3 be added to the numerator its value will be $\frac{1}{3}$, and if 1 be subtracted from the denominator its value will be $\frac{1}{5}$. What is the fraction?

9. A man has two horses, and a saddle worth $10. The value of the saddle and the first horse is double that of the second horse, but the value of the saddle and the second horse lacks $13 of being equal to the value of the first horse. What is the value of each horse?

10. Two purses contain together $300. If $30 is taken from the first and put into the second, there will be the same amount in each. How much money is there in each?

11. A and B have $570. If A's money were three times, and B's were five times as great as it really is, they would have $2350. How much has each?

12. There is a fraction such that if 4 be added to the numerator, its value will be $\frac{1}{2}$, and if 7 be added to the denominator its value will be $\frac{1}{5}$. What is the fraction?

13. There is a number of two digits, which is equal to 4 times the sum of the digits, and if 18 be added to the number, the result will be expressed by the digits inverted. What is the number?

14. A person had two kinds of money, such that it took 10 pieces of one kind to make a dollar, and two pieces of the other to make a dollar. He paid a man a dollar, giving him 6 pieces. How many of each kind were used?

15. A party which had hired a coach, found that if there had been 3 more persons, they would each have had to pay $1 less than they did; and if there had been 2 less they would each have had to pay $1 more. How many persons were there? How much did each pay?

16. A wine-merchant sold at one time 20 dozen bottles of port wine and 30 dozen of sherry for £120. At another time he sold 30 dozen bottles of port and 25 dozen of sherry for £140. What was the price per dozen bottles of each?

17. There is a number expressed by two figures. If to the sum of the digits 7 be added, the result will be 3 times the left-hand digit, and if 18 be subtracted from the number, the digits will be inverted. What is the number?

18. A and B had together a capital of $9800. A invested $\frac{1}{6}$ of his capital and B $\frac{1}{5}$ of his, whereupon each had the same sum left. How much had each before the investment?

19. A farmer purchased 100 acres of land for $2450. For a part of it he paid $20 an acre and for the rest $30 an acre. How many acres were there in each part?

20. The sum of the ages of a father and a son is 80 years. If the age of the son is doubled, it will exceed the age of the father by 10 years. What is the age of each?

21. A said to B: "Give me 20 cents of your money and I will have 4 times as much as you." B said to A: "Give me 20 cents of your money and I will have $1\frac{1}{2}$ times as much as you." How much had each?

22. A farmer bought 100 acres of land, part at $37 and part at $45 an acre, paying for the whole $4220. How much land was there in each part?

23. A boy expended 30 cents for apples and pears, buying the apples at 4 for a cent and the pears at 5 for a cent. He

SIMULTANEOUS EQUATIONS. 157

then sold $\frac{1}{2}$ of his apples and $\frac{1}{3}$ of his pears for 13 cents, which was what they cost him. How many of each did he buy?

24. A farmer sold 47 bushels of corn and 18 bushels of wheat to one man for $45.26, and 3 bushels of corn and 63 bushels of wheat to another at the same price, for $64.74. Find the cost of each per bushel.

25. The sum of $\frac{1}{2}$ of one number and $\frac{2}{3}$ of another is 38; and, if 3 be added to the first, the sum will be equal to $\frac{3}{8}$ of the difference between the second and 8. Find the numbers.

26. Said one boy to another: "If you will give me $\frac{1}{2}$ of your money and 50 cents, I shall have 4 times as much as you; but if I give you 50 cents, you will have $2.50 more than I." How much had each?

27. One quarter of the number of sheep in one field equaled $\frac{1}{6}$ of the number in an adjoining field. Ten sheep jumped from the first into the second field, and there were then 4 times as many in the second as in the first. How many were there then in each?

28. Three years ago Paul was 3 times as old as his sister Mabel, and 3 years hence 3 times his age will equal 5 times hers. Find their ages now.

29. The sum of two fractions whose numerators are 3, is 3 times the smaller, and 3 times the smaller subtracted from twice the larger gives $\frac{3}{8}$. What are the fractions?

30. If I loan my money at 6% for a given time, I shall receive $720 interest; but if I loan it for three years longer, I shall receive $1800. Find amount of money and the time.

31. If a certain rectangle were 1 foot longer and 1 foot broader, it would contain 14 square feet more area. If it were 1 foot shorter and $\frac{1}{2}$ foot wider, its total area would be unchanged. Find the dimensions of the rectangle.

32. The number of marbles a boy has in one pocket is 9 more than $\frac{2}{3}$ of what he has in the other. How many has he in each, if there are 39 marbles in both?

33. A number consists of two figures. The number is 2 more than 8 times the sum of the digits, and if 54 be subtracted from the number, the digits will be inverted. Find the number.

34. A and C can do a piece of work in 6 days; B and C can do it in 8 days. In what time can they all do it working together, if A can do $\frac{3}{2}$ as much as B?

35. A man invested $4400, part of it in railroad bonds bearing 3% interest, and the remainder in state bonds bearing $2\frac{1}{2}$% interest. He received the same income from each. How much did he invest in each?

SUGGESTION. — Since the rate of interest upon the railroad bonds was 3%, the income was $\frac{3}{100}$ of the investment. The income from the state bonds was $\frac{2\frac{1}{2}}{100}$, or $\frac{5}{200}$, of the investment.

36. A gives B 100 yards start and overtakes him in 4 minutes. He also gains 750 feet on B in running 9000 feet. Find the rate at which each runs.

37. A railway train, after traveling an hour, is detained 30 minutes. It then proceeds at $\frac{6}{5}$ of its former rate, and arrives 10 minutes late. If the detention had occurred 12 miles further on the train would have arrived 4 minutes later than it did. At what rate did the train travel before the detention, and what was the whole distance traveled?

THREE OR MORE UNKNOWN QUANTITIES.

191. 1. In the equations $2x + 3y + 4z = 26$ and $x + 4y + 5z = 18$, how may z be eliminated?

2. If one of the quantities in the above equations is eliminated, how many quantities will be left?

3. How many independent equations are necessary before the values of two unknown quantities can be found?

4. How many independent equations containing the same two unknown quantities can be formed by combining the equations in (1)?

SIMULTANEOUS EQUATIONS.

5. Since in order to find the values of two unknown quantities we must have two independent equations, and since from the two equations given in (1) only one derived equation can be formed containing the same unknown quantities, how many independent equations must be given so that the value of any of those unknown quantities may be found?

192. Since it is necessary to have *two* independent equations to find the values of *two* unknown quantities, and *three* independent equations to find the values of *three* unknown quantities, etc., a general law may be expressed as follows:

193. PRINCIPLE. — *To solve equations containing unknown quantities, there must be as many independent equations as there are unknown quantities.*

EXAMPLES.

1. Given $\begin{cases} x + 2y + 3z = 14 \\ 2x + y + 2z = 10 \\ 3x + 4y - 3z = 2 \end{cases}$, to find x, y, and z.

SOLUTION.

$$x + 2y + 3z = 14 \quad (1)$$
$$2x + y + 2z = 10 \quad (2)$$
$$3x + 4y - 3z = 2 \quad (3)$$

$(1) \times 2$
$$2x + 4y + 6z = 28 \quad (4)$$
(2)
$$2x + y + 2z = 10$$

$(4) - (2)$
$$3y + 4z = 18 \quad (5)$$

$(1) \times 3$
$$3x + 6y + 9z = 42 \quad (6)$$
(3)
$$3x + 4y - 3z = 2$$

$(6) - (3)$
$$2y + 12z = 40 \quad (7)$$
$(5) \times 3$
$$9y + 12z = 54 \quad (8)$$

$$7y = 14 \quad (9)$$
$$y = 2 \quad (10)$$
$$4 + 12z = 40 \quad (11)$$
$$12z = 36 \quad (12)$$
$$z = 3 \quad (13)$$
$$x + 4 + 9 = 14 \quad (14)$$
$$x = 1 \quad (15)$$

HIGH SCHOOL ALGEBRA.

Find the value of each unknown quantity in the following:

2. $\begin{cases} x - 2y + 2z = 5. \\ 5x + 3y + 6z = 57. \\ x + 2y + 2z = 21. \end{cases}$

10. $\begin{cases} x + 2y + 3z = 34. \\ 2x - 3y + 4z = 19. \\ 3x + 4y - 5z = -6. \end{cases}$

3. $\begin{cases} 7x - 4y + 3z = 35. \\ 4x - 5y + 2z = 6. \\ 2x + 3y - z = 20. \end{cases}$

11. $\begin{cases} x + y + z = 90. \\ 2x - 3y = -20. \\ 2x + 3z = 145. \end{cases}$

4. $\begin{cases} x + y + z = 6. \\ 5x + 4y + 3z = 22. \\ 3x + 4y - 3z = 2. \end{cases}$

12. $\begin{cases} x + y = 35. \\ x + z = 40. \\ y + z = 45. \end{cases}$

5. $\begin{cases} x - 4y + 3z = 2. \\ 4x - 3y + z = 9. \\ 2x + 6y - 4z = 14. \end{cases}$

13. $\begin{cases} 2x + 4y - 3z = 22. \\ 4x - 2y + 5z = 18. \\ 6x + 7y - z = 63. \end{cases}$

6. $\begin{cases} x + y + z = 35. \\ x - 2y + 3z = 15. \\ y - x + z = -5. \end{cases}$

14. $\begin{cases} 8x - 4y = 24 - z. \\ 6x + y = z + 84. \\ x + 80 = 3y + 4z. \end{cases}$

7. $\begin{cases} x + 22 = y + z. \\ y + 22 = 2x + 2z. \\ z + 22 = 3x + 3y. \end{cases}$

15. $\begin{cases} x + 3y - z = 10. \\ 5x - 2y + 2z = 6. \\ 3x + 2y + z = 13. \end{cases}$

8. $\begin{cases} x + y + z = 12. \\ x - y = 2. \\ x - z = 4. \end{cases}$ *

16. $\begin{cases} x + y + z = 18. \\ x - y + z = 6. \\ x + y - z = 4. \end{cases}$

9. $\begin{cases} u + y + z = 2x. \\ u + x + z = 3y. \\ u + x + y = 4z. \\ u + x = y + 36. \end{cases}$

17. $\begin{cases} 2x - 4y + 3z = 10. \\ 3x + y - 2z = 6. \\ x + 3y - z = 20. \end{cases}$

---

* The sum of the first members of the three equations is 3 x.

SIMULTANEOUS EQUATIONS.

18. $\begin{cases} 5x - 3y + z = 16. \\ 9x + 2y - 3z = 14. \\ x - 4y - 5z = 10. \end{cases}$

19. $\begin{cases} x + a = y + z. \\ y + a = 2x + 2z. \\ z + a = 3x + 3y. \end{cases}$

20. $\begin{cases} x + y + 2z = 2(b+c). \\ x + z + 2y = 2(a+c). \\ y + z + 2x = 2(a+b). \end{cases}$

21. $\begin{cases} ax + by = r. \\ by + cz = s. \\ ax + cz = v. \end{cases}$

22. $\begin{cases} x + y + 2z + w = 18. \\ x + 2y + z + w = 17. \\ x + y + z + 2w = 19. \\ 2x + y + z + w = 16. \end{cases}$ *

23. $\begin{cases} u + v + x + y = 14. \\ u + v + x + z = 15. \\ u + v + y + z = 16. \\ u + x + y + z = 17. \\ v + x + y + z = 18. \end{cases}$

24. $\begin{cases} \dfrac{1}{x} + \dfrac{1}{y} = 5. \\ \dfrac{1}{y} + \dfrac{1}{z} = 7. \\ \dfrac{1}{x} + \dfrac{1}{z} = 6. \end{cases}$ †

25. $\begin{cases} x + \dfrac{1}{3}y = 5. \\ x + \dfrac{1}{3}z = 6. \\ y + \dfrac{1}{3}z = 9. \end{cases}$

26. $\begin{cases} \dfrac{x}{a} + \dfrac{y}{b} = 2. \\ \dfrac{x}{a} + \dfrac{z}{c} = 2. \\ \dfrac{y}{b} + \dfrac{z}{c} = 2. \end{cases}$

27. $\begin{cases} x + y = a. \\ x + z = b. \\ y + z = c. \end{cases}$

28. $\begin{cases} x + y = 9. \\ y + z = 11. \\ z + w = 13. \\ w + u = 15. \\ u + x = 12. \end{cases}$

29. $\begin{cases} \dfrac{xy}{x+y} = \dfrac{1}{5}. \\ \dfrac{yz}{x+z} = \dfrac{1}{6}. \\ \dfrac{xz}{x+z} = \dfrac{1}{7}. \end{cases}$

---

* The sum of the four equations divided by 5, is the sum of the unknown quantities.
† This example may be solved without clearing of fractions.

ALGEBRA. — 11.

30. $\begin{cases} \dfrac{1}{x}+\dfrac{1}{y}=a. \\ \dfrac{1}{x}+\dfrac{1}{z}=b. \\ \dfrac{1}{y}+\dfrac{1}{z}=c. \end{cases}$

31. $\begin{cases} 3x+4y+z=35. \\ 3z+2y-3t=4. \\ 2x-y+2t=17. \\ 3z-2t+u=9. \\ t+y=13. \end{cases}$

32. $\begin{cases} x-y=a. \\ y+z=3a. \\ 5z-x=2a. \end{cases}$

33. $\begin{cases} y+z-x=a. \\ x+y-z=b. \\ z+x-y=c. \end{cases}$

34. $\begin{cases} x-z+y=a. \\ z+y-v=b. \\ v-x+z=c. \\ x-y+v=d. \end{cases}$

35. $\begin{cases} \dfrac{a}{x}+\dfrac{b}{y}+\dfrac{c}{z}=a. \\ \dfrac{b}{y}-\dfrac{c}{z}+\dfrac{a}{x}=b. \\ \dfrac{a}{x}-\dfrac{b}{y}-\dfrac{c}{z}=c. \end{cases}$

36. $\begin{cases} x+2y=a. \\ y+2z=b. \\ 2z+x=c. \end{cases}$

37. $\begin{cases} 3x+4y+z=a. \\ 3y+4z+x=b. \\ 3z+4x+y=c. \end{cases}$

38. $\begin{cases} x+y=a+b. \\ y+z=b+c. \\ z+v=a-b. \\ v-x=c-b. \end{cases}$

39. $\begin{cases} \dfrac{a}{x}-\dfrac{c}{z}+\dfrac{b}{y}=ab. \\ \dfrac{b}{y}-\dfrac{a}{x}+\dfrac{c}{z}=bc. \\ \dfrac{a}{x}+\dfrac{c}{z}-\dfrac{b}{y}=ac. \end{cases}$

PROBLEMS.

194. 1. Find three numbers such that their sum is 60; $\frac{1}{2}$ of the first plus $\frac{1}{3}$ of the second, and $\frac{1}{5}$ of the third is 19; and twice the first with three times the remainder, when the third is subtracted from the second, is 50.

2. Find three numbers such that the first with $\frac{1}{2}$ of the sum of the second and third is 119; the second with $\frac{1}{3}$ of the remainder, when the first is subtracted from the third, is 68; and $\frac{1}{2}$ the sum of the three numbers is 94.

SIMULTANEOUS EQUATIONS.

3. A, B, and C together possess $1500. If B gives A $200 of his money, A will have $280 more than B; but if B should receive $180 from C, B and C would have equal amounts. How much has each?

4. Three persons purchased sugar, coffee, and tea at the same rates. A paid $4.20 for 7 pounds of sugar, 5 pounds of coffee, and 3 pounds of tea; B paid $3.40 for 9 pounds of sugar, 4 pounds of coffee, and 2 pounds of tea; C paid $3.25 for 5 pounds of sugar, 2 pounds of coffee, and 3 pounds of tea. What was the price of each per pound?

5. Divide 125 into four such parts that, if the first be increased by 4, the second diminished by 4, the third multiplied by 4, and the fourth divided by 4, the sum, product, difference, and quotient will all be equal.

6. A and B can perform a piece of work in 8 days; A and C can do it in 9 days, and B and C in 10 days. In how many days can each do the same work alone?

7. A certain number is expressed by three digits whose sum is 10. The sum of the first and last digits is $\frac{2}{3}$ of the second digit; and, if 198 be subtracted from the number, the digits will be inverted. What is the number?

Let $x =$ the first digit, or hundreds; y, the second digit, or tens; z, the third digit, or units. Then, $100x + 10y + z =$ the number.

8. A farmer has 80 horses, cows, and sheep; $\frac{1}{4}$ of the number of horses, $\frac{1}{6}$ of the sheep, and $\frac{1}{2}$ the cows equal 19. If the number of horses and cows be subtracted from $\frac{1}{2}$ the number of sheep, the remainder will be 10. How many animals of each kind has he?

9. A merchant found, on counting his cash, that he had 84 pieces of silver, in dollars, half-dollars, and quarters, worth $42. He also found that $\frac{1}{3}$ of his half-dollars and $\frac{1}{4}$ of his quarters were worth $6.50. How many pieces of each kind had he?

10. A, B, and C were discussing their ages. A said: "If you add to twice my age, once B's, the sum will be 20 years more than C's. If you take $\frac{1}{2}$ of my age and twice B's, the sum will be 20 years less than C's. But the sum of my age and C's is equal to 6 times as much as B's." What were their ages?

11. A jeweler has a fine gold watch, a plain gold one, and an open face one, and one gold chain. If he puts the chain on the fine watch, that watch and chain will be worth $40 more than both the other watches. If he puts the chain on the plain watch, that watch and chain will be worth $20 more than twice the value of the open face watch. If the chain is placed on the open face watch, the value of that watch and chain will be $\frac{1}{5}$ of that of the fine watch, and $\frac{1}{3}$ of the plain watch. Find the value of each watch and of the chain, if they all cost $400.

12. Bought 8 horses, a number of cows, and 100 sheep for $2500. The number of cows was equal numerically to 4 times the price of a sheep, and a sheep and a horse cost $5 less than $\frac{1}{5}$ the cost of all the cows. Find the cost of a horse, and a sheep, and the number of cows, if a cow cost $40.

13. Three boys, A, B, and C, had each a bag of nuts. Each boy gave to each of the others $\frac{1}{5}$ of the number of nuts he had in his bag. They then counted their nuts, and A had 740, B 580, and C 380. How many had each at first?

14. There are two fractions which have the same denominator. If 1 be subtracted from the numerator of the smaller, its value will be $\frac{1}{3}$ of the larger fraction; but if 1 be subtracted from the numerator of the larger, its value will be twice that of the smaller. The difference between the fractions is $\frac{1}{3}$. What are the fractions?

15. A man divided a sum of money amongst his four sons, so that the share of the eldest was $\frac{1}{2}$ of the shares of the other three, the share of the second $\frac{1}{3}$ of the shares of the other three, and the share of the third $\frac{1}{4}$ of the shares of the other three. The eldest had $14 more than the youngest. What was the share of each?

SIMULTANEOUS EQUATIONS. 165

16. A farmer found that the number of his sheep was 26 more than the number of his cows and horses together; that $\frac{1}{5}$ of the number of sheep was equal to the number of horses together with $\frac{1}{4}$ of his cows; and that $\frac{1}{8}$ of his cows, $\frac{1}{2}$ of his horses, and $\frac{1}{5}$ of his sheep amounted to 12. How many had he of each?

17. There are three purses such that if $20 be taken out of the first and put into the second, the second will contain 4 times as much as remains in the first; if $60 be taken from the second and put into the third, the third will contain $1\frac{3}{4}$ times as much as remains in the second; if $40 be taken from the third and put into the first, the third will contain $2\frac{1}{4}$ times as much as the first. How much is there in each purse?

18. Three men, A, B, and C, jointly purchased a quantity of lumber which cost $900. One half of what A paid added to $\frac{1}{4}$ of what B paid and $\frac{1}{5}$ of what C paid, will make $279; and the sum that A paid increased by $\frac{2}{3}$ of what B paid, and diminished by $\frac{1}{2}$ of what C paid, will make $320. How much did each pay?

19. There are three numbers such that $\frac{1}{2}$ the first, $\frac{1}{3}$ the second, and $\frac{1}{4}$ the third, together make 115; $\frac{1}{3}$ the first, $\frac{1}{4}$ the second, and $\frac{1}{5}$ the third, together make 86; and $\frac{1}{4}$ the first, $\frac{1}{5}$ the second, and $\frac{1}{6}$ the third, together make 69. What are the numbers?

20. A gives to B and C as much as each of them has; B gives to A and C as much as each of them then has; and C gives to A and B as much as each of them then has, after which each has $8. How much had each at first?

21. A, B, and C have certain sums of money. If B should give A $350, A would have twice as much as B would have left; if C should give B $700, C would have left $\frac{1}{3}$ as much money as B would then have; if A should give C $210, C would then have 5 times as much as A would have remaining. How much money has each?

INVOLUTION.

195. 1. How many times is a used as a factor in producing a^2? a^3? a^4? a^5? a^6? a^n?

2. How many times is a quantity used as a factor in producing the *second* power? The *third* power? The *fourth* power? The *fifth* power? The nth power?

3. What sign has the second power of $+a$? The third power? The fourth power? Any power?

4. What sign has the second power of $-a$? The third power? The fourth power? The fifth power? The sixth power?

5. Which powers of a negative quantity are *positive?* Which are *negative?*

6. What is the third power of a^3? How many times is a^3 used as a factor?

7. What is the fourth power of a^3? Of a^4? Of a^5?

8. What is the fifth power of a^3? Of a^4? Of a^5?

9. How is the exponent of a power of a quantity determined?

10. What is the mth power of a^n?

196. Exponent (Art. 17). Power (Art. 18). **Names of Powers** (Art. 18).

197. Involution is the process of finding **a power of a quantity.**

INVOLUTION.

198. PRINCIPLES.—1. *All powers of a positive quantity are positive.*

2. *All even powers of a negative quantity are positive, and all odd powers are negative.*

3. *The exponent of a power of a quantity is equal to the exponent of the quantity multiplied by the exponent of the power to which the quantity is to be raised.*

199. Involution of monomials.

1. What is the third power of $6a^2b$?

SOLUTION.
$$(6a^2b)^3 = 6a^2b \times 6a^2b \times 6a^2b = 216a^6b^3$$

2. What is the fifth power of $-2a^2b^3$?

SOLUTION.
$$(-2a^2b^3)^5 = -2a^2b^3 \times -2a^2b^3 \times -2a^2b^3 \times -2a^2b^3 \times -2a^2b^3 = -32a^{10}b^{15}$$

RULE. — *Raise the numerical coëfficient to the required power; multiply the exponent of each literal quantity by the exponent of the power to which it is to be raised, and prefix the proper sign to the result.*

Find the values of the following:

3. $(6x^2y)^2$.
4. $(-4a^2b^2)^3$.
5. $(-3ab^3)^3$.
6. $(-3c^2d^2)^2$.
7. $(2ax^3y^2)^5$.
8. $(2x^2yz^3)^6$.
9. $(4ab^4d^3)^4$.
10. $(-a^5b^2c^6)^7$.
11. $(-4a^2b^2c^4)^4$.
12. $(2x^2yz^4)^5$.
13. $(-5a^3bc^2)^3$.
14. $(a^4b^2c^3)^8$.
15. $(-4a^2b^3c^4)^4$.
16. $(2a^2b^2c^3)^9$.
17. $(2x^2y^3z)^4$.
18. $(3x^2y^4)^3$.
19. $(-4a^3z^2y)^3$.
20. $(2x^3y^4z^3)^5$.
21. $(-2a^2y^3)^5$.
22. $(2ax^2y^3)^4$.
23. $(x^4y^2z^n)^3$.
24. $(x^my^nz^m)^5$.
25. $(a^{2n}z^nw^m)^4$.
26. $(a^4y^{5n}z^{3n})^2$.
27. $(-x^4y^3z^n)^n$.
28. $(-a^4b^2c^nd^{2n})^{5n}$.
29. $(a^3b^4c^3d^{n-2})^{n-2}$.

HIGH SCHOOL ALGEBRA.

30. What is the third power of $\dfrac{2\,x^2y}{3\,a^2b}$?

PROCESS.

$$\left(\dfrac{2\,x^2y}{3\,a^2b}\right)^3 = \dfrac{2\,x^2y}{3\,a^2b} \times \dfrac{2\,x^2y}{3\,a^2b} \times \dfrac{2\,x^2y}{3\,a^2b} = \dfrac{8\,x^6y^3}{27\,a^6b^3}$$

EXPLANATION. — In raising a fraction to a power, both numerator and denominator must be raised to the required power.

Raise to the required power:

31. $\left(\dfrac{2\,a}{3\,b}\right)^2$.

32. $\left(\dfrac{2\,x^2}{3\,y}\right)^2$.

33. $\left(-\dfrac{6\,xy}{5\,ab}\right)^3$.

34. $\left(\dfrac{8\,a^2b}{7\,x^2y}\right)^4$.

35. $\left(\dfrac{a^2b^n}{c^4d^3}\right)^6$.

36. $\left(\dfrac{x^ny^{2n}}{a^nz^4}\right)^7$.

37. $\left(\dfrac{a^2b^2c^{n-1}}{x^{n-2}y^{n-3}}\right)^n$.

38. $\left(\dfrac{a^2b^3c^4}{x^4y^4z^4}\right)^{2n}$.

39. $\left(\dfrac{a^3x^4}{b^4y^3}\right)^{3n}$.

40. $\left(\dfrac{ab^2c^{n+1}}{x^my^{n-1}}\right)^n$.

41. $\left(\dfrac{a^4b^3c^2}{x^2y^4}\right)^4$.

42. $\left(\dfrac{ab^3c^n}{x^5y^2z^{3m}}\right)^{2n}$.

200. Involution of polynomials.

$$(x+y)^2 = x^2 + 2xy + y^2 \qquad \text{(Art. 73)}$$
$$(x-y)^2 = x^2 - 2xy + y^2 \qquad \text{(Art. 75)}$$
$$(x+y-z)^2 = x^2 + y^2 + z^2 + 2xy - 2xz - 2yz \qquad \text{(Art. 81)}$$

Raise to the required power:

1. $(2a+b)^2$.

2. $(3a-2c)^2$.

3. $(2a+3b)^2$.

4. $(4a-2b)^2$.

5. $(3a+4c)^2$.

6. $(5a-4c)^2$.

7. $(2x^2-y)^2$.

8. $(3x^2+2y)^2$.

9. $(3x^2-3y^2)^2$.

10. $(4x^2+5y)^2$.

11. $(5x^3-3y^2)^2$.

12. $(3x^2+6y^2)^2$.

13. $(a+b+c-d)^2$.

14. $(2x-3y+2z)^2$.

15. $(x^2-y^2+2z^2)^2$.

16. $(x^2+2y+3z^2)^2$.

17. $(2x+3y^2-z^2)^2$.

18. $(2x^2-3y^2-2z^2)^2$.

INVOLUTION.

Quantities may be raised to higher powers by multiplication. The quantity must be used as many times as a factor as there are units in the index of the required power.

Raise to the required power:

19. $(x+y)^3$.
20. $(a-b)^5$.
21. $(a+b)^4$.
22. $(x+y)^6$.
23. $(a-c)^7$.
24. $(2b+c)^4$.
25. $(2a-b)^3$.
26. $(3a+b)^4$.
27. $(4x-y)^4$.
28. $(5x+z)^5$.
29. $(2x^2-3y^2)^3$.
30. $(3x^2+2y^2)^3$.
31. $(4x^2+3y)^4$.
32. $(5x^2-2y^2)^3$.
33. $(4x^2-2y^2)^4$.

201. Involution of binomials by the Binomial Theorem.

$(a+b)^2 = a^2 + 2ab + b^2$.

$(a+b)^3 = a^3 + 3a^2b + 3ab^2 + b^3$.

$(a+b)^4 = a^4 + 4a^3b + 6a^2b^2 + 4ab^3 + b^4$.

$(a+b)^5 = a^5 + 5a^4b + 10a^3b^2 + 10a^2b^3 + 5ab^4 + b^5$.

$(a-b)^2 = a^2 - 2ab + b^2$.

$(a-b)^3 = a^3 - 3a^2b + 3ab^2 - b^3$.

$(a-b)^4 = a^4 - 4a^3b + 6a^2b^2 - 4ab^3 + b^4$.

$(a-b)^5 = a^5 - 5a^4b + 10a^3b^2 - 10a^2b^3 + 5ab^4 - b^5$.

Examine carefully the above powers of $(a+b)$ and $(a-b)$.

1. How does the number of terms in any power of these binomials compare with the exponent of the power?

2. In what terms of any power of these binomials is the *first* letter of the binomial found?

3. In what terms of any power of these binomials is the *second* letter of the binomial found?

4. What is the exponent of the first letter of the binomials in the first term of any power of the binomials? In the second? In the third, etc.?

5. What is the exponent of the second letter of the binomials in the second term of any power of the binomials? In the third term? In the fourth term, etc.?

6. What is the coëfficient of the first and the last terms in any power of the binomials?

7. How does the coëfficient of the second term of the power compare with the exponent of the power to which the binomial is to be raised?

8. If the coëfficient of the second term of the power is multiplied by the exponent of the first quantity in that term, and the result divided by the number of the term, or by the exponent of the second quantity in that term increased by 1, what is the result? Of what term is it the coëfficient?

9. If the same thing is done to the coëfficient of each succeeding term of the power, what coëfficient is obtained?

10. What are the signs of the terms in all the powers of $(a+b)$?

11. What terms are positive and what are negative in any power of $(a-b)$?

202. PRINCIPLES.—1. *The number of terms in a positive integral power of any binomial is one more than the exponent of the required power.*

2. *The letter of the first term of the binomial is found in all the terms of the power except the last; the letter of the second term in all the terms of the power except the first; both letters are found in all the terms of the power except the first and last.*

3. *The exponent of the first term of the binomial in the first term of the power is the same as the index of the required power, and decreases by 1 in each term at the right. The exponent of the second term of the binomial is 1 in the second term of the power, and increases by 1 in each term at the right.*

4. *The coëfficient of the first term of the power is 1. The coëfficient of the second term is the same as the index of the required power.*

INVOLUTION.

5. *The coëfficient of any term of the power, multiplied by the exponent of the first letter in that term, and the result divided by the number of the term, or by the exponent of the second letter increased by 1, will be the coëfficient of the next term.*

6. *If both terms of the binomial are positive, all the terms of any power of the binomial will be positive.*

7. *If the second term of the binomial is negative, all the odd terms of any power counting from the left will be positive, and all the even terms will be negative.*

The sum of the exponents in any term is always equal to the exponent of the required power.

EXAMPLES.

1. Find the fifth power of $(x-y)$ by the binomial theorem.

SOLUTION.

Letters,	x	xy	xy	xy	xy	y
Letters and exponents,	x^5	x^4y	x^3y^2	x^2y^3	xy^4	y^5
Coëfficients,	1	5	10	10	5	1
Signs,		$-$	$+$	$-$	$+$	$-$

Combined, $x^5 - 5x^4y + 10x^3y^2 - 10x^2y^3 + 5xy^4 - y^5$

Expand:

2. $(x+y)^3$.
3. $(a-b)^3$.
4. $(a+c)^3$.
5. $(a+x)^4$.
6. $(a-x)^3$.
7. $(a+b)^3$.
8. $(x-y)^3$.
9. $(b+c)^4$.
10. $(x+1)^3$.
11. $(x-1)^3$.
12. $(1+a)^3$.
13. $(1-a)^4$.
14. $(x+a)^4$.
15. $(x+b)^3$.
16. $(x-c)^4$.
17. $(x-y)^4$.
18. $(a+b)^4$.
19. $(a-c)^5$.
20. $(a-x)^4$.
21. $(a+x)^6$.
22. $(a-c)^9$.
23. $(x-y)^7$.
24. $(x+y)^{10}$.
25. $(x+1)^5$.
26. $(x-1)^6$.
27. $(1+a)^5$.
28. $(1-a)^7$.
29. $(x+ac)^4$.
30. $(x+bc)^5$.
31. $(x-ac)^6$.

172   HIGH SCHOOL ALGEBRA.

When the terms of the binomial have coëfficients, the binomial may be expanded as follows:

32. Find the third power of $2a^2 - 3b$.

SOLUTION.

Let $\qquad 2a^2 = x$ and $3b = y$
Then, $\qquad 2a^2 - 3b = x - y$

$$(x-y)^3 = x^3 - 3x^2y + 3xy^2 - y^3$$

Restoring $2a^2$ for x,		$8a^6$	$4a^4$	$2a^2$	
Restoring $3b$ for y,			$3b$	$9b^2$	$27b^3$
Coëfficients,	1	3	3	1
Signs,	+	−	+	−
Products,		$8a^6 - 36a^4b + 54a^2b^2 - 27b^3$			

Expand the following:

33. $(a+2b)^3$.   **37.** $(3a+2c)^4$.   **41.** $(4a^2-c)^4$.

34. $(3a-b)^3$.   **38.** $(2a^2-c)^3$.   **42.** $(3a^2+4c^2)^3$.

35. $(2a+3b)^3$.   **39.** $(2a+c^2)^3$.   **43.** $(3x^2-5c^2)^4$.

36. $(3a-3c)^3$.   **40.** $(3x^2+2y^2)^3$.   **44.** $(5a^2+4c^2)^4$.

45. Expand $(a+b-c)^3$.

SOLUTION.

$(a + b - c)^3 = [(a + b) - c]^3$, a binomial form.

$[(a + b) - c]^3 = (a + b)^3 - 3(a + b)^2c + 3(a + b)c^2 - c^3$
$\qquad = a^3 + 3a^2b + 3ab^2 + b^3 - 3(a^2 + 2ab + b^2)c + 3ac^2 + 3bc^2 - c^3$
$\qquad = a^3 + 3a^2b + 3ab^2 + b^3 - 3a^2c - 6abc - 3b^2c + 3ac^2 + 3bc^2 - c^3$

46. Expand $(a + b + c - d)^3$.

SUGGESTION. — $(a + b + c - d)^3 = [(a + b) + (c - d)]^3$, a binomial form

Expand:

47. $(x-y+z)^3$.   **50.** $(a-b-c-d)^3$.

48. $(x-y-z)^3$.   **51.** $(a+b-c+d)^3$.

49. $(a+b-c)^3$.   **52.** $(1+x-y-z)^3$.

INVOLUTION. 173

203. The following method of expanding binomials whose terms have coëfficients will be found of great value.

1. Find the third power of $5a + 4b$.

PROCESS.

$(5a + 4b)^3 = 125\,a^3 + 300\,a^2b + 240\,ab^2 + 64\,b^3$

$(5a)^3 = 125\,a^3$, the first term.

$\dfrac{125 \times 4 \times 3}{5} = 300$, the coëfficient of the second term.

$\dfrac{300 \times 4 \times 2}{5 \times 2} = 240$, the coëfficient of the third term.

$\dfrac{240 \times 4 \times 1}{5 \times 3} = 64$, the coëfficient of the fourth term.

EXPLANATION. — The first term is the cube of $5a$ or $125\,a^3$.

Since the second term of the power contains the second power of the first term of the binomial multiplied by the second term of the binomial, the coëfficient of the second term of the power has as two of its factors the second power of 5, the coëfficient of the first term of the binomial, and 4, the coëfficient of the second term. To save computations, the second power of the coëfficient of the first term of the binomial is obtained by dividing the coëfficient of the third power of that quantity by the coëfficient of the quantity. Therefore, $\frac{125}{5} \times 4$ are two of the factors of the coëfficient of the second term of the power.

When the coëfficients of the terms of a binomial are each 1, the coëfficient of each succeeding term of the power is obtained from the preceding term by multiplying the coëfficient of that term by the exponent of the first quantity in that term and then dividing the product by the exponent of the second quantity increased by 1. Consequently, to obtain the *entire* coëfficient of the second term of the power, the expression $\dfrac{125 \times 4}{5}$ must be multiplied by 3, the exponent of the first quantity, and divided by 1, the exponent of the second quantity increased by 1. (The exponent of the second quantity in the first term is 0.) Indicating the operations, the second coëfficient is equal to $\dfrac{125 \times 4 \times 3}{5 \times 1}$ or 300.

The coëfficients of the succeeding terms are found in the same manner, and the method of finding the coëfficients may be seen to be expressed by the following general rule:

RULE. — *The first term of the power is the power of the first term of the binomial corresponding to the exponent of the binomial.*

The coëfficient of any term may be found from the preceding term by dividing the coëfficient of the preceding term by the coëfficient of the first term of the binomial and then multiplying the quotient by the coëfficient of the second term of the binomial and by the exponent of the first quantity in that term, and dividing the product by the exponent of the second quantity increased by 1.

1. Since the number of terms in a power of a binomial is one more than the exponent of the power to which it is to be raised, the exponent of the first quantity in any term of a power is equal to the *number of terms that term is from the last.*

2. For the same reason the exponent of the second quantity in any term is equal to the number of terms that term is from the first.

3. The laws of letters, signs, and exponents are the same as have been learned heretofore.

2. Find the fourth power of $2a + 3b$.

SOLUTION.

$(2a)^4 = 16a^4$, the first term.

$\dfrac{16 \times 3 \times 4}{2} = 96$, the coëfficient of the second term, and the second term is $96\,a^3b$.

$\dfrac{96 \times 3 \times 3}{2 \times 2} = 216$, the coëfficient of the third term, and the third term is $216\,a^2b^2$.

$\dfrac{216 \times 3 \times 2}{2 \times 3} = 216$, the coëfficient of the fourth term, and the fourth term is $216\,ab^3$.

$\dfrac{216 \times 3 \times 1}{2 \times 4} = 81$, the coëfficient of the fifth term, and the fifth term is $81\,b^4$.

$\therefore (2a + 3b)^4 = 16a^4 + 96a^3b + 216a^2b^2 + 216ab^3 + 81b^4.$

Expand the following:

3. $(2a + 3c)^3$.
4. $(2a - 4b)^3$.
5. $(3a + 2x)^4$.
6. $(2a - 5x)^5$.
7. $(5a + 3c)^4$.
8. $(4a + 3x)^5$.
9. $(3a - 5x)^5$.
10. $(7a - 4c)^4$.
11. $(6a + 5b)^6$.
12. $(3x - 1)^5$.
13. $(2x + 5)^4$.
14. $\left(a + \dfrac{1}{2a}\right)^5$
15. $(1 - \tfrac{3}{2}x)^5$.
16. $(\tfrac{3}{2} + \tfrac{5}{3}c)^4$.
17. $(a^2 + 2c^2)^5$.

EVOLUTION.

204. 1. What are the two equal factors of 16? 36? 49? 81? 121?

2. What is one of the two equal factors of a^2, or what is the second root of a^2? $4a^2$? $16a^2$? $25a^4$? $36a^4$?

3. What is one of the three equal factors of a^3, or what is the third root of a^3? $8a^3$? $27a^3$? $8a^6$? $64a^6$?

4. What sign has the second power of $+2$? The third power? The fourth power? The fifth power? Any power?

5. What sign has the second power of -2? The third power? The fourth power? The fifth power? The sixth power?

6. What powers of a positive quantity are positive? What powers of a negative quantity are positive? What powers of a negative quantity are negative?

7. Since a power which has the negative sign is the product of an odd number of equal negative factors, what is the sign of an *odd* root of a *negative* quantity?

8. What is the sign of an *odd* root of a *positive* quantity?

9. Since a power having the plus sign may be the product of an even number of either positive or negative factors, what is the sign of an *even* root of a positive quantity?

10. What quantity used twice as a factor will give a product with the negative sign? What quantity used four times as a factor? What quantity used six times as a factor?

11. What may be said, then, regarding the sign of an *even* root of a negative quantity?

205. Root (Art. 19). Names of Roots (Art. 20). Sign of Evolution or Radical Sign (Art. 21).

206. Evolution is the process of finding a root of a quantity.

207. PRINCIPLES. — 1. *An odd root of a quantity has the same sign as the quantity itself.*

2. *An even root of a positive quantity is either positive or negative.*

3. *An even root of a negative quantity is impossible or imaginary.*

208. Evolution of any monomial.

1. What is the square root of $25\,x^8y^6$?

PROCESS.

$$\sqrt{25\,x^8y^6} = \pm\,5\,x^4y^3$$

EXPLANATION. — Since, in squaring a monomial, we square the coëfficient and multiply the exponents of the letters by 2, to extract the square root we must extract the square root of the coëfficient and divide the exponents of the letters by 2.

Since the even root of a positive quantity is either positive or negative (Prin. 2), the sign of the root is either plus or minus. Hence, the square root of the quantity is $\pm\,5\,x^4y^3$.

RULE. — *Extract the required root of the numerical coëfficient; divide the exponent of each letter by the index of the root sought, and prefix the proper sign to the result.*

The root of a fraction is found by taking the root of the numerator and of the denominator.

If the root of a number is not readily discovered, it may be separated into its prime factors, and the factors grouped into as many equal sets as are required by the root sought. The product of one set of the factors will be the root.

Thus, to find the fourth root of 1296, the number is separated into its prime factors, 2, 2, 2, 2, 3, 3, 3, 3, and 2×3, or 6 is its fourth root.

Find the values of the following:

2. $\sqrt{16\,a^2b^4c^4}$.

3. $\sqrt[3]{-8\,a^3b^6c^3}$.

4. $\sqrt{4\,a^4c^4x^2}$.

5. $\sqrt[3]{27\,x^3y^6z^3}$.

6. $\sqrt[4]{16\,a^4b^8c^8}$.

7. $\sqrt[3]{-8\,ab^3c^2}$.

8. $\sqrt[5]{-a^5c^{10}x^2y}$.

9. $\sqrt[m]{a^{4}x^{3m}y^{m}}$.

10. $\sqrt[3]{a^3x^3y^2z^5}$.

11. $\sqrt[4]{x^4y^8z^3w^2}$.

12. $\sqrt[4]{16\,x^{4n}y^{2n}z}$.

13. $\sqrt{a^4x^2y^4z^8}$.

14. $\sqrt[n]{a^{2n}x^{4n}y^{3n}}$.

15. $\sqrt{\dfrac{16\,a^2}{25\,y^4}}$.

16. $\sqrt[3]{\dfrac{8\,x^3}{27\,y^6}}$.

17. $\sqrt[3]{\dfrac{125\,x^3y^3}{216\,a^6y^3}}$.

18. $\sqrt{\dfrac{81\,x^{10}}{289\,y^4z^6}}$.

19. $\sqrt[3]{-\dfrac{27\,a^6b^9}{343\,x^{12}y^{18}}}$.

20. $\sqrt[5]{x^8y^5z^n}$.

21. $\sqrt[7]{x^7y^{14}z^{21}}$.

22. $\sqrt[m]{a^m x^{2m} y^{3m} z^{4m}}$.

23. $\sqrt[5]{-32\,a^{10}x^{15}y^{20}}$.

209. To extract the square root of a polynomial,

1. Since $a^2 + 2ab + b^2$ is the square of $(a+b)$, what is the square root of $a^2 + 2ab + b^2$?

2. How may the first term of the square root be found from $a^2 + 2ab + b^2$?

3. How may the second term of the square root be found from $2ab$, the second term of the power?

4. What are the factors of $2ab + b^2$?

5. Since $2ab + b^2$ is equal to $b(2a + b)$, what are the factors of the last two terms of the square of a binomial?

6. By what quantity, then, must the last two terms of $a^2 + 2ab + b^2$ be divided so that the quotient may be the second term of the square root?

ALGEBRA. — 12.

HIGH SCHOOL ALGEBRA.

EXAMPLES.

1. Find the square root of $a^2 + 2ab + b^2$.

PROCESS.

$$
\begin{array}{r|l}
a^2 + 2ab + b^2 & \underline{\,a + b\,} \\
a^2 & \\
\end{array}
$$

Trial divisor, $2a$ $\quad\overline{|\;2ab + b^2}$
Complete divisor, $2a + b$ $\;\underline{|\;2ab + b^2}$

EXPLANATION. — The quantities are arranged according to the descending powers of a.

Since the first term of the quantity is a^2, its square root a is the first term of the root sought. Subtracting a^2 from the quantity, there is a remainder of $2ab + b^2$.

Since the first term of the remainder is $2ab$, if it is divided by $2a$, the quotient b will be the second term of the root. $2a$, or twice the root found is called the *trial divisor*. The complete divisor, or the divisor which multiplied by b will produce the remainder $2ab + b^2$, is, therefore, $2a + b$. It is formed by adding to the trial divisor the second term of the root. Multiplying by b and subtracting, there is no remainder.

Therefore $a + b$ is the square root of the quantity.

Since, in squaring $a + b + c$, $a + b$ may be represented by x, and the square of the quantity by $x^2 + 2xc + c^2$, it is obvious that the square root of a quantity, whose root consists of *more than two terms*, may be extracted in the same way as in example 1, *by considering the terms already found as one term.*

2. Find the square root of $x^4 + 4x^3 - 6x^2 - 20x + 25$.

PROCESS.

$$
\begin{array}{r|l}
x^4 + 4x^3 - 6x^2 - 20x + 25 & \underline{\,x^2 + 2x - 5\,} \\
x^4 & \\
\end{array}
$$

$\underline{2x^2 + 2x}\;\;|\;\;4x^3 - 6x^2$
$4x^3 + 4x^2$
$\underline{2x^2 + 4x - 5}\;\;\overline{|\;-10x^2 - 20x + 25}$
$-10x^2 - 20x + 25$

EXPLANATION. — By proceeding as in the previous example, the first two terms of the root are found to be $x^2 + 2x$.

EVOLUTION. 179

To find the next term of the root, we consider $x^2 + 2x$ as one quantity, which we multiply by 2 for the trial divisor. Dividing the first term of the remainder by the first term of the divisor, the third term of the root is obtained, which is -5. Annexing this, as before, to the trial divisor already found, the entire divisor is $2x^2 + 4x - 5$. This multiplied by -5, and the product subtracted from $-10x^2 - 20x + 25$, leaves no remainder.

Hence, the square root of the quantity is $x^2 + 2x - 5$.

RULE. — *Arrange the terms of the polynomial with reference to the consecutive powers of some letter.*

Extract the square root of the first term, write the result as the first term of the root, and subtract its square from the given polynomial.

Divide the first term of the remainder by twice the root already found, as a trial divisor, and the quotient will be the next term of the root. Write this result in the root, and annex it to the trial divisor, to form a complete divisor.

Multiply the complete divisor by this term of the root, and subtract the product from the first remainder.

Continue in this manner until all the terms of the root are found.

Find the square root of

3. $x^2 + 4x + 4$.
4. $b^2 + 2bx + x^2$.
5. $4x^2 + 4x + 1$.
6. $a^2 + ab + \frac{1}{4}b^2$.
7. $9a^2 - 12ab + 4b^2$.
8. $a^2 + 2ab + b^2 - 2ac - 2bc + c^2$.
9. $4x^4 - 12x^3 + 13x^2 - 6x + 1$.
10. $4a^4 + 4a^3 - 7a^2 - 4a + 4$.
11. $x^6 - 4x^5 + 10x^4 - 12x^3 + 9x^2$.
12. $16a^4 - 24a^3x + 49a^2x^2 - 30ax^3 + 25x^4$.
13. $a^2 + b^2 + c^2 - 2ab - 2ac + 2bc$.
14. $40x^2 - 12x^3 + 9x^4 - 24x + 36$.
15. $4x^6 + 5x^4 + 12x^5 - 5x^2 - 10x^3 + 2x + 1$.
16. $49x^4 - 28x^3 - 17x^2 + 6x + \frac{9}{4}$.
17. $9a^2 + 6ac + c^2 + 1 - 2c - 6a$.

180   HIGH SCHOOL ALGEBRA.

18. $4m^2 + 9x^2 - 12mx + 8m - 12x + 4$.

19. $a^2 + b^2 - 12ac + 12bc - 2ab + 36c^2$.

20. $a^{2m} - 2a^m b^2 + b^4 + 2a^m c^{3n} - 2b^2 c^{3n} + c^{6n}$.

21. $1 - a + \dfrac{a^2}{4} + \dfrac{2a}{3} - \dfrac{a^2}{3} + \dfrac{a^2}{9}$.

22. $m^6 - 2m^5 - m^4 + 3m^2 + 2m + 1$.

23. $a^4 + a^3 + \tfrac{3}{4} a^2 + \tfrac{1}{4} a + \tfrac{1}{16}$.

24. $c^4 - 2c^3 + \dfrac{3c^2}{2} - \dfrac{c}{2} + \dfrac{1}{16}$.

25. $\dfrac{c^4}{4} + \dfrac{x^2}{c^2} + \dfrac{c^3}{x} - cx + \dfrac{c^2}{x^2} - 2$.

26. Find four terms of the square root of $1 + x$.

Solution.

$$\begin{array}{r|l}
1+x & 1 + \tfrac{1}{2}x - \tfrac{1}{8}x^2 + \tfrac{1}{16}x^3 \\
1 & \\
\hline
2 + \tfrac{1}{2}x \;\big|\; x & \\
\phantom{2 + \tfrac{1}{2}x\;\big|\;} x + \tfrac{1}{4}x^2 & \\
\hline
2 + x - \tfrac{1}{8}x^2 \;\big|\; -\tfrac{1}{4}x^2 & \\
\phantom{2 + x - \tfrac{1}{8}x^2 \;\big|\;} -\tfrac{1}{4}x^2 - \tfrac{1}{8}x^3 + \tfrac{1}{64}x^4 & \\
\hline
2 + x - \tfrac{1}{4}x^2 + \tfrac{1}{16}x^3 \;\big|\; +\tfrac{1}{8}x^3 - \tfrac{1}{64}x^4 &
\end{array}$$

Find the square root of the following to three terms:

27. $1 - x$.   30. $a^2 + b$.   33. $4x^2 - 1$.

28. $1 - 2a$.   31. $4 + 3a$.   34. $1 - 2y^2$.

29. $a^2 - 2b$.   32. $4a^2 + 3$.   35. $a^2 + \tfrac{1}{2}b$.

SQUARE ROOT OF NUMBERS.

$1^2 = 1$ $10^2 = 100$ $100^2 = 10000$

$9^2 = 81$ $99^2 = 9801$ $999^2 = 998001$

210. 1. How many figures are required to express the square of any number of *units?*

2. How does the number of figures required to express the second power of any number of *tens* compare with the number of figures expressing the tens? How does the number of figures expressing the second power of *hundreds* compare with the number of figures expressing hundreds?

3. If the second power of a number is expressed by 3 figures, how many orders of units are there in the number? If by 4, how many? By 5? By 7?

4. How, therefore, may the number of figures in the square root of any number be found?

211. Principles. — 1. *The square of a number is expressed by twice as many figures as is the number itself, or by one less than twice as many.*

2. *The orders of units in the square root of a number correspond to the number of periods of two figures each into which the number can be separated, beginning at units.*

212. If the tens of a number are represented by t and the units by u, the square of a number consisting of tens and units will be the square of $(t + u)$ or $t^2 + 2tu + u^2$.

Thus, $35 = 3$ tens plus 5, or $30+5$, and $35^2 = 30^2 + 2(30 \times 5) + 5^2 = 1225$.

EXAMPLES.

1. What is the square root of 1225?

FIRST PROCESS.

$$\begin{array}{r|l} 12{\cdot}25 & 30 + 5 \\ 900 & \\ \hline 325 & \\ & \\ & \\ \hline 325 & \end{array}$$

$t^2 = $

$2t = 60$
$u\ \ = 5$
$\overline{2t + u = 65}$

Explanation. — According to Prin. 2, Art. 211, the orders of units in the square root of a number may be determined by separating the number into periods of two figures each, beginning at units. Separating 1225 thus, there are found to be two orders of units in the root, — that is, it is composed of tens and units. Since the square of tens is hundreds, and the hundreds of the power are less than 16 or 4^2, and more than 9 or 3^2, the tens' figure of the root must be 3. 3 tens, or 30 squared, is 900, and 900 subtracted from 1225 leaves 325, which is equal to 2 times the tens × the units + the units ^{2}.

182 HIGH SCHOOL ALGEBRA.

Since two times the tens multiplied by the units is much greater than the square of the units, 325 is *nearly* 2 times the tens multiplied by the units. Therefore, if 325 is divided by 2 times the tens, or 60, the quotient will be approximately the units of the root. Dividing by 60, the *trial divisor*, the units are found to be 5. And since the complete divisor is found by adding the units to twice the tens, the complete divisor is $60 + 5$, or 65. This multiplied by 5 gives as a product 325, which, subtracted from 325, leaves no remainder. Therefore, the square root of 1225 is 35.

SECOND PROCESS.

$$\begin{array}{ll} & 12\cdot 25 \ \underline{|\ 35} \\ t^2 = & 9 \\ 2t\ \ \ = 60 & \overline{325} \\ u\ \ \ = 5 & \\ 2t + u = 65 & \overline{325} \end{array}$$

EXPLANATION. — In practice it is usual to place the figures of the same order in a column, and to disregard the ciphers on the right of the products.

Since any number may be regarded as composed of tens and units, the processes given above have a general application.

Thus, $325 = 32$ tens $+ 5$ units; $4685 = 468$ tens $+ 5$ units.

2. Find the square root of 137641.

SOLUTION.

$$\begin{array}{ll} & 13\cdot 76\cdot 41\ \underline{|\ 371} \\ & \ \ 9 \\ \text{Trial divisor}\ \ \ \ \ \ \ = 2\ \times 30 = \ \ 60 & 476 \\ \text{Complete divisor} = 60 + 7 = \ \ 67 & 469 \\ \text{Trial divisor}\ \ \ \ \ \ \ = 2 \times 370 = 740 & 741 \\ \text{Complete divisor} = 740 + 1 = 741 & 741 \end{array}$$

RULE. — *Separate the number into periods of two figures each, beginning at units.*

Find the greatest square in the left-hand period, and write its root for the first figure of the required root.

Square this root, subtract the result from the left-hand period, and annex to the remainder the next period for a new dividend.

Double the root already found, with a cipher annexed, for a trial divisor, and by it divide the dividend. The quotient, or quotient diminished, will be the second figure of the root. Add **to the trial divisor the figure last found, multiply this complete**

EVOLUTION.

divisor by the figure of the root found, subtract the product from the dividend, and to the remainder annex the next period for the next dividend.

Proceed in this manner until all the periods have been used thus. The result will be the square root sought.

1. When the number is not a perfect square, annex periods of decimal ciphers and continue the process.
2. Decimals are pointed off into periods of two figures each, by beginning with tenths and passing to the right.
3. The square root of a common fraction may be found by extracting the square root of both numerator and denominator separately, or by reducing it to a decimal and then extracting its root.

Extract the square root of the following:

3. 2809.
4. 3969.
5. 4356.
6. 9216.
7. 2209.
8. 1681.
9. 65536.
10. 54289.
11. 42849.
12. 70756.
13. 118336.
14. 674041.
15. 784996.
16. 776161.
17. 107584.
18. 234.09.
19. 17.3056.
20. 576.4801.
21. 938961.
22. 5875776.
23. 12574116.
24. 30858025.
25. 1.338649.
26. .000729.
27. 1034.2656.
28. 29635.6225.
29. .00720801.

Find the square root to four decimal places:

30. 5.
31. 11.
32. 13.
33. 15.
34. 7.2
35. 5.3.
36. .07.
37. .007.
38. $\frac{4}{7}$.
39. $\frac{3}{8}$.
40. $\frac{5}{9}$.
41. $\frac{6}{11}$.
42. $\frac{4}{3}$.
43. $\frac{3}{5}$.
44. $\frac{7}{15}$.
45. $\frac{10}{15}$.

213. To extract the cube root of a polynomial.

1. What is the cube of $(a+b)$?
2. Since $a^3 + 3a^2b + 3ab^2 + b^3$ is the cube of $(a+b)$, what is the cube root of $a^3 + 3a^2b + 3ab^2 + b^3$?

3. How may the first term of the root be found from $a^3 + 3a^2b + 3ab^2 + b^3$?

4. How may the second term of the root be found from the second term of the power, $3a^2b$?

5. What are the factors of $3a^2b + 3ab^2 + b^3$?

6. Since $3a^2b + 3ab^2 + b^3$ is equal to $b(3a^2 + 3ab + b^2)$, what are the factors of the last three terms of the cube of a binomial?

7. By what quantity, then, must the last three terms of $a^3 + 3a^2b + 3ab^2 + b^3$ be divided so that the quotient may be the second term of the cube root?

EXAMPLES.

1. Find the cube root of $a^3 + 3a^2b + 3ab^2 + b^3$.

PROCESS.

$$\begin{array}{l} a^3 + 3a^2b + 3ab^2 + b^3 \ \underline{\vert\ a+b} \\ a^3 \end{array}$$

Trial divisor, $\quad 3a^2 \qquad \vert\ 3a^2b + 3ab^2 + b^3$
Complete divisor, $3a^2 + 3ab + b^2\ \vert\ 3a^2b + 3ab^2 + b^3$

EXPLANATION.—Since, if the quantity is a cube, one of the terms is a cube, the first term of the root is the cube root of a^3, which is a. Subtracting a from the entire quantity, there is left $3a^2b + 3ab^2 + b^3$.

Since the second term of the root can be obtained from the first term of the remainder, by dividing it by three times the square of the root already found, the second term of the root of the quantity will be found by dividing $3a^2b$ by $3a^2$, the *trial divisor*, which gives b for the second term of the root. And since the last three terms of the cube of a binomial consist of the product of the second term of the root with 3 times the square of the first, 3 times the product of the first and second, and the square of the second, $3a^2 + 3ab + b^2$, is the entire quantity, or *complete divisor*, which is to be multiplied by b.

Multiplying by b, and subtracting, there is no remainder. Therefore, $a + b$ is the cube root of $a^3 + 3a^2b + 3ab^2 + b^3$.

Since, in cubing $a + b + c$, $a + b$ may be expressed by x, its cube will be $x^3 + 3x^2c + 3xc^2 + c^3$. Hence, it is obvious that the cube root of a quantity, whose root consists of *more than two terms*, may be extracted in the same way as in example **1**, by considering the terms already found as one term.

EVOLUTION.

2. Find the cube root of $x^6 - 3x^5 + 5x^3 - 3x - 1$.

PROCESS.

$$x^6 - 3x^5 + 5x^3 - 3x - 1 \;\underline{|\,x^2 - x - 1}$$
$$x^6$$

Trial divisor, $\quad 3x^4 \qquad\qquad | \;\; -3x^5 + 5x^3$
Complete divisor, $\; 3x^4 - 3x^3 + x^2 \;|\; -3x^5 + 3x^4 - x^3$
Trial divisor, $\qquad 3x^4 - 6x^3 + 3x^2 \qquad|\; -3x^4 + 6x^3 - 3x - 1$
Complete divisor, $\; 3x^4 - 6x^3 + 3x + 1 \;\underline{|\; -3x^4 + 6x^3 - 3x - 1}$

EXPLANATION. — The first two terms are found in the same manner as in the previous example. In finding the next term, $x^2 - x$ is considered as one quantity, which we square and multiply by 3 for a trial divisor. Dividing the remainder by this trial divisor, the next term of the root is found to be -1. Adding to this trial divisor 3 times $(x^2 - x)$, multiplied by -1 and the square of -1, we obtain the complete divisor. This, multiplied by -1, and subtracted from $-3x^4 + 6x^3 - 3x - 1$, leaves no remainder. Hence, the cube root of the quantity is $x^2 - x - 1$.

RULE. — *Arrange the polynomial with reference to the consecutive powers of some letter.*

Extract the cube root of the first term, write the result as the first term of the root, and subtract its cube from the given polynomial.

Divide the first term of the remainder by 3 times the square of the root already found, as a trial divisor, and the quotient will be the next term of the root.

Add to this trial divisor 3 times the product of the first and second terms of the root, and the square of the second term. The result will be the complete divisor.

Multiply the complete divisor by the last term of the root found, and subtract this product from the dividend. Continue in this manner until all the terms of the root are found.

Each succeeding *term of the root* may be obtained by dividing the first term of each successive remainder by 3 times the square of the first term of the root.

Find the cube root of:

3. $x^3 + 6x^2y + 12xy^2 + 8y^3$.   5. $8x^3 - 36x^2 + 54x - 27$.
4. $27a^3 + 27a^2 + 9a + 1$.   6. $27x^3 + 108x^2 + 144x + 64$.

7. $a^3 + 3a + \dfrac{3}{a} + \dfrac{1}{a^3}$.

8. $a^3 - 12a^2 + 48a - 64$.

9. $8a^3 + 12a^2 + 6a + 1$.

10. $27x^6 - 54x^5 + 63x^4 - 44x^3 + 21x^2 - 6x + 1$.

11. $8m^6 + 36m^5 + 66m^4 + 63m^3 + 33m^2 + 9m + 1$.

12. $1 - 3a + 6a^2 - 7a^3 + 6a^4 - 3a^5 + a^6$.

13. $m^3 - 3m^2 + 5 - \dfrac{3}{m^2} - \dfrac{1}{m^3}$.

14. $x^3 - 3x^2y - y^3 + 8z^3 + 6x^2z - 12xyz + 6y^2z + 12xz^2 - 12yz^2 + 3xy^2$.

15. $8a^3 - 84a^2b + 294ab^2 - 343b^3$.

16. $y^6 - 6y^5 + 21y^4 - 44y^3 + 63y^2 - 54y + 27$.

17. $1 - 9a + 39a^2 - 99a^3 + 156a^4 - 144a^5 + 64a^6$.

18. $66c^4 + 1 - 63c^3 - 9c + 8c^6 - 36c^5 + 33c^2$.

19. $a^6 + 3a^4 + 6a^2 + 7 + \dfrac{6}{a^2} + \dfrac{3}{a^4} + \dfrac{1}{a^6}$.

20. $x^6 - 12x^5y + 60x^4y^2 - 192xy^5 + 240x^2y^4 - 160x^3y^3 + 64y^6$.

214. To extract any root of a polynomial.

To find a formula for obtaining the complete divisor in extracting the *fourth, fifth, sixth,* or *any* required root of a polynomial, raise $(a + b)$ to the required power, and separate all the terms after the first into two factors, one of which shall be the first power of the second term of the root. The other factor will be the formula for the complete divisor. Thus,

$$(a + b)^5 = a^5 + 5a^4b + 10a^3b^2 + 10a^2b^3 + 5ab^4 + b^5.$$

Trial divisor, $5a^4$.

Complete divisor, $(5a^4 + 10a^3b + 10a^2b^2 + 5ab^3 + b^4)$.

$(a + b)^7 = a^7 + 7a^6b + 21a^5b^2 + 35a^4b^3 + 35a^3b^4 + 21a^2b^5 + 7ab^6 + b^7$.

Trial divisor, $7a^6$.

Complete divisor,

$(7a^6 + 21a^5b + 35a^4b^2 + 35a^3b^3 + 21a^2b^4 + 7ab^5 + b^6)$.

EVOLUTION. 187

Since the *fourth* power is the *square* of the *second* power, the *sixth* power the *cube* of the *second* power, etc., any root whose index is composed of the factors 2 or 3, or 2 and 3, may be found by extracting successively the roots corresponding to the factors of the index.

Thus, the fourth root may be obtained by extracting the square root of the square root; the sixth root, by extracting the cube root of the square root; and the eighth root, by extracting the square root of the square root of the square root.

1. Find the fourth root of $1 - 8a + 24a^2 - 32a^3 + 16a^4$.

2. Find the fifth root of $x^5 + 5x^4 + 10x^3 + 10x^2 + 5x + 1$.

3. Find the sixth root of $x^6 + 6x^5 + 15x^4 + 20x^3 + 15x^2 + 6x + 1$.

4. Find the fourth root of $a^4 + 8a^3b + 24a^2b^2 + 32ab^3 + 16b^4$.

5. Find the fifth root of $x^5 - 5x^4y + 10x^3y^2 - 10x^2y^3 + 5xy^4 - y^5$.

6. Find the fifth root of $a^5 - \dfrac{5a^4x}{y} + \dfrac{10a^3x^2}{y^2} - \dfrac{10a^2x^3}{y^3} + \dfrac{5ax^4}{y^4} - \dfrac{x^5}{y^5}$.

7. Find the sixth root of $c^6 + 15c^4z^2 + 15c^2z^4 + 6cz^5 + 6c^5z + 20c^3z^3 + z^6$.

CUBE ROOT OF NUMBERS.

$1^3 = 1$ $10^3 = 1000$ $100^3 = 1000000$

$3^3 = 27$ $36^3 = 46656$ $361^3 = 47045881$

$9^3 = 729$ $99^3 = 970299$ $999^3 = 997002999$

215. 1. How many figures are required to express the cube of any number of *units*?

2. How does the number of figures required to express the cube of any number of *tens* compare with the number of figures expressing the tens? How does the number of figures expressing the cube of any number of hundreds compare with the number of figures expressing hundreds?

3. If, then, the cube of a number is expressed by 4 figures, how many orders of units are there in the root? If by 5 figures, how many? If by 6 figures, how many? If by 8 figures, how many?

4. How may the number of figures in the cube root of a number be found?

216. PRINCIPLES. — 1. *The cube of a number is expressed by three times as many figures as the number itself, or by one or two less than three times as many.*

2. *The orders of units in the cube root of a number correspond to the number of periods of three figures each into which the number can be separated, beginning at units.*

217. If the tens of a number are represented by t and the units by u, the cube of a number consisting of tens and units will be the cube of $(t + u)$ or $t^3 + 3t^2u + 3tu^2 + u^3$.

Thus, $35 = 3$ tens $+ 5$ units, or $30 + 5$, and $35^3 = 30^3 + 3(30^2 \times 5) + 3(30 + 5^2) + 5^3 = 42875$.

EXAMPLES.

1. What is the cube root of 13824?

FIRST PROCESS.

$$\begin{array}{rrr|l}
 & & 13\cdot 824 &)20 + 4 \\
t^3 = & & 8000 & \\
\text{Trial divisor,} \quad 3t^2 = & 1200 & \overline{5824} & \\
3tu = & 240 & & \\
u^2 = & 16 & & \\
\text{Complete divisor,} \quad = & 1456 & \overline{5824} &
\end{array}$$

EXPLANATION. — According to Prin. 2, Art. 216, the orders of units may be determined by separating the number into periods of three figures each, beginning at units. Separating 13824 thus, there are found to be two orders of units in the root; that is, it is composed of tens and units.

Since the cube of tens is thousands, and the thousands of the power are less than 27 or 3^3, and more than 8 or 2^3, the tens figure of the root must be 2. 2 tens, or 20 cubed, is 8000, and 8000 subtracted from 13824 leaves 5824, which is equal to three times the tens ^{2} × the units $+$ 3 times the tens × the units ^{2} + the units ^{3}.

Since three times the tens ^{2} is much greater than three times the tens × he units ^{2}, 5824 is a little more than three times the tens ^{2} × the units.

EVOLUTION.

If, then, 5824 is divided by 3 times the tens ^{2}, or 1200, the trial divisor, the quotient 4 will be the units of the root, provided proper allowance has been made for the additions necessary to obtain the complete divisor.

Since the complete divisor is found by adding to 3 times the tens ^{2}, 3 times the tens × the units and the units ^{2} (Art. 217), the complete divisor will be $1200 + 240 + 16$, or 1456. This multiplied by 4 gives as a product, 5824, which, subtracted from 5824, leaves no remainder. Therefore, the cube root of 13824 is 24.

SECOND PROCESS.

$$\begin{array}{r|l} 13\cdot 824 & 24 \\ \hline \end{array}$$

$t^3 =$ 8
$3t^2 = 1200$ | 5824
$3tu = 240$
$u^2 = 16$
$\overline{1456}$ | 5824

EXPLANATION. — In practice it is usual to place figures of the same order in a column, and to disregard the ciphers on the right of the products.

Since any number may be regarded as composed of tens and units, the processes given have a general application.

Thus, $468 = 46$ tens $+ 8$ units ; $3829 = 382$ tens $+ 9$ units.

2. What is the cube root of 48228544?

SOLUTION.

$$\begin{array}{r|l} 48\cdot 228\cdot 544 & 364 \\ \hline \end{array}$$

27 | 21228

Trial divisor $= 3(30)^2 = 2700$
$3(30 \times 6) = 540$
$6^2 = 36$
Complete divisor $= 3276$ | 19656
Trial divisor $= 3(360)^2 = 388800$ | 1572544
$3(360 \times 4) = 4320$
$4^2 = 16$
Complete divisor $= 393136$ | 1572544

RULE. — *Separate the number into periods of three figures each, beginning at units.*

Find the greatest cube in the left-hand period, and write its root for the first figure of the required root. Cube the root, subtract the result from the left-hand period, and annex to the remainder the next period for a dividend.

Take 3 times the square of the root already found with a cipher annexed, for a trial divisor, and by it divide the dividend.

The quotient, or quotient diminished, will be the second figure of the root.

To this trial divisor add 3 times the product of the first part of the root with a cipher annexed, multiplied by the second part, and also the square of the second part. Their sum will be the complete divisor.

Multiply the complete divisor by the second part of the root, and subtract the product from the dividend. Continue thus until all the figures of the root have been found.

1. When there is a remainder after subtracting the last product, annex decimal ciphers and continue the process.
2. Decimals are pointed off into periods of three figures each, by beginning at tenths and passing to the right.
3. The cube root of a common fraction may be found by extracting the cube root of both numerator and denominator separately, or by reducing it to a decimal and then extracting its root.
4. A rule for the extraction of any root may be formed from the general formulas, as is shown on page 188.

Extract the cube root of the following:

3. 74088.
4. 262144.
5. 166375.
6. 34965783.
7. 130323843.
8. 704969.
9. 185193.
10. 250047.
11. .015625.
12. 12.812904.
13. 5545233.
14. 2000376.
15. 153990656.
16. 60236.288.
17. .000064.

Extract the cube root to three decimal places.

18. 3.
19. .27.
20. 6.4.
21. .00465.
22. $\frac{5}{6}$.
23. $\frac{7}{8}$.
24. $\frac{3}{11}$.
25. $\frac{4}{15}$.

218. To extract any root of numbers.

By following the method for finding the trial and complete divisors exemplified in Art. 214, any root of a number may be found.

Roots of numbers higher than the cube are, however, rarely extracted, and when it is necessary to obtain them, they are usually found by the aid of Logarithms, as will be shown hereafter.

THEORY OF EXPONENTS.

219. Thus far the exponents used have been positive integers. It is, however, found desirable to use fractional and negative exponents, even though they seem to be meaningless or inconsistent with the definition which has been given of the term exponent. By definition, an exponent shows how many times a quantity is used as a factor, but the exponent in the expression $a^{\frac{3}{4}}$ cannot indicate any such fact, nor can the exponent in the expression a^{-2}.

But since the laws governing exponents that are positive integers have been already established, to avoid complexity and confusion, fractional and negative exponents must conform to the same fundamental laws, so that the laws may be universal in their application. Consequently it is necessary to interpret or explain the meaning of fractional and negative exponents.

220. From the definition of an exponent the following laws for positive integral exponents have been established:

1. $a^m \times a^n = a^{m+n}$.
2. $a^m \div a^n = a^{m-n}$.
3. $(a^m)^n = a^{mn}$.
4. $\sqrt[n]{a^{mn}} = a^m$.

221. The meaning of fractional exponents.

1. What is the cube root of x^6? How is it obtained? Express the root by indicating the division of the exponent.

2. In the expression $x^{\frac{6}{3}}$, what does 6 express? What does 3 express?

3. What is the fourth root of x^8? Express the root by indicating the division of the exponent.

4. In the expression $x^{\frac{8}{4}}$, what does 8 express? What does 4 express?

5. What does the numerator of a fractional exponent indicate? What, the denominator?

222. Principle. — *When the terms of the fractional exponent are positive integers, the numerator indicates a power, and the denominator a root.*

223. The truth of the principle may be shown as follows: Since the laws of exponents already established must apply to fractional as well as integral exponents,

1. $a^{\frac{3}{4}} \times a^{\frac{3}{4}} \times a^{\frac{3}{4}} \times a^{\frac{3}{4}} = a^{\frac{12}{4}} = a^3$.

2. $a^{\frac{1}{n}} \times a^{\frac{1}{n}} \times a^{\frac{1}{n}}$ to n factors $= a^{\frac{n}{n}} = a$, n being a positive integer.

3. $a^{\frac{m}{n}} \times a^{\frac{m}{n}} \times a^{\frac{m}{n}}$ to n factors $= a^{\frac{nm}{n}} = a^m$, m and n being positive integers.

That is, $a^{\frac{3}{4}}$ or $\sqrt[4]{a^3}$ is one of the four equal factors of a^3; $a^{\frac{1}{n}}$ or $\sqrt[n]{a}$ is one of the n equal factors of a; $a^{\frac{m}{n}}$ or $\sqrt[n]{a^m}$ is one of the n equal factors of a^m.

Therefore, since the laws for positive integral exponents are to apply to positive fractional exponents, the numerator indicates a power of a quantity, and the denominator its root.

EXAMPLES.

Express with radical signs:

1. $a^{\frac{1}{2}} b^{\frac{1}{3}}$.
2. $3 a^{\frac{1}{3}}$.
3. $(3a)^{\frac{1}{3}}$.
4. $2 a^{\frac{1}{1}} b^{\frac{2}{5}}$.
5. $3 x^{\frac{1}{2}} y^{\frac{2}{3}}$.
6. $a x^{\frac{1}{n}}$.
7. $4 a^{\frac{1}{n}} b^{\frac{1}{m}}$.
8. $a^{\frac{1}{3}} b^{\frac{1}{4}} c^{\frac{1}{n}}$.

Express the following with fractional exponents:

9. $\sqrt{ab}$.
10. $\sqrt[3]{a^2}$.
11. $(\sqrt[3]{a})^4$.
12. $\sqrt[7]{x^6}$.
13. $\sqrt[6]{a^2 b^3}$.
14. $\sqrt[3]{x^2 y^3}$.
15. $\sqrt[5]{a^2 b^3}$.
16. $3 \sqrt[3]{a^2 x^3 y^4}$.

Find the values of the following:

17. $16^{\frac{3}{2}}$.
18. $25^{\frac{1}{2}}$.
19. $-27^{\frac{4}{3}}$.
20. $8^{\frac{2}{3}}$.
21. $36^{\frac{3}{2}}$.
22. $9^{\frac{5}{2}}$.
23. $\left(\frac{9}{16}\right)^{\frac{3}{2}}$.
24. $\left(\frac{81}{256}\right)^{\frac{3}{4}}$.

The process is shortened by extracting the root first.

THEORY OF EXPONENTS.

224. The meaning of any finite quantity having the exponent 0.

It has been shown in Art. 92, that any finite quantity having 0 for an exponent is equal to 1. The truth of the principle may be shown as follows:

By the laws of exponents already established,
$$a^m \div a^m = a^0$$
also
$$a^m \div a^m = 1$$
$$\therefore a^0 = 1$$

225. The meaning of negative exponents.

It was shown in Art. 92, that any quantity with a negative exponent is equal to its reciprocal with an equal positive exponent. The truth of the principle may be shown as follows:

Assuming the laws relating to exponents to be of general application,
$$a^n \times a^{-n} = a^0 = 1$$
Dividing by a^n,
$$a^{-n} = \frac{a^0}{a^n} = \frac{1}{a^n}$$

226. PRINCIPLE. — *Any factor may be transferred from one term of a fraction to the other, if the sign of its exponent is changed.*

EXAMPLES.

Write the following with positive exponents:

1. $3a^{-2}$.
2. $4b^{-3}$.
3. $a^{-2}b^{-3}$.
4. $x^{-1}y^{-2}$.
5. $6a^{-3}y$.
6. x^3y^{-2}.
7. a^2b^{-5}.
8. $4x^3y^{-7}$.
9. $3c^{-1}d^{-3}x$.
10. $2ab^{-n}c$.
11. $7a^{-1}b^2c^{-3}$.
12. $5a^{-3}b^2c^n$.

Transfer the literal factors from the denominator to the numerator:

13. $\dfrac{2x}{y^2}$.
14. $\dfrac{a^2x}{y^3}$.
15. $\dfrac{3ax}{y^{-4}}$.
16. $\dfrac{c}{xy}$.
17. $\dfrac{abc}{x^{-1}y^2}$.
18. $\dfrac{5x^2}{yz^2}$.
19. $\dfrac{3a^2x^3}{y^nz^3}$.
20. $\dfrac{2bc}{d^{-n}}$.

HIGH SCHOOL ALGEBRA.

227. Since the meanings of fractional and negative exponents were obtained upon the theory that the principles and laws already established, relating to exponents, were of general application, it follows that $a^m \times a^n = a^{m+n}$, whether m and n are integral or fractional, positive or negative.

$$a^m \times a^n = a^{m+n}$$
$$a^3 \times a^{-6} = a^{3-6} = a^{-3}$$
$$a^{-\frac{2}{3}} \times a^{-\frac{3}{4}} = a^{-\frac{2}{3}-\frac{3}{4}} = a^{-1\frac{5}{12}}, \text{ etc.}$$

EXAMPLES.

Find the product of the following:

1. $2a^5 \times 3a^{\frac{2}{3}}$.
2. $5x^{-1} \times 3x^4$.
3. $3a^3 \times 2a^{\frac{1}{2}}$.
4. $4b^{-2} \times 2b^3$.
5. $5c^4 \times 2c^{\frac{2}{3}}$.
6. $7a^5 \times a^{-4}$.
7. $3b \times ab^{-\frac{3}{4}}$.
8. $2a \times a^{-\frac{2}{3}}b^2$.
9. $3x \times ax^{-\frac{2}{3}}$.
10. $4ax \times a^{\frac{2}{3}}x$.
11. $2c^3 \times 4c^{-\frac{3}{5}}x$.
12. $5ad \times 2d^{-n}$.
13. $2a^{-\frac{2}{3}} \times 3b\sqrt[4]{a^2}$.
14. $b^{-\frac{3}{4}} \times a\sqrt{b^2}$.
15. $3c^{-\frac{2}{3}} \times 5ab\sqrt{c}$.
16. $ab^{\frac{1}{2}} \times c\sqrt[4]{b^3}$.
17. $a^{\frac{2}{3}}b^{\frac{1}{4}}c^{-\frac{1}{2}} \times a^{\frac{1}{3}}b^{-\frac{3}{4}}c^2$.
18. $x^{\frac{5}{8}}y^{\frac{2}{3}}z \times x^{-\frac{1}{8}}y\sqrt[5]{z^3}$.

19. Multiply $x^{\frac{2}{3}} - x^{\frac{1}{3}}y^{\frac{1}{3}} + y^{\frac{2}{3}}$ by $x^{\frac{1}{3}} + y^{\frac{1}{3}}$.

$$\begin{array}{l}
x^{\frac{2}{3}} - x^{\frac{1}{3}}y^{\frac{1}{3}} + y^{\frac{2}{3}} \\
x^{\frac{1}{3}} + y^{\frac{1}{3}} \\
\hline
x\phantom{^{\frac{2}{3}}} - x^{\frac{2}{3}}y^{\frac{1}{3}} + x^{\frac{1}{3}}y^{\frac{2}{3}} \\
 + x^{\frac{2}{3}}y^{\frac{1}{3}} - x^{\frac{1}{3}}y^{\frac{2}{3}} + y \\
\hline
x\phantom{^{\frac{2}{3}} - x^{\frac{2}{3}}y^{\frac{1}{3}} + x^{\frac{1}{3}}y^{\frac{2}{3}}\ } + y
\end{array}$$

THEORY OF EXPONENTS. 195

20. Multiply $a^2b^{-2} - 2 + a^{-2}b^2$ by $a^2b^{-2} + 2 + a^{-2}b^2$.

$$
\begin{array}{l}
a^2b^{-2} - 2 + a^{-2}b^2 \\
a^2b^{-2} + 2 + a^{-2}b^2 \\
\hline
a^4b^{-4} - 2a^2b^{-2} + a^0b^0 \\
\qquad + 2a^2b^{-2} \qquad\quad - 4 + 2a^{-2}b^2 \\
\qquad\qquad\qquad + a^0b^0 \quad - 2a^{-2}b^2 + a^{-4}b^4 \\
\hline
a^4b^{-4} \qquad\qquad + 2a^0b^0 - 4 \qquad\qquad + a^{-4}b^4
\end{array}
$$

$$= a^4b^{-4} \qquad\qquad + 2 \quad - 4 \qquad\qquad + a^{-4}b^4$$

$$= a^4b^{-4} - 2 + a^{-4}b^4.$$

21. Multiply $a^{\frac{1}{2}} + b^{\frac{1}{2}}$ by $a^{\frac{1}{2}} + b^{\frac{1}{2}}$.

22. Multiply $x^{\frac{1}{3}} + y^{\frac{1}{3}}$ by $x^{\frac{2}{3}} + y^{\frac{2}{3}}$.

23. Multiply $x^{\frac{1}{3}} + x^{\frac{1}{6}}y^{\frac{1}{6}} + y^{\frac{1}{3}}$ by $x^{\frac{1}{3}} + y^{\frac{1}{3}}$.

24. Multiply $2x^{\frac{1}{3}} + x + 3x^{\frac{2}{3}}$ by $x^{\frac{2}{3}} + 2$.

25. Multiply $x^{\frac{1}{4}} + x^{\frac{1}{2}}y^{\frac{1}{2}} + y^{\frac{1}{4}}$ by $x^{\frac{1}{4}} - y^{\frac{1}{4}}$.

26. Multiply $a^{\frac{2}{3}} + a^{\frac{1}{3}}b^{\frac{1}{3}} + b^{\frac{2}{3}}$ by $a^{-\frac{1}{3}} - b^{-\frac{1}{3}}$.

27. Multiply $x^2 + x + 1$ by $x - x^{-1} - 1$.

28. Multiply $x^3y^{-3} - 3 + x^{-3}y^3$ by $x^3y^{-3} + 3 + x^{-3}y^3$.

29. Multiply $x^2y^{-2} - xy + x^{-2}y^2$ by $x + xy - y$.

30. Multiply $x^{-3} - y^{-3} + x^{-\frac{3}{2}}y^{-\frac{3}{2}}$ by $x^3 + y^3$.

228. Since the law relating to exponents in division is of general application, $a^m \div a^n = a^{m-n}$, whether m and n are positive or negative, integral or fractional.

$$a^m \div a^n = a^{m-n}$$

$$a^3 \div a^{-6} = a^{3+6} = a^9$$

$$a^{-\frac{2}{3}} \div a^{-\frac{3}{4}} = a^{-\frac{2}{3}+\frac{3}{4}} = a^{\frac{1}{12}}, \text{ etc.}$$

EXAMPLES.

Find the quotients of:

1. $a^5 \div a^3$.
2. $a^3 \div a^{-2}$.
3. $a^{-3} \div a^{-1}$.
4. $a^{-5} \div a^{-9}$.
5. $a^{\frac{3}{4}} \div a^{-\frac{2}{3}}$.
6. $c^{-\frac{4}{5}} \div c^{-\frac{2}{3}}$.
7. $x^{-\frac{3}{4}} \div x^{\frac{1}{3}}$.
8. $z^{\frac{3}{7}} \div z^{\frac{2}{3}}$.
9. $b^{\frac{5}{6}} \div b^{\frac{1}{3}}$.
10. $b^2 c^{-\frac{2}{3}} \div bc^{\frac{1}{4}}$.
11. $a^{\frac{1}{2}} x^{\frac{3}{4}} \div a^{\frac{1}{4}} x^{\frac{2}{3}}$.
12. $x^{-\frac{3}{5}} y^{\frac{3}{4}} \div x^{\frac{3}{4}} y^{\frac{3}{5}}$.

13. Divide $a - b$ by $a^{\frac{1}{3}} - b^{\frac{1}{3}}$.

$$\begin{array}{r|l} a - b & a^{\frac{1}{3}} - b^{\frac{1}{3}} \\ \underline{a - a^{\frac{2}{3}} b^{\frac{1}{3}}} & \overline{a^{\frac{2}{3}} + a^{\frac{1}{3}} b^{\frac{1}{3}} + b^{\frac{2}{3}}} \\ a^{\frac{2}{3}} b^{\frac{1}{3}} - b & \\ \underline{a^{\frac{2}{3}} b^{\frac{1}{3}} - a^{\frac{1}{3}} b^{\frac{2}{3}}} & \\ a^{\frac{1}{3}} b^{\frac{2}{3}} - b & \\ \underline{a^{\frac{1}{3}} b^{\frac{2}{3}} - b} & \end{array}$$

14. Divide $2 x^{\frac{2}{3}} y^{-1} - 20 + 18 x^{-\frac{2}{3}} y$ by $x^{\frac{2}{3}} y^{-\frac{1}{2}} + 2 x^{\frac{1}{3}} - 3 y^{\frac{1}{2}}$.

$$\begin{array}{r|l} 2 x^{\frac{2}{3}} y^{-1} - 20 + 18 x^{-\frac{2}{3}} y & x^{\frac{2}{3}} y^{-\frac{1}{2}} + 2 x^{\frac{1}{3}} - 3 y^{\frac{1}{2}} \\ \underline{2 x^{\frac{2}{3}} y^{-1} + 4 x^{\frac{1}{3}} y^{-\frac{1}{2}} - 6} & \overline{2 y^{-\frac{1}{2}} - 4 x^{-\frac{1}{3}} - 6 x^{-\frac{2}{3}} y^{\frac{1}{2}}} \\ - 4 x^{\frac{1}{3}} y^{-\frac{1}{2}} - 14 + 18 x^{-\frac{2}{3}} y & \\ \underline{- 4 x^{\frac{1}{3}} y^{-\frac{1}{2}} - 8 + 12 x^{-\frac{1}{3}} y^{\frac{1}{2}}} & \\ - 6 - 12 x^{-\frac{1}{3}} y^{\frac{1}{2}} + 18 x^{-\frac{2}{3}} y & \\ \underline{- 6 - 12 x^{-\frac{1}{3}} y^{\frac{1}{2}} + 18 x^{-\frac{2}{3}} y} & \end{array}$$

Divide:

15. $x + 2 x^{\frac{1}{2}} y^{\frac{1}{2}} + y$ by $x^{\frac{1}{2}} + y^{\frac{1}{2}}$.

16. $x - y$ by $x^{\frac{1}{3}} - y^{\frac{1}{3}}$.

17. $a^{\frac{4}{3}} + 2 a^{\frac{2}{3}} b^{\frac{2}{3}} + b^{\frac{4}{3}}$ by $a^{\frac{2}{3}} + b^{\frac{2}{3}}$.

18. $a + 3 a^{\frac{2}{3}} x^{\frac{1}{3}} + 3 a^{\frac{1}{3}} x^{\frac{2}{3}} + x$ by $a^{\frac{1}{3}} + x^{\frac{1}{3}}$.

19. $x - y$ by $x^{\frac{2}{3}} + x^{\frac{1}{3}} y^{\frac{1}{3}} + y^{\frac{2}{3}}$.

THEORY OF EXPONENTS. 197

20. $x^{-3}+x^{-2}y^{-1}+x^{-1}y^{-2}+y^{-3}$ by $x^{-1}+y^{-1}$.
21. $x^{-6}-3x^{-4}y^{-2}+3x^{-2}y^{-4}-y^{-6}$ by $x^{-2}-y^{-2}$.
22. $a^{-2}b^2+a^{-1}b+ab^{-1}-a^2b^{-2}$ by a^2+b^2.
23. $c^{\frac{4}{3}}+c+c^{\frac{1}{3}}+1$ by $c^{\frac{2}{3}}+2c^{\frac{1}{3}}+1$.

229. Since the law relating to exponents in involution is of general application, $(a^m)^n = a^{mn}$ whether m and n be positive or negative, integral or fractional.

According to the general law $(a^m)^{-n} = a^{-mn}$.

It can be shown to be true in the following manner:

1. To show that $(a^m)^{-n} = a^{-mn}$.

$$(a^m)^{-n} = \frac{1}{(a^m)^n} = \frac{1}{a^{mn}} = a^{-mn}$$

2. To show that $(a^m)^{-\frac{p}{q}} = a^{-\frac{mp}{q}}$.

$$(a^m)^{-\frac{p}{q}} = \frac{1}{(a^m)^{\frac{p}{q}}} = \frac{1}{(a^{mp})^{\frac{1}{q}}} = \frac{1}{\sqrt[q]{a^{mp}}} = \frac{1}{a^{\frac{mp}{q}}} = a^{-\frac{mp}{q}}$$

The other instances may be shown by similar processes.

EXAMPLES.

Find the values of:

1. $(a^{\frac{1}{2}})^4$.
2. $(a^{-\frac{1}{2}})^{\frac{2}{3}}$.
3. $(c^{\frac{1}{3}})^{-3}$.
4. $(b^{\frac{2}{3}})^{\frac{3}{2}}$.
5. $(d^{\frac{2}{5}})^{\frac{1}{2}}$.
6. $(x^{\frac{2}{3}})^{-\frac{3}{4}}$.
7. $(x^{-\frac{3}{5}})^{-\frac{2}{5}}$.
8. $(z^{-\frac{1}{2}})^4$.
9. $(c^{-\frac{2}{3}})^6$.
10. $(9c^{-\frac{2}{3}})^{-\frac{3}{2}}$.
11. $(8a^{-3})^{-\frac{2}{3}}$.
12. $(36x^4)^{-\frac{3}{2}}$.
13. $(z^{\frac{2}{5}})^{-\frac{1}{2}}$.
14. $(x^{\frac{3}{4}}y^{\frac{1}{2}})^{-\frac{2}{3}}$.
15. $(16x^2)^{-\frac{7}{2}}$.
16. $\left(\dfrac{9x^4}{25y^{-3}}\right)^{-\frac{3}{2}}$.

230. MISCELLANEOUS EXAMPLES.

Expand:

1. $(a^{\frac{1}{3}}+b^{\frac{1}{3}})^2$.
2. $(x^{\frac{2}{3}}-y^{\frac{2}{3}})^2$.
3. $(c^{\frac{5}{6}}+x^{\frac{5}{6}})^2$.
4. $(x^{\frac{2}{3}}+y^{-\frac{2}{3}})(x^{\frac{2}{3}}-y^{-\frac{2}{3}})$.
5. $(x^{-\frac{1}{2}}+y^{\frac{1}{2}})^2(x-y)$.
6. $(x^{\frac{3}{4}}+z^{\frac{3}{4}})^2(x^{\frac{3}{2}}-z^{\frac{3}{2}})$.

HIGH SCHOOL ALGEBRA.

Multiply:

7. $a^{\frac{2}{3}} + a^{\frac{1}{3}}b^{\frac{1}{3}} + b^{\frac{2}{3}}$ by $a^{\frac{1}{3}} - b^{\frac{1}{3}}$.

8. $c^{\frac{3}{4}} - 2c^{\frac{1}{4}} + 4$ by $c^{\frac{1}{4}} - 1$.

9. $4a^{\frac{3}{5}} + 2a^{\frac{2}{5}}b^{\frac{1}{5}} + 3a^{\frac{1}{5}}b^{\frac{2}{5}} + 9$ by $2a^{\frac{2}{5}} - b$.

10. $x^{2p} - x^p y^p + y^{2p}$ by $x^p + y^p$.

Divide:

11. $21c + c^{\frac{2}{3}} + c^{\frac{1}{3}} + 1$ by $3c^{\frac{1}{3}} + 1$.

12. $a^2 b^{-2} + 2 + a^{-2} b^2$ by $ab^{-1} + a^{-1} b$.

13. $x^{5n} - y^{5n}$ by $x^n - y^n$.

14. $a^{\frac{4}{3}} + a + a^{\frac{1}{3}} + 1$ by $a^{\frac{2}{3}} + 2a^{\frac{1}{3}} + 1$.

Extract the square root of:

15. $9c - 4c^{-\frac{1}{2}} + 10 - 12c^{\frac{1}{2}} + c^{-1}$.

16. $4a + 12a^{\frac{1}{2}}b^{\frac{1}{3}} + 4a^{\frac{1}{2}}c + 9b^{\frac{2}{3}} + 6b^{\frac{1}{3}}c + c^2$.

17. $x^{-\frac{2}{3}} + 4x^{\frac{1}{3}} - 10x^{-\frac{1}{3}} + 4x^{\frac{4}{3}} - 20x^{\frac{2}{3}} + 25$.

18. $x + 2x^{\frac{3}{4}} + 3x^{\frac{1}{2}} + 2x^{\frac{1}{4}} + 1$.

19. $4x^{\frac{1}{2}} - 20x^{\frac{3}{8}}y^{\frac{1}{8}} + 37x^{\frac{1}{4}}y^{\frac{1}{4}} - 30x^{\frac{1}{8}}y^{\frac{3}{8}} + 9y^{\frac{1}{2}}$.

Extract the cube root of:

20. $x + 12x^{\frac{2}{3}} + 48x^{\frac{1}{3}} + 64$.

21. $x^2 - 3x^{\frac{5}{3}} + 5x - 3x^{\frac{1}{3}} - 1$.

22. $\frac{1}{8}x^3 - \frac{3}{2}x^2 y^{\frac{1}{2}} + 6xy - 8y^{\frac{3}{2}}$.

23. $8a^3 + 12a^2 - 30a - 35 + 45a^{-1} + 27a^{-2} - 27a^{-3}$.

RADICAL QUANTITIES.

231. 1. In what two ways may the root of a quantity be indicated?

2. What is indicated by $\sqrt{x^2}$? By $(3a)^{\frac{1}{3}}$? By $5b^{\frac{1}{4}}$?

3. In the expression $5\sqrt{x^3}$, what shows how many times $\sqrt{x^3}$ is taken?

4. What is that called which shows how many times a quantity is taken?

5. To what quantity without the radical sign is $\sqrt{9x^2}$ exactly equal? $\sqrt{36b^2}$? $\sqrt{49m^2}$? $\sqrt[3]{27a^3}$?

6. To what quantity without the radical sign is $\sqrt{7a}$ exactly equal? $\sqrt{8x}$? $\sqrt{12x^3}$? $\sqrt[3]{15a^2}$? $\sqrt[3]{10x^2}$?

7. What is the square root of 9? Of 4? Of 9×4? What is the square root of a^2? Of b^2? Of $a^2 \times b^2$? What is the cube root of 8? Of 27? Of 8×27?

8. How does the product of the roots of two quantities compare with the root of the product of the quantities?

9. How, then, does the root of a quantity compare with the product of the roots of its factors?

232. A **Radical Quantity,** or **Radical,** is an indicated root of a quantity.

The root may be indicated by the *radical sign* or by a *fractional exponent.*

Thus, $\sqrt{7x^2}$, $\sqrt{25a^2}$, $\sqrt[3]{4x}$, $x^{\frac{1}{3}}$, $5x^{\frac{1}{4}}$, and $4a^{\frac{2}{3}}$ are radical quantities.

233. The Coëfficient of a radical is the quantity prefixed to the radical part to show how many times it is taken.

Thus, a is the coëfficient of $a\sqrt{x^2y}$, and bc of $bcx^{\frac{1}{4}}$.

234. The Degree of a radical is indicated by the index of the radical or the denominator of the fractional exponent.

Thus, $\sqrt{x}$, $ax^{\frac{1}{2}}$, $(x+y)^{\frac{1}{2}}$, are radicals of the *second* degree.

$\sqrt[3]{y^2}$, $\sqrt[3]{a+b}$, $(c+d)^{\frac{1}{3}}$, are radicals of the *third* degree.

235. Similar Radicals are those which have the same quantity under the radical sign, with the same index, or which have the same fractional exponent.

Thus, $5\sqrt[3]{x^2y}$, $a\sqrt[3]{x^2y}$, $5\,b(x^2y)^{\frac{1}{3}}$, are similar radicals.

236. A Rational Quantity is a quantity without a radical sign, or one whose indicated root can be extracted exactly.

Thus, $\sqrt{49\,x^2}$, $\sqrt[3]{27\,y^3}$, $(a^2+2\,ab+b^2)^{\frac{1}{2}}$, are rational quantities.

237. A Surd, or Irrational Quantity, is a quantity whose indicated root cannot be extracted exactly.

Thus, $\sqrt{3\,x}$, $\sqrt[3]{7\,x^2}$, $(a^2+b^2)^{\frac{1}{2}}$, are irrational quantities.

A surd of the second degree is called a Quadratic Surd.

238. Principle. — *The root of any quantity is equal to the product of the same roots of its factors.*

REDUCTION OF RADICALS.

239. To reduce a radical to its simplest form.

When a radical is integral and contains no factor which is a perfect power corresponding to the degree of the radical, it is in its *simplest form*.

1. Reduce $\sqrt{16\,a^2x}$ to its simplest form.

PROCESS.

$$\sqrt{16\,a^2x} = \sqrt{16\,a^2} \times \sqrt{x} = 4\,a\sqrt{x}$$

Explanation. — Since a radical is in its simplest form when it contains no factor which is a perfect power corresponding to the degree of the

RADICAL QUANTITIES.

radical, we separate $16\,a^2x$ into two factors, one of which is a perfect square, since the radical is of the second degree. The factors are $16\,a^2$ and x. Then, by Principle (Art. 238), $\sqrt{16\,a^2x} = \sqrt{16\,a^2} \times \sqrt{x}$; or, extracting the root of the perfect square, the original expression becomes $4\,a\sqrt{x}$.

2. Reduce $\sqrt[3]{432}$ to its simplest form.

Solution.

$$\sqrt[3]{432} = \sqrt[3]{216} \times \sqrt[3]{2} = 6\sqrt[3]{2}$$
$$\therefore \sqrt[3]{432} = 6\sqrt[3]{2}$$

Rule. — *Separate the quantity under the radical sign into two factors, one of which is the highest factor which is a perfect power of the radical corresponding in degree with the radical. Extract the required root of the rational factor, multiply the result by the coëfficient of the given radical, and place the product as the coëfficient of the surd.*

Reduce the following to their simplest form:

3. $\sqrt{45}$.
4. $\sqrt{75}$.
5. $\sqrt[3]{54}$.
6. $\sqrt{108}$.
7. $\sqrt{192}$.
8. $\sqrt[3]{162}$.
9. $\sqrt[3]{-56}$.
10. $\sqrt[3]{375}$.
11. $\sqrt[3]{-686}$.
12. $\sqrt[5]{486}$.
13. $\sqrt{18\,x^2}$.
14. $\sqrt{36\,a^2b}$.
15. $\sqrt{75\,x^3y^4}$.
16. $\sqrt{100\,a^6b^3}$.
17. $\sqrt{432\,a^3x^2y}$.
18. $\sqrt{81\,a^2xy}$.
19. $3\sqrt{4\,xy^2z^2}$.
20. $5\sqrt{18\,x^3y^2z^2}$.
21. $\sqrt{a^2 - a^2x}$.
22. $\sqrt[3]{x^4 - x^3y^2}$.
23. $\sqrt{a^3 - a^2b}$.
24. $\sqrt[3]{x^5 + x^3y^2}$.
25. $\sqrt{ax^2 - x^3}$.
26. $\sqrt{2\,b^3 + b^4y}$.
27. $\sqrt[3]{3\,b^3 + b^4x}$.
28. $\sqrt[3]{ax^4 - 2\,x^3y}$.
29. $\sqrt{4\,ax - 8}$.
30. $\sqrt{9\,a^3 + 18\,y}$.
31. $\sqrt[3]{16\,x^3 + 54\,x^4y^2}$.
32. $\sqrt[3]{(a-x)^5y^4}$.

33. $a\sqrt[3]{a^4 - a^3x^2}$.
34. $a\sqrt{a^3 + 2\,a^2b + ab^2}$.
35. $(a+b)\sqrt{a^4 - a^2b^2}$.
36. $(x+y)\sqrt{x^3 - 2\,x^2y + xy^2}$.
37. $(x^2 + x^2y + x^2y^2)^{\frac{1}{2}}$.
38. $4\,a(x + 2\,xy + xy^2)^{\frac{1}{2}}$.

240. A Fractional Radical Quantity is in its *simplest form*, when only a surd in its simplest form without a denominator is left under the radical.

39. Reduce $\sqrt{\dfrac{3a^2}{5b}}$ to its simplest form.

PROCESS.

$$\sqrt{\dfrac{3a^2}{5b}} = \sqrt{\dfrac{3a^2 \times 5b}{5b \times 5b}} = \sqrt{\dfrac{15a^2 b}{25b^2}} = \sqrt{15b \times \dfrac{a^2}{25b^2}} = \dfrac{a}{5b}\sqrt{15b}$$

EXPLANATION.—Since the denominator is to be removed, it must be made a perfect square. This can be done by multiplying the terms of the fraction by $5b$. Then factoring, as in previous examples, and extracting the root of the rational part, the result is $\dfrac{a}{5b}\sqrt{15b}$, which is the simplest form of the original expression.

Reduce the following to their simplest form:

40. $\sqrt{\dfrac{2}{3}}.$ **43.** $\sqrt[3]{\dfrac{5}{9}}.$ **46.** $\sqrt{\dfrac{5xy}{6ab}}.$ **49.** $\sqrt{\dfrac{3x^3}{5a}}.$

41. $\sqrt{\dfrac{3}{7}}.$ **44.** $\sqrt{\dfrac{5a}{7b}}.$ **47.** $\sqrt[3]{\dfrac{7a}{8b}}.$ **50.** $2\sqrt{\dfrac{5}{8}}.$

42. $\sqrt{\dfrac{5}{8}}.$ **45.** $\sqrt{\dfrac{7x}{9y}}.$ **48.** $\sqrt[3]{\dfrac{2c}{3x}}.$ **51.** $3\sqrt{\dfrac{7a}{12b}}.$

52. $(x-2)\sqrt{\dfrac{x-2}{x+2}}.$ **55.** $(a^2-b^2)\sqrt{\dfrac{x}{a^2+b^2}}.$

53. $(a+b)\sqrt{\dfrac{a}{a-b}}.$ **56.** $3a\left(\dfrac{3x}{7y}\right)^{\frac{1}{2}}.$

54. $(x+y)\sqrt[3]{\dfrac{x}{(x+y)^2}}.$ **57.** $5xy\left(\dfrac{3x^2y}{4x^2y^2}\right)^{\frac{1}{3}}.$

241. To reduce a rational quantity to the form of a radical.

1. What is the root of $\sqrt{4a^2}$? To what quantity under the radical sign of the second degree is $2a$ equal?

2. What is the value of $\sqrt[3]{27a^3}$? To what quantity under the radical sign of the third degree is $3a$ equal?

RADICAL QUANTITIES.

EXAMPLES.

1. Reduce $6a^2xy$ to the form of the cube root.

PROCESS.

$$6a^2xy = \sqrt[3]{(6a^2xy)^3} = \sqrt[3]{216\,a^6x^3y^3}$$

EXPLANATION.— Since any quantity is equal to the cube root of its cube, $6a^2xy$ is equal to the cube root of $(6a^2xy)^3$, or $\sqrt[3]{216\,a^6x^3y^3}$.

RULE. — *Raise the quantity to the power corresponding to the index of the given root, and place the result under the appropriate radical sign.*

The coëfficient of a radical quantity may be placed under the radical sign by raising the coëfficient to the power indicated by the index of the radical, and multiplying the quantity under the radical by the result.

Reduce to the form of the square root:

2. $3ax^2$.   4. $4a^2xy$.   6. $3x^3y^4$.   8. $a+b$.

3. $2xy$.   5. $5x^2y$.   7. $5a^{-2}x^{\frac{1}{2}}$.   9. $3a^2x^{-3}$.

Reduce to the form of the cube root:

10. $2a^2x$.   12. $2b^4c$.   14. $3a^{\frac{1}{2}}b^2$.   16. $a-x$.

11. $3ax^3$.   13. $4x^{-2}y$.   15. $a^{\frac{1}{3}}x^{\frac{1}{4}}$.   17. $(a+b)^{\frac{1}{2}}$.

Express entirely under the radical:

18. $2x\sqrt{x^2+y^2}$.

19. $(x+y)\sqrt{2x}$.

20. $3a\sqrt{a^2-b^2}$.

21. $(a-b)\sqrt{a+b}$.

22. $(x-y)\sqrt{\dfrac{a}{x-y}}$.

23. $(x^2-y^2)\sqrt{\dfrac{x}{(x-y)^2}}$.

242. To reduce radicals to equivalent radicals of the same degree.

1. To what fraction in lower terms is $\frac{3}{6}$ equal? By what equivalent expression having an exponent in lower terms can $a^{\frac{3}{6}}$ be expressed? $a^{\frac{2}{6}}$? $a^{\frac{2}{8}}$? $a^{\frac{4}{8}}$?

2. Express $a^{\frac{1}{2}}$ by an equivalent expression, with an exponent in higher terms. Express, in like manner, $x^{\frac{1}{3}}$ and $y^{\frac{1}{5}}$.

3. By what equivalent expressions with a common denominator, may $a^{\frac{1}{2}}$ and $a^{\frac{1}{3}}$ be expressed?

EXAMPLES.

1. Reduce $\sqrt{xy}$ and $\sqrt[3]{x^2y}$ to equivalent radicals of the same degree.

PROCESS.

$$\sqrt{xy} = (xy)^{\frac{1}{2}} = (xy)^{\frac{3}{6}} = \sqrt[6]{x^3y^3}$$
$$\sqrt[3]{x^2y} = (x^2y)^{\frac{1}{3}} = (x^2y)^{\frac{2}{6}} = \sqrt[6]{x^4y^2}$$

EXPLANATION. — Expressing the quantities with their fractional exponents, they are $(xy)^{\frac{1}{2}}$ and $(x^2y)^{\frac{1}{3}}$. These expressed as radicals of the same degree, are $(xy)^{\frac{3}{6}}$ and $(x^2y)^{\frac{2}{6}}$. Raising each quantity to the power indicated by the numerator of its exponent, and placing the result under the radical sign, with an index equal to the denominator of the exponent, we have as a result, $\sqrt[6]{x^3y^3}$ and $\sqrt[6]{x^4y^2}$.

RULE. — *If necessary, express the given radicals with fractional exponents. Reduce the fractional exponents to equivalent exponents having a common denominator.*

Raise each quantity to the power indicated by the numerator of its exponent, and indicate the root which is expressed by the common denominator.

Reduce the following to radicals of the same degree:

2. $\sqrt{10}$ and $\sqrt[3]{3}$.

3. $\sqrt{a^2x}$ and $\sqrt[3]{x^2y}$.

4. $\sqrt{x^2y^3}$ and $\sqrt[4]{xy^2}$.

5. $\sqrt[4]{a^2xy}$ and $\sqrt[3]{xy^2z}$.

6. $\sqrt{axy^4}$ and $\sqrt[5]{a^2xy^2}$.

7. $\sqrt[n]{ax}$ and $\sqrt[m]{by}$.

8. $\sqrt{ax}$, $\sqrt[3]{by}$, and $\sqrt{az}$.

9. $\sqrt{a^2x}$, $\sqrt[3]{ay^2}$, and $\sqrt{by^3}$.

10. $\sqrt{\frac{2}{3}}$, $\sqrt[3]{\frac{2}{5}}$, and $\sqrt{2}$.

11. $\sqrt{\frac{x}{3}}$, $\sqrt[3]{\frac{x}{4}}$, and $\sqrt{x^3}$.

12. $\sqrt{a+b}$ and $\sqrt[3]{a+b}$.

ADDITION AND SUBTRACTION OF RADICALS.

243. Principle.—*Only similar radicals can be combined into one term by addition or subtraction.*

1. Find the sum of $\sqrt{27}$, $2\sqrt{48}$, and $3\sqrt{108}$.

PROCESS.

$\sqrt{27} = 3\sqrt{3}$
$2\sqrt{48} = 8\sqrt{3}$
$3\sqrt{108} = 18\sqrt{3}$
Sum $= 29\sqrt{3}$

EXPLANATION.—Since the radical quantities are not similar, they must be reduced to similar radicals before adding. Expressing the radicals in their simplest form, $\sqrt{27} = 3\sqrt{3}$, $2\sqrt{48} = 8\sqrt{3}$, and $3\sqrt{108} = 18\sqrt{3}$. Therefore, since they all have the same radical factor with the same index, they are similar, and their sum is the sum of the coëfficients prefixed to the common radical factor.

Find the sum of:

2. $\sqrt{18}$, $\sqrt{128}$, and $\sqrt{32}$.
3. $\sqrt{75}$, $\sqrt{12}$, and $\sqrt{48}$.
4. $\sqrt[3]{32}$, $\sqrt[3]{108}$, and $\sqrt[3]{256}$.
5. $3\sqrt{200}$, $4\sqrt{50}$, and $\sqrt{72}$.
6. $\sqrt[3]{162}$, $\sqrt[3]{384}$, and $\sqrt[3]{750}$.
7. $3\sqrt{\frac{2}{3}}$ and $7\sqrt{\frac{27}{50}}$.
8. $\sqrt{24}$, $\sqrt{54}$, and $\sqrt{294}$.
9. $\sqrt{\frac{3}{4}}$, $\sqrt{\frac{1}{3}}$, and $\frac{1}{2}\sqrt{3}$.
10. $\frac{1}{2}\sqrt{\frac{1}{2}}$, $\frac{3}{4}\sqrt{2}$, and $\sqrt{\frac{1}{8}}$.
11. $3\sqrt{242\,x^5y^5}$ and $11\,xy\sqrt{2\,x^3y^3}$.
12. $\sqrt{108}$, $\sqrt{243}$, and $\sqrt{300}$.
13. $\sqrt{98\,x^2y^2}$, $\sqrt{128\,x^2y^2}$, and $\sqrt{162\,x^2y^2}$.
14. $\sqrt[3]{40}$, $\sqrt[3]{135}$, and $\sqrt[3]{320}$.

Simplify:

15. $\sqrt{24} + \sqrt{54} - \sqrt{96}$.
16. $\sqrt{12} + 3\sqrt{75} - 2\sqrt{27}$.
17. $5\sqrt{72} + 3\sqrt{18} - \sqrt{50}$.
18. $\sqrt{45a^3} + 5\sqrt{20a^3} - \sqrt{80a^3}$.
19. $\sqrt{8\,a^2y} + \sqrt{8\,b^2y} - \sqrt{50\,c^2y}$.
20. $\sqrt[3]{128} + \sqrt[3]{686} - \sqrt[3]{54}$.
21. $\sqrt{192\,a} - \sqrt{75\,a} + \sqrt{12\,a}$.
22. $4\sqrt{3} + 7\sqrt{1\frac{1}{3}} - 5\sqrt{1\frac{1}{3}}$.

23. $\sqrt{\dfrac{a}{x^2}} + \sqrt{\dfrac{a}{y^2}} - \sqrt{\dfrac{a}{z^2}}.$

24. $\sqrt{\dfrac{a^5 b^2}{c^2 x^3}} - \sqrt{\dfrac{a^3 c^2}{d^4 x}} + \sqrt{\dfrac{a d^2 y^4}{p^2 x}}.$

25. $\sqrt[4]{48} + \sqrt[4]{768} - \sqrt[4]{243}.$

26. $\sqrt{(x+y)^2 z} + \sqrt{(x-y)^2 z}.$

27. $\sqrt{5} + \sqrt{20} - 2\sqrt{\tfrac{1}{5}}.$

28. $\sqrt{\dfrac{x+y}{x-y}} + \sqrt{x^2 - y^2}.$

29. $\sqrt[5]{x^3 y^6 z^7} + \sqrt[5]{x^8 y z^2} - \sqrt[5]{x^{13} y^{11} z^{12}}.$

30. $\sqrt[3]{256} + 3\sqrt[3]{500} - 2\sqrt[3]{108}.$

31. $\sqrt{288} - \sqrt{392} + \sqrt{450}.$

32. $\sqrt[3]{-24} + 3\sqrt[3]{-375} + \sqrt[3]{-81}.$

33. $6\sqrt{\tfrac{3}{4}} + \tfrac{7}{4}\sqrt{\tfrac{1}{3}} - \sqrt{27}.$

34. $2\sqrt[3]{3x} + 5\sqrt[6]{9 x^2} - \sqrt[9]{27 x^3}.$

35. $\sqrt{2x^2 + 4xy + 2y^2} - \sqrt{2x^2 - 4xy + 2y^2}.$

36. $\sqrt[4]{32 x^4 y^8} + \sqrt[4]{162 x^4 y^8} - \sqrt[4]{512 x^4 y^8} + \sqrt[4]{1250 x^4 y^8}.$

37. $3\sqrt{\dfrac{3a}{4b^2}} + 2\sqrt{\dfrac{a}{3b^2}} - \sqrt{\dfrac{a}{27 b^2}}.$

38. $\sqrt{72} + \tfrac{3}{4}\sqrt{50} - \dfrac{1}{\sqrt{2}}.$

MULTIPLICATION OF RADICALS.

244. 1. When similar quantities have exponents, how is their product obtained?

2. What is the product of $a^2 \times a^3$? $a^m \times a^n$? $a^{\frac{1}{2}} \times a^{\frac{1}{2}}$? $a^{\frac{2}{3}} \times a^{\frac{1}{3}}$? $a^{\frac{1}{2}} \times a^{\frac{1}{4}}$? $a^{\frac{1}{5}} \times a^{\frac{1}{10}}$? $a^{\frac{1}{m}} \times a^{\frac{1}{m}}$? $a^{\frac{1}{m}} \times a^{\frac{1}{n}}$? $\sqrt[n]{a} \times \sqrt[n]{b}$?

3. Express with fractional exponents $\sqrt{a^3}$; $\sqrt[3]{a^2}$; $\sqrt[4]{a^5}$; $\sqrt[n]{a^m}$.

EXAMPLES.

1. $3\sqrt{4} \times 2\sqrt{12} = 6\sqrt{48} = 6\sqrt{3} \times \sqrt{16} = 24\sqrt{3}$

2. $3a\sqrt{6xy} \times 2b\sqrt{3xy} = 6ab\sqrt{18 x^2 y^2} = 6ab\sqrt{9 x^2 y^2} \times \sqrt{2}$
$\qquad = 18\, abxy\sqrt{2}$

RADICAL QUANTITIES. 207

3. $\sqrt[3]{axy} \times \sqrt{ax^3y^3} = (axy)^{\frac{1}{3}} \times (ax^3y^3)^{\frac{1}{2}} = (axy)^{\frac{2}{6}} \times (ax^3y^3)^{\frac{3}{6}}$
$= \sqrt[6]{a^2x^2y^2} \times \sqrt[6]{a^3x^9y^9} = \sqrt[6]{a^5x^{11}y^{11}} = xy\sqrt[6]{a^5x^5y^5}$

RULE. — *Reduce the radical factors to the same degree, if necessary.*

Multiply the coëfficients together for the coëfficient of the product, and the factors under the radical for the radical factor of the product, and simplify the result.

Multiply:

4. $2\sqrt{3}$ by $3\sqrt{8}$.
5. $2\sqrt{8}$ by $3\sqrt{6}$.
6. $3\sqrt{5}$ by $2\sqrt{15}$.
7. $3\sqrt{9}$ by $2\sqrt{8}$.
8. $3\sqrt{6}$ by $\sqrt{9}$.
9. $3\sqrt{10}$ by $4\sqrt{14}$.
10. $2\sqrt{10}$ by $3\sqrt{125}$.
11. $3\sqrt[3]{8}$ by $2\sqrt[3]{16}$.
12. $4\sqrt[3]{3}$ by $5\sqrt[3]{45}$.
13. $2\sqrt[3]{10}$ by $3\sqrt[3]{50}$.
14. $2\sqrt{3}$ by $3\sqrt[3]{2}$.
15. $3\sqrt{2}$ by $\sqrt[3]{3}$.
16. $\sqrt{4}$ by $\sqrt[4]{5}$.
17. $3\sqrt{3}$ by $2\sqrt[3]{5}$.
18. $2\sqrt[3]{3}$ by $3\sqrt[4]{4}$.
19. $3\sqrt{32}$ by $4\sqrt[3]{108}$.

Find the value of:

20. $\sqrt[3]{3a^2b} \times \sqrt[3]{9a^2b} \times \sqrt[3]{2ab^2}$.
21. $\sqrt{3a} \times \sqrt{2ab} \times \sqrt{c}$.
22. $2\sqrt{xy} \times \sqrt{3xy^2} \times \sqrt{6xy}$.
23. $\sqrt[3]{x^2y} \times \sqrt[3]{2x^2y^2} \times \sqrt{3x^3y^2}$.
24. $\sqrt[3]{3axy^2} \times \sqrt{axy} \times \sqrt{ax^2y}$.
25. $\sqrt{ab} \times 2\sqrt{ac} \times \sqrt[3]{a^2b^2}$.
26. $2\sqrt{2xy} \times 3\sqrt[3]{x^2y} \times 3\sqrt[3]{8xy^2}$.
27. $\sqrt{\frac{5}{8}} \times \sqrt{\frac{3}{4}} \times \sqrt{\frac{8}{5}}$.

28. $\sqrt{\frac{2}{3}} \times \sqrt{\frac{4}{5}} \times \sqrt{\frac{5}{6}}.$ 31. $\sqrt[3]{\frac{2}{5}} \times \sqrt{\frac{2}{3}} \times \sqrt[3]{3}.$

29. $\sqrt{\frac{3}{7}} \times \sqrt{\frac{8}{9}} \times \sqrt{\frac{7}{9}}.$ 32. $\sqrt[3]{\frac{5}{6}} \times \sqrt[3]{\frac{4}{3}} \times \sqrt{\frac{3}{2}}.$

30. $\sqrt[3]{\frac{2}{3}} \times \sqrt{\frac{1}{2}} \times \sqrt[3]{\frac{3}{4}}.$ 33. $\sqrt[3]{\frac{7}{2}} \times \sqrt{\frac{5}{3}} \times \sqrt[3]{\frac{3}{10}}.$

34. Multiply $8 + 4\sqrt{2}$ by $2 - \sqrt{2}.$

$$\begin{array}{r} 8 + 4\sqrt{2} \\ 2 - \sqrt{2} \\ \hline 16 + 8\sqrt{2} \\ -8\sqrt{2} - 4\sqrt{4} \\ \hline 16 - 4\sqrt{4} \end{array}$$

$$= 16 - 8 = 8.$$

Multiply:

    35. $3 + 4\sqrt{2}$ by $4 - 3\sqrt{2}.$

    36. $4 + 3\sqrt{3}$ by $5 + 2\sqrt{3}.$

    37. $5 + 4\sqrt{5}$ by $5 - 4\sqrt{5}.$

    38. $\sqrt{3} + 4\sqrt{2}$ by $\sqrt{3} - 3\sqrt{2}.$

    39. $2\sqrt{6} - \sqrt{3}$ by $\sqrt{6} - 5\sqrt{3}.$

    40. $4\sqrt{3} - 3\sqrt{5}$ by $2\sqrt{3} + \sqrt{5}.$

    41. $7\sqrt{7} - 2\sqrt{5}$ by $2\sqrt{7} + 7\sqrt{5}.$

    42. $4\sqrt{a} + 2\sqrt{b}$ by $4\sqrt{a} + 2\sqrt{b}.$

    43. $9\sqrt{3a} + 2$ by $2 - 9\sqrt{3a}.$

    44. $a + \sqrt{ab} + b$ by $\sqrt{a} - \sqrt{b}.$

    45. $3\sqrt{x} + \sqrt{2 - 3x}$ by $3\sqrt{x} - \sqrt{2 - 3x}.$

    46. $x + x\sqrt{3} + 1$ by $x\sqrt{3} + 1.$

DIVISION OF RADICALS.

245. 1. When similar quantities have exponents, how is their quotient obtained?

2. What is the quotient of $a^3 \div a^2$? $a^m \div a^n$? $a^{\frac{3}{4}} \div a^{\frac{1}{4}}$? $a^{\frac{3}{5}} \div a^{\frac{1}{5}}$? $a^{\frac{1}{2}} \div a^{\frac{1}{4}}$? $a^{\frac{2}{m}} \div a^{\frac{1}{m}}$? $\sqrt[n]{a^4} \div \sqrt[n]{a^2}$?

EXAMPLES.

1. $3\sqrt{405} \div 6\sqrt{5} = \frac{1}{2}\sqrt{81} = \frac{9}{2}$

2. $\sqrt{54} \div \sqrt[4]{36} = \sqrt{54} \div \sqrt[4]{6^2} = \sqrt{54} \div \sqrt{6} = \sqrt{9} = 3$

3. $\sqrt{ax^3y^3} \div \sqrt[3]{axy} = (ax^3y^3)^{\frac{1}{2}} \div (axy)^{\frac{1}{3}} = (ax^3y^3)^{\frac{3}{6}} \div (axy)^{\frac{2}{6}}$
$= \sqrt[6]{a^3x^9y^9} \div \sqrt[6]{a^2x^2y^2} = \sqrt[6]{ax^7y^7} = xy\sqrt[6]{axy}$

Find the quotients of:

4. $2\sqrt{27} \div 6\sqrt{3}$.
5. $15\sqrt{48} \div 5\sqrt{12}$.
6. $3\sqrt{75} \div 9\sqrt{3}$.
7. $5\sqrt{60} \div 6\sqrt{5}$.
8. $4\sqrt{72} \div 8\sqrt{32}$.
9. $7\sqrt{225} \div 5\sqrt{5}$.
10. $9\sqrt{256} \div 18\sqrt{2}$.
11. $3\sqrt{2} \div \sqrt[3]{2}$.
12. $\sqrt{3} \div \sqrt[3]{4}$.
13. $2\sqrt{5} \div \sqrt[3]{5}$.
14. $3\sqrt{x^2z} \div 4\sqrt[3]{x^2z^2}$.
15. $\sqrt{4} \div \sqrt[3]{6}$.
16. $\sqrt{72} \div \sqrt[4]{64}$.
17. $\sqrt[3]{12} \div \sqrt{8}$.
18. $3\sqrt[3]{18} \div \sqrt{6}$.
19. $6\sqrt[3]{ax} \div 2\sqrt{xy}$.
20. $\sqrt[3]{4x^2} \div \sqrt{2ax}$.
21. $\sqrt[3]{4xz} \div \sqrt[4]{2xz}$.
22. $\sqrt[3]{3x^2y} \div \sqrt[5]{6x^3y^2}$.
23. $2\sqrt[4]{x^2y^3z} \div 3\sqrt[3]{xy^2z^2}$.
24. $\sqrt[6]{12a^3b^2c^5} \div \sqrt[4]{2a^2b^3c}$.
25. $3\sqrt[3]{a^4x^7y} \div \sqrt[4]{ax}$.

26. Divide $\sqrt{20} + \sqrt{12}$ by $\sqrt{5} + \sqrt{3}$.

27. Divide $6 + 11\sqrt{2} + 6$ by $3 + \sqrt{2}$.

28. Divide $8 + 10\sqrt{5} + 15$ by $2 + \sqrt{5}$.

HIGH SCHOOL ALGEBRA.

29. Divide $12 + 17\sqrt{6} + 36$ by $3 + 2\sqrt{6}$.

30. $(12 - 2\sqrt{4} - \sqrt{48}) \div \sqrt{12}$.

31. $(a\sqrt[3]{a} - \sqrt[3]{a^2b} - b\sqrt[3]{a^2b} + b\sqrt[3]{b^2}) \div (\sqrt[3]{a^2} - \sqrt[3]{b})$.

32. $(x + y - z + 2\sqrt{xy}) \div (\sqrt{x} + \sqrt{y} - \sqrt{z})$.

INVOLUTION AND EVOLUTION OF RADICALS.

246. In finding the powers and roots of radicals it is commonly convenient to use fractional exponents.

EXAMPLES.

1. What is the cube of $3\sqrt{x}$?

$$(3\sqrt{x})^3 = (3x^{\frac{1}{2}})^3 = 3^3 x^{\frac{3}{2}} = 27 x^{\frac{3}{2}} = 27\sqrt{x^3} = 27x\sqrt{x}$$

2. What is the square of $3\sqrt[4]{16}$?

$$(3\sqrt[4]{16})^2 = (3 \times 16^{\frac{1}{4}})^2 = 3^2 \times 16^{\frac{1}{2}} = 9 \times 4 = 36$$

3. What is the square of $2 + 3\sqrt{x}$?

$$(2 + 3\sqrt{x})^2 = 4 + 12\sqrt{x} + 9\sqrt{x^2} = 4 + 12\sqrt{x} + 9x$$

In such cases use the binomial formula in expanding.

Square: Cube: Square:

4. $2\sqrt{x}$.   **8.** $3\sqrt{4x}$.   **12.** $4 + 5\sqrt{x}$.

5. $3\sqrt[3]{2x^2}$.   **9.** $2a\sqrt{3a}$.   **13.** $3 - 7\sqrt{x}$.

6. $4\sqrt[3]{3b^2}$.   **10.** $3\sqrt[3]{12ax}$.   **14.** $\sqrt{3} + 2\sqrt{5}$.

7. $3a\sqrt[3]{4b^2}$.   **11.** $4x\sqrt[3]{ax}$.   **15.** $\sqrt{7} + 5x\sqrt{ax}$.

16. What is the cube root of $8\sqrt{ab}$?

$$\sqrt[3]{8\sqrt{ab}} = \sqrt[3]{8} \times \sqrt[3]{\sqrt{ab}} = 2[(ab)^{\frac{1}{2}}]^{\frac{1}{3}} = 2(ab)^{\frac{1}{6}} = 2\sqrt[6]{ab}.$$

17. What is the square root of $\sqrt[3]{16x}$?

$$\sqrt{\sqrt[3]{16x}} = [(16x)^{\frac{1}{3}}]^{\frac{1}{2}} = (16x)^{\frac{1}{6}} = \sqrt[6]{16x}.$$

Observe that the root of the quantity under the radical may be found by multiplying the index of the radical by the index of the required root.

RADICAL QUANTITIES.

Find the square root of: Find the cube root of:

18. $\sqrt{3}$.   21. $\sqrt{ax}$.   24. $\sqrt{48}$.   27. $\sqrt[5]{x^8y^7}$.

19. $3\sqrt{5}$.   22. $\sqrt[3]{a^2x}$.   25. $\sqrt[3]{x^2y^5}$.   28. $\sqrt[3]{36}$.

20. $4\sqrt[3]{10}$.   23. $\sqrt[4]{xy^2z^3}$.   26. $\sqrt{125}$.   29. $\sqrt[7]{a^5x^9y^3}$.

RATIONALIZATION.

247. 1. What is a rational quantity?

2. By what quantity may $\sqrt{a}$ be multiplied to produce a rational quantity? $\sqrt{3a}$? $\sqrt{5x}$? $\sqrt[3]{a}$? $\sqrt[3]{a^2}$?

248. Rationalization is the process of changing a quantity under the radical sign so that the indicated root may be extracted.

249. To rationalize a monomial surd.

EXAMPLES.

1. What factor will rationalize $\sqrt{2a}$?

PROCESS.

$$\sqrt{2a} \times \sqrt{2a} = \sqrt{4a^2} = 2a$$

$\therefore \sqrt{2a}$ is the rationalizing factor.

EXPLANATION. — Since the quantity under the radical sign is to be changed so that the indicated root may be extracted, $\sqrt{2a}$ must be multiplied by a factor which will render the radical a perfect square. Any quantity multiplied by itself gives a perfect square, hence $\sqrt{2a}$ is multiplied by $\sqrt{2a}$.

2. What factor will rationalize $\sqrt[3]{3a^2}$?

$$\sqrt[3]{3a^2} \times \sqrt[3]{9a} = \sqrt[3]{27a^3} = 3a$$

$\therefore \sqrt[3]{9a}$ is the rationalizing factor.

3. Express the fraction $\dfrac{2a}{\sqrt{5ab}}$ with a rational denominator.

$$\frac{2a}{\sqrt{5ab}} = \frac{2a\sqrt{5ab}}{\sqrt{5ab}\sqrt{5ab}} = \frac{2a\sqrt{5ab}}{\sqrt{25a^2b^2}} = \frac{2a\sqrt{5ab}}{5ab} = \frac{2\sqrt{5ab}}{5b}$$

HIGH SCHOOL ALGEBRA.

Express the following with rational denominators:

4. $\dfrac{3}{\sqrt{2a}}$.

5. $\dfrac{2}{\sqrt[3]{4}}$.

6. $\dfrac{2\sqrt{a}}{\sqrt{x}}$.

7. $\dfrac{5}{\sqrt[3]{2a^2}}$.

8. $\dfrac{3a^2}{\sqrt{2a}}$.

9. $\dfrac{3ab}{\sqrt{5abx}}$.

10. $\dfrac{7}{\sqrt[3]{9z^2}}$.

11. $\dfrac{3ax}{\sqrt[5]{16x^3}}$.

12. $\dfrac{2z}{\sqrt[4]{27a^2}}$.

250. To rationalize a binomial surd of the second degree.

1. By what quantity must $a + b$ be multiplied so that the terms of the product will be squares?

2. By what quantity must $\sqrt{a} + \sqrt{b}$ be multiplied so that the terms of the product may be rational?

3. What is the product of $(\sqrt{x} + \sqrt{y})(\sqrt{x} - \sqrt{y})$?

EXAMPLES.

1. What factor will rationalize $\sqrt{2} - \sqrt{3}$?

PROCESS.

$$(\sqrt{2} - \sqrt{3})(\sqrt{2} + \sqrt{3}) = \sqrt{4} - \sqrt{9} = 2 - 3 = -1$$

$\therefore \sqrt{2} + \sqrt{3}$ is the rationalizing factor.

EXPLANATION. — Since the binomial will be rationalized when both terms are made perfect squares, it must be multiplied by $\sqrt{2} + \sqrt{3}$, since the product of the sum and difference of two quantities is the difference of their squares.

2. Express with a rational denominator $\dfrac{5}{2 + \sqrt{3}}$.

SOLUTION.

$$\dfrac{5}{2 + \sqrt{3}} = \dfrac{5(2 - \sqrt{3})}{(2 + \sqrt{3})(2 - \sqrt{3})} = \dfrac{10 - .5\sqrt{3}}{4 - \sqrt{9}} = \dfrac{10 - 5\sqrt{3}}{1}$$

$\therefore 2 - \sqrt{3}$ is the factor which rationalizes the denominator.

RADICAL QUANTITIES.

3. Express with a rational denominator $\dfrac{\sqrt{a+x}-\sqrt{a}}{\sqrt{a+x}+\sqrt{a}}$.

SOLUTION.

$$\frac{\sqrt{a+x}-\sqrt{a}}{\sqrt{a+x}+\sqrt{a}} = \frac{(\sqrt{a+x}-\sqrt{a})^2}{(\sqrt{a+x}+\sqrt{a})(\sqrt{a+x}-\sqrt{a})}$$

$$= \frac{a+x-2\sqrt{a^2+ax}+a}{a+x-a}$$

$$= \frac{2a+x-2\sqrt{a^2+ax}}{x}$$

Express the following with rational denominators:

4. $\dfrac{4-\sqrt{5}}{2+\sqrt{3}}$.

5 $\dfrac{3}{2+\sqrt{2}}$.

6. $\dfrac{x}{x-\sqrt{y}}$.

7. $\dfrac{c}{\sqrt{x}+\sqrt{z}}$.

8. $\dfrac{\sqrt{3}-\sqrt{2}}{\sqrt{3}+\sqrt{2}}$.

9. $\dfrac{x+\sqrt{z}}{x-\sqrt{z}}$.

10. $\dfrac{\sqrt{a}+\sqrt{x}}{\sqrt{a}-\sqrt{x}}$.

11. $\dfrac{\sqrt{x+1}+3}{\sqrt{x+1}+1}$.

12. $\dfrac{x-\sqrt{x^2-1}}{x+\sqrt{x^2-1}}$.

13. $\dfrac{\sqrt{a+3}-\sqrt{a}}{\sqrt{a+3}+\sqrt{a}}$.

14. $\dfrac{\sqrt{b+c}+\sqrt{b-c}}{\sqrt{b+c}-\sqrt{b-c}}$.

15. $\dfrac{\sqrt{a^2-2}-\sqrt{x^2+2}}{\sqrt{a^2-2}+\sqrt{x^2+2}}$.

16. Find the approximate value of $\dfrac{5}{\sqrt{3}}$.

SOLUTION.

$$\frac{5}{\sqrt{3}} = \frac{5\sqrt{3}}{\sqrt{3}\sqrt{3}} = \frac{5\sqrt{3}}{3} = \frac{5 \times 1.73205}{3} = 2.88675$$

Find the approximate values of:

17. $\dfrac{7}{\sqrt{2}}.$

18. $\dfrac{16}{\sqrt{3}}.$

19. $\dfrac{4}{\sqrt{2}}.$

20. $\dfrac{4}{\sqrt{300}}.$

21. $\dfrac{5}{\sqrt{147}}.$

22. $\dfrac{1}{3\sqrt{2}}.$

23. $\dfrac{\sqrt{3}-\sqrt{2}}{\sqrt{3}-\sqrt{2}}.$

24. $\dfrac{\sqrt{8}+\sqrt{5}}{\sqrt{8}-\sqrt{5}}.$

25. $\dfrac{4}{\sqrt{3}+\sqrt{2}}.$

251. To find the square root of a binomial surd.

$$(2+\sqrt{3})^2 = 4+4\sqrt{3}+3 = 7+4\sqrt{3}$$
$$(\sqrt{3}+\sqrt{5})^2 = 3+2\sqrt{15}+5 = 8+2\sqrt{15}$$
$$(2\sqrt{5}+\sqrt{7})^2 = 20+4\sqrt{35}+7 = 27+4\sqrt{35}$$

1. What is the square of $2+\sqrt{3}$? Since $7+4\sqrt{3}$ is the square of $2+\sqrt{3}$, what is the square root of $7+4\sqrt{3}$?

2. What is the square root of $8+2\sqrt{15}$, given in the second example? Of $27+4\sqrt{35}$, given in the third example?

3. How may 7 in the first example be obtained from the terms of the binomial? 8 in the second? 27 in the third?

4. How is each of the surd quantities in the power obtained from the terms of the binomial?

252. A **Binomial Surd** is a binomial, either or both of whose terms are surds.

Thus $2+\sqrt{3}$, $\sqrt{3}+\sqrt{5}$, etc., are binomial surds.

253. PRINCIPLE. — *A binomial surd whose square root can be extracted is composed of two parts, one of which is rational and equal to the sum of the squares of the terms of the root, and the other of which is irrational and composed of twice the product of the terms of the root.*

RADICAL QUANTITIES.

EXAMPLES.

1. Find the square root of $12 + 2\sqrt{35}$.

SOLUTION.

Let	$x + y =$ the square root of $12 + 2\sqrt{35}$	
Then,	$x^2 + y^2 = 12$	(1)
and	$2xy = 2\sqrt{35}$	(2)
(1) + (2)	$x^2 + 2xy + y^2 = 12 + 2\sqrt{35}$	(3)
(1) − (2)	$x^2 - 2xy + y^2 = 12 - 2\sqrt{35}$	(4)
(3)	$(x + y)^2 = 12 + 2\sqrt{35}$	
(4)	$(x - y)^2 = 12 - 2\sqrt{35}$	
(3) × (4)	$(x^2 - y^2)^2 = 144 - 140 = 4$	(5)
Square root,	$x^2 - y^2 = 2$	(6)
(6) + (1)	$2x^2 = 14$	
	$x = \sqrt{7}$	
	$y = \sqrt{5}$	
	$\therefore \sqrt{12 + 2\sqrt{35}} = \sqrt{7} + \sqrt{5}$	

EXPLANATION. — Since the rational quantity is the sum of the squares of the terms of the root of the binomial, it may be assumed to be equal to $x^2 + y^2$, and since the surd is twice the product of the two terms of the root, it will be represented by $2xy$. Consequently, $x + y$ will be the square root of the quantity.

2. Find the square root of $7 + 4\sqrt{3}$ by inspection.

EXPLANATION. — Since $4\sqrt{3}$ is twice the product of the terms of the root, their product is $2\sqrt{3}$. Since their product is $2\sqrt{3}$ or $\sqrt{12}$, the factors of $\sqrt{12}$, the sum of whose squares is 7, must be the terms of the root. They are readily seen to be $\sqrt{4}$ and $\sqrt{3}$, hence the square root of the quantity is $2 + \sqrt{3}$.

3. Find the square root of $31 + 4\sqrt{21}$ by inspection.

SOLUTION.

$$\tfrac{1}{2} \text{ of } 4\sqrt{21} = 2\sqrt{21} \text{ or } \sqrt{84}$$
$$\sqrt{84} = \sqrt{3} \times \sqrt{28}$$
$$\therefore \sqrt{3} + \sqrt{28} \text{ or } \sqrt{3} + 2\sqrt{7} \text{ is the square root.}$$

HIGH SCHOOL ALGEBRA.

Find the square root of the following:

4. $5 + 2\sqrt{6}$.
5. $9 + 2\sqrt{8}$.
6. $8 - 2\sqrt{15}$.
7. $11 - 6\sqrt{2}$.
8. $6a + 2a\sqrt{5}$.
9. $5 - \sqrt{24}$.
10. $15 + 4\sqrt{14}$.
11. $28 + 5\sqrt{12}$.
12. $13 + 2\sqrt{30}$.
13. $7 - 2\sqrt{10}$.
14. $11 + 2\sqrt{30}$.
15. $2m - 2\sqrt{m^2 - n^2}$.
16. $15 - 2\sqrt{56}$.
17. $14 + 3\sqrt{20}$.
18. $6 + \sqrt{35}$.
19. $28 + 10\sqrt{3}$.
20. $m^2 + n + 2m\sqrt{n}$.
21. $25 + 2\sqrt{156}$.

254. If $x + \sqrt{y} = a + \sqrt{b}$, then $x = a$ and $\sqrt{y} = \sqrt{b}$.

If x does not equal a, x must equal $a \pm m$.

Then, $\qquad a \pm m + \sqrt{y} = a + \sqrt{b}$

Reducing, $\qquad \pm m + \sqrt{y} = \sqrt{b}$

Squaring, $\qquad m^2 \pm 2m\sqrt{y} + y = b$

Transposing, etc., $\qquad \sqrt{y} = \dfrac{b - m^2 - y}{\pm 2m}$

This conclusion is absurd, for $\sqrt{y}$, a surd, cannot be equal to a rational quantity. Hence, x cannot be equal to $a \pm m$, and must be equal to a, and $\sqrt{y} = \sqrt{b}$. Therefore,

In any equation consisting of rational quantities and quadratic surds, the rational parts are equal and also the irrational parts.

REVIEW EXERCISES.

255. Expand:

1. $(a + b)^7$.
2. $\left(\dfrac{2x}{5} + \dfrac{3y}{7}\right)^3$.
3. $(2a - 3b)^4$.
4. $\left(\dfrac{x}{4} + \dfrac{y}{3}\right)^5$.
5. $(3a + 4b)^5$.
6. $\left(\dfrac{a}{2} - \dfrac{b}{3}\right)^3$.
7. $(a + b)^n$ to 6 terms.
8. $(a + b)^{n-2}$ to 5 terms.
9. $(x + y)^r$ to 6 terms.
10. $(1 + 2x + 3y + z)^2$.

RADICAL QUANTITIES.

Simplify:

11. $\sqrt{128} - 2\sqrt{50} + \sqrt{72} - \sqrt{78}.$

12. $6\sqrt[6]{4a^2} + 2\sqrt[3]{2a} + \sqrt[9]{2a^3}.$

13. $2\sqrt{3} - \tfrac{1}{2}\sqrt{12} + 4\sqrt{27} - 2\sqrt{\tfrac{3}{16}}.$

14. $\sqrt[4]{16} + \sqrt[3]{81} - \sqrt[3]{-512} + \sqrt[3]{192} - 7\sqrt[6]{9}.$

15. $\sqrt{48ab^2} + b\sqrt{75a} + \sqrt{3a(a-9b)^2}.$

Extract the square root of:

16. $9x - 24x^{\frac{1}{2}}y^{\frac{2}{3}} + 12x^{\frac{1}{2}} + 16y^{\frac{4}{3}} - 16y^{\frac{2}{3}} + 4.$

17. $4a^{\frac{4}{3}} + 4a^{\frac{5}{3}} - 15a^2 - 8a^{\frac{7}{3}} + 16a^{\frac{8}{3}}.$

18. $x^{\frac{4}{3}} - 4xy^{\frac{1}{3}} + 6x^{\frac{2}{3}}y^{\frac{2}{3}} - 4x^{\frac{1}{3}}y + y^{\frac{4}{3}}.$

19. $9x - 12x^{\frac{5}{6}} + 10x^{\frac{2}{3}} - 28x^{\frac{1}{2}} + 17x^{\frac{1}{3}} - 8x^{\frac{1}{6}} + 16.$

Extract the cube root of:

20. $5x^3 - 1 - 3x^5 + x^6 - 3x.$

21. $x^{\frac{6}{5}} - 6x + 15x^{\frac{4}{5}} - 20x^{\frac{3}{5}} + 15x^{\frac{2}{5}} - 6x^{\frac{1}{5}} + 1.$

22. $x^{\frac{1}{2}} + 3x^{\frac{5}{12}}y^{\frac{1}{12}} - 3x^{\frac{1}{3}}y^{\frac{1}{6}} - 11x^{\frac{1}{4}}y^{\frac{1}{4}} + 6x^{\frac{1}{6}}y^{\frac{1}{3}} + 12x^{\frac{1}{12}}y^{\frac{5}{12}} - 8y^{\frac{1}{2}}$

23. $a^{\frac{1}{3}}x^{-1} + a^{-\frac{1}{3}}x.$

Multiply:

24. $\sqrt{a} + \sqrt{a-x} + \sqrt{x}$ by $\sqrt{a} - \sqrt{a-x} + \sqrt{x}.$

25. $\sqrt{a} + \sqrt{a-x}$ by $\sqrt{a} - \sqrt{a-x}.$

26. $2\sqrt[3]{x^5} - x^{\frac{1}{3}} - \dfrac{3}{x}$ by $2x - 3\sqrt[3]{\dfrac{1}{x}} - x^{-\frac{5}{3}}.$

Divide:

27. $m^2 + mn + n^2$ by $m + \sqrt{mn} + n$.
28. $x^4 + y^4$ by $x^2 + xy\sqrt{2} + y^2$.
29. $\dfrac{a - 7a^{\frac{1}{2}}}{a - 5\sqrt{a} - 14}$ by $\left(1 + \dfrac{2}{\sqrt{a}}\right)^{-1}$.

Expand:

30. $(\sqrt[3]{2y^2} + \sqrt[6]{x^3y})^2$.
31. $(2 + a^2\sqrt{2x+y})^2$.
32. $(3\sqrt{8} + 5\sqrt{3})^2$.
33. $(\sqrt{1-x^2} + x)^2$.

Find the values of the following:

34. $(\sqrt[6]{12})^3$.
35. $(\sqrt[3]{a^2x})^5$
36. $(a\sqrt[3]{a})^2$.
37. $(3\sqrt[6]{3})^3$.
38. $\sqrt[5]{\sqrt{64}}$.
39. $\sqrt[3]{\sqrt[5]{54x^3}}$.

Rationalize the denominators of:

40. $\dfrac{2 + \sqrt{3}}{3 - \sqrt{3}}$.
41. $\dfrac{3 + \sqrt{5}}{\sqrt{3} - \sqrt{2}}$.
42. $\dfrac{\sqrt{a}}{\sqrt{a} - \sqrt{b}}$.
43. $\dfrac{4 + 3\sqrt{2}}{3 - 2\sqrt{2}}$.
44. $\dfrac{2\sqrt{3} + 3\sqrt{2}}{5 + 2\sqrt{6}}$.
45. $\dfrac{10\sqrt{6} - 2\sqrt{7}}{3\sqrt{6} + 2\sqrt{7}}$.
46. $\dfrac{a + \sqrt{a^2 - x^2}}{a - \sqrt{a^2 - x^2}}$.
47. $\dfrac{\sqrt{m^2 + 1} - \sqrt{m^2 - 1}}{\sqrt{m^2 + 1} + \sqrt{m^2 - 1}}$.

Find the approximate values of the following to three decimal places:

48. $\dfrac{3}{\sqrt{5}}$.
49. $\dfrac{5\sqrt{2}}{\sqrt{3}}$.
50. $\dfrac{\sqrt{7} + \sqrt{3}}{\sqrt{7} - \sqrt{3}}$.
51. $\dfrac{2\sqrt{5} - \sqrt{3}}{3\sqrt{5} + 2\sqrt{3}}$.

Extract the square root of:

52. $9 - 4\sqrt{5}$.
53. $43 - 15\sqrt{8}$.
54. $24 - 2\sqrt{63}$.
55. $22 - 4\sqrt{30}$.
56. $47 - 6\sqrt{10}$.
57. $23 - \sqrt{528}$.

RADICAL EQUATIONS.

256. 1. In the equation $\sqrt{x} = 2$, what is the value of x? How is it obtained?

2. In the equation $\sqrt{2x} = 4$, how may the value of x be obtained? How in the equation $\sqrt{x} + 2 = 4$?

257. A **Radical Equation** is an equation containing a radical quantity.

Thus, $\sqrt{x} = 5$, $\sqrt{x} + 3 = 6$, $\sqrt{x} + 5 = \sqrt{3x}$, are radical equations.

EXAMPLES.

1. Given $\sqrt{x} + 6 = 9$, to find the value of x.

SOLUTION.
$$\sqrt{x} + 6 = 9$$
Transposing, $\quad \sqrt{x} = 3$
Squaring, $\quad x = 9$

2. Given $\sqrt{4+x} = 4 - \sqrt{x}$, to find the value of x

SOLUTION.
$$\sqrt{4+x} = 4 - \sqrt{x}$$
Squaring, $\quad 4 + x = 16 - 8\sqrt{x} + x$
Transposing, $\quad 8\sqrt{x} = 12$
Dividing by 4, $\quad 2\sqrt{x} = 3$
Squaring, $\quad 4x = 9$
$$x = 2\tfrac{1}{4}$$

HIGH SCHOOL ALGEBRA.

3. Given $\dfrac{x-ax}{\sqrt{x}} = \dfrac{\sqrt{x}}{x}$, to find the value of x.

SOLUTION.

$$\frac{x-ax}{\sqrt{x}} = \frac{\sqrt{x}}{x}$$

Clearing of fractions, $\quad x^2 - ax^2 = x$
Dividing by x, $\quad x - ax = 1$
Factoring, $\quad (1-a)x = 1$

$$x = \frac{1}{1-a}$$

4. Given $\dfrac{2\sqrt{4x}+2}{2\sqrt{4x}-7} = \dfrac{\sqrt{4x}+6}{\sqrt{4x}-1}$.

SOLUTION.

$$\frac{2\sqrt{4x}+2}{2\sqrt{4x}-7} = \frac{\sqrt{4x}+6}{\sqrt{4x}-1}$$

Reducing to mixed quantities,

$$1 + \frac{9}{2\sqrt{4x}-7} = 1 + \frac{7}{\sqrt{4x}-1}$$

Transposing, etc., $\quad \dfrac{9}{2\sqrt{4x}-7} = \dfrac{7}{\sqrt{4x}-1}$

Clearing of fractions, $\quad 9\sqrt{4x} - 9 = 14\sqrt{4x} - 49$
Transposing, etc., $\quad 5\sqrt{4x} = 40$
Dividing by 5, $\quad \sqrt{4x} = 8$
Squaring, $\quad 4x = 64$

$$x = 16$$

SUGGESTIONS. —1. *Transpose the terms so that the radical quantity, if there be but one, or the more complex radical, if there be more than one, may constitute one member of the equation; then raise each member to a power of the same degree as the radical.*

2. *When the equation is not freed from radicals by the first involution, proceed again as indicated in Suggestion 1.*

3. *Simplify the equation as much as possible before performing the involution.*

RADICAL EQUATIONS.

4. *Sometimes it may be advantageous to clear a radical equation of fractions, only in part. This will be especially convenient when a radical denominator and another numerator are similar.*

5. *It is sometimes convenient to rationalize the denominator before clearing of fractions or involving.*

Solve the following equations:

5. $\sqrt{x} + 7 = 9$.

6. $\sqrt{2x} - 5 = 3$.

7. $\sqrt{x-7} = 7$.

8. $\sqrt[3]{4x-16} = 2$.

9. $\sqrt{x+9} = 6$.

10. $\sqrt[3]{2x+3} + 4 = 7$.

11. $\sqrt{x^2-9} + x = 9$.

12. $\sqrt{x^2-11} + 1 = x$.

13. $\sqrt{16+x} + \sqrt{x} = 8$.

14. $\sqrt{x-16} = 8 - \sqrt{x}$.

15. $\sqrt{x-21} = \sqrt{x} - 1$.

16. $\sqrt{x+5} + \sqrt{x} = 5$.

17. $\sqrt{x-15} + \sqrt{x} = 15$.

18. $2\sqrt{x} - \sqrt{4x-22} = \sqrt{2}$.

19. $\sqrt{x} + \sqrt{x-9} = \dfrac{36}{\sqrt{x-9}}$.

20. $\sqrt{x} + \sqrt{x-4} = \dfrac{4}{\sqrt{x-4}}$.

21. $\sqrt{x} + \sqrt{8+x} = \dfrac{12}{\sqrt{8+x}}$.

22. $\dfrac{3x-1}{\sqrt{3x}+1} - \dfrac{\sqrt{3x}-1}{2} = 4$.

23. $x + \sqrt{x^2 - \sqrt{1-x}} = 1$.

24. $\sqrt{1 + x\sqrt{x^2+12}} = 1 + x$.

25. $\sqrt{a-x} = \dfrac{a}{\sqrt{a-x}} - x$.

26. $\dfrac{ax-1}{\sqrt{ax}+1} = \dfrac{\sqrt{ax}-1}{2} + 4$.

27. $\dfrac{\sqrt{x}+28}{\sqrt{x}+4} = \dfrac{32}{\sqrt{x}+6} + 1$.

28. $\dfrac{\sqrt{x}+16}{\sqrt{x}+4} = \dfrac{\sqrt{x}+32}{\sqrt{x}+12}$.

29. $\dfrac{\sqrt{x}-1}{\sqrt{x}-5} = \dfrac{\sqrt{x}-3}{\sqrt{x}-1}$.

30. $\dfrac{\sqrt{x}-3}{\sqrt{x}+7} = \dfrac{\sqrt{x}-4}{\sqrt{x}+1}$.

31. $\dfrac{\sqrt{x}-8}{\sqrt{x}-6} = \dfrac{\sqrt{x}-4}{\sqrt{x}+2}$.

32. $\dfrac{3\sqrt{2x}+10}{3\sqrt{2x}-10} = \dfrac{\sqrt{2x}+16}{\sqrt{2x}-4}.$

33. $\sqrt{x+8} - \sqrt{x-8} = 2\sqrt{2}.$

34. $\dfrac{a}{x} + \dfrac{\sqrt{a^2-x^2}}{x} = \dfrac{1}{b}.$

35. $\sqrt{\dfrac{1}{x}} - \sqrt{x} = \sqrt{1+x}.$

36. $\dfrac{\sqrt{a+x}+\sqrt{x}}{\sqrt{a+x}-\sqrt{x}} = b.$

37. $\dfrac{\sqrt{4x+1}+2\sqrt{x}}{\sqrt{4x+1}-2\sqrt{x}} = 9.$

38. $\dfrac{\sqrt{5+x}+\sqrt{5-x}}{\sqrt{5+x}-\sqrt{5-x}} = 2.$

39. $\dfrac{\sqrt{a}+\sqrt{a-x}}{\sqrt{a}-\sqrt{a-x}} = a.$

40. $\dfrac{a+\sqrt{x}}{a-\sqrt{x}} = b.$

41. $\dfrac{b-\sqrt{b^2-x}}{b+\sqrt{b^2-x}} = a.$

42. $\dfrac{2x-\sqrt{2x-4x^2}}{2x+\sqrt{2x-4x^2}} = \dfrac{1}{3}.$

43. $\sqrt{2x+1} = \dfrac{2x-1}{\sqrt{2x+1}} - 1.$

44. $\dfrac{\sqrt{4x+3}+2\sqrt{x-1}}{\sqrt{4x+3}-2\sqrt{x-1}} = 5.$

45. $\dfrac{x-3}{\sqrt{x}-\sqrt{3}} = \dfrac{\sqrt{x}+\sqrt{3}}{2} + 2\sqrt{3}.$

46. $\dfrac{(\sqrt{4x}-3)(\sqrt{x}+4)}{\sqrt{x}-2} = 2\sqrt{x}+15.$

47. $\sqrt{x^2+4x+12} + \sqrt{x^2-12x-20} = 8.$

48. $\dfrac{x-a}{\sqrt{x}+\sqrt{a}} = \dfrac{\sqrt{x}-\sqrt{a}}{3} + 2\sqrt{a}.$

49. $\sqrt{x+a} + \sqrt{x+b} = \sqrt{4x+a+3b}.$

50. $\dfrac{\sqrt{x^2+1}-\sqrt{x^2-1}}{\sqrt{x^2+1}+\sqrt{x^2-1}} = \dfrac{1}{2}.$

51. $\dfrac{\sqrt{a^2+x^2}+x}{\sqrt{a^2+x^2}-x} = \dfrac{b}{c}.$

52. $\dfrac{a+x+\sqrt{2ax+x^2}}{a+x-\sqrt{2ax+x^2}} = b.$

QUADRATIC EQUATIONS.

258. 1. How is the degree of an equation determined? (Art. 170.)

2. What is the degree of the equation $x+2=7$? Of $x^2+4=8$? Of $x^2+4x=15$? Of $x+xy=12$?

3. What are the equations of the second degree called?

4. When $x^2=4$, what is the value of x? What, when $x^2=9$? What, when $x^2=a^2$?

5. How many values has x in these equations? How do the values of x compare numerically? What are the signs of the values of x?

259. A **Quadratic Equation** is an equation of the second degree.

Thus, $x^2=9$, and $x^2+bx=c$, are quadratic equations.

260. A **Pure Quadratic Equation** is an equation which contains only the second power of the unknown quantity.

Thus, $4x^2=16$, and $ax^2=b$, are pure quadratics.

261. A Pure Quadratic Equation is sometimes called an *Incomplete Quadratic Equation* because the first power of the unknown quantity is wanting.

262. A **Root** of an equation is a value of the unknown quantity.

When a root of an equation is substituted for the unknown quantity it *satisfies* the equation, or renders the members of the equation identical.

PURE QUADRATICS.

263. Since pure quadratic equations contain only the second power of the unknown quantity, they may always be reduced to the general form of $ax^2 = b$, in which a represents the coëfficient of x^2, and b the other terms.

264. PRINCIPLE. — *Every pure quadratic equation has two roots numerically equal, but having opposite signs.*

EXAMPLES.

1. Given $3x^2 + \dfrac{x^2}{2} = 14$, to find the value of x.

SOLUTION.

$$3x^2 + \frac{x^2}{2} = 14$$
$$6x^2 + x^2 = 28$$
$$7x^2 = 28$$
$$x^2 = 4$$

Extracting the square root, $\quad x = \pm 2$

2. Given $ax^2 + c = bx^2 + d$, to find x.

SOLUTION.

$$ax^2 + c = bx^2 + d$$
$$ax^2 - bx^2 = d - c$$
$$(a - b)x^2 = d - c$$
$$x^2 = \frac{d - c}{a - b}$$

Extracting the square root, $\quad x = \pm\sqrt{\dfrac{d-c}{a-b}}$

Solve the following equations:

3. $x^2 - 3 = 46$.

4. $3x^2 + 7 = x^2 + 15$.

5. $2x^2 - 6 = 66$.

6. $\dfrac{x^2 - 12}{3} = \dfrac{x^2 - 4}{4}$.

7. $3x^2 + 7 = \dfrac{3x^2}{4} + 43$.

8. $2x^2 = \dfrac{3x^2}{5} + 35$.

9. $\dfrac{x^2 + 2}{3} + 8 = 42$.

QUADRATIC EQUATIONS.

10. $\dfrac{x-3}{4} = \dfrac{4}{x+3}.$

11. $\dfrac{x^2}{3} + 9 = \dfrac{3x^2-3}{5}.$

12. $5x^2 - 3 = 2x^2 + 24.$

13. $7x^2 + 4 = 3x^2 + 40.$

14. $(x+2)^2 = 4x + 5.$

15. $\dfrac{x+4}{x-4} + \dfrac{x-4}{x+4} = 3\tfrac{1}{3}.$

16. $\dfrac{1}{1-x} + \dfrac{1}{1+x} = 2\tfrac{2}{3}.$

17. $\dfrac{x^2+9x}{15} = \dfrac{3(x+2)}{5}.$

18. $(x+2)(x-2) = \dfrac{x^2}{4} - 1.$

19. $\sqrt{a+x} = \dfrac{a}{\sqrt{x-a}}.$

20. $x + \sqrt{x^2+3} = \dfrac{6}{\sqrt{x^2+3}}.$

21. $\dfrac{15}{\sqrt{x^2+5}} = x + \sqrt{x^2+5}.$

22. $x + \sqrt{x^2 - 2\sqrt{1-x}} = 1$

23. $\dfrac{x-m}{x+m} + \dfrac{x+m}{x-m} = 4.$

24. $\dfrac{x+a}{x-a} + \dfrac{x-a}{x+a} = \dfrac{2a}{1-a}$

25. $\dfrac{\sqrt{25+16x}}{\sqrt{8}} + \dfrac{\sqrt{25-16x}}{\sqrt{8}} = \sqrt{8}.$

26. $x - \sqrt{x^2-14} = \dfrac{5}{\sqrt{x^2-14}}.$

27. $(x+2)^2 - 6 = 2(2x+17).$

28. $\dfrac{3}{2x+1} - \dfrac{4x-7}{4x^2-1} = \dfrac{2x}{2x-1}.$

29. $(x^2+4)(x+1) = (x-1)^3 + x + 13.$

30. $\dfrac{4x^2+6}{14} - \dfrac{3x^2-2}{2x^2-1} = \dfrac{2x^2-3}{7}.$

31. $\sqrt{a+x} = \sqrt{x + \sqrt{x^2-b^2}}.$

32. $\dfrac{\sqrt{x^2+1} - \sqrt{x^2-1}}{\sqrt{x^2+1} + \sqrt{x^2-1}} = \dfrac{1}{2}.$

33. $\dfrac{2}{x+\sqrt{2-x^2}} + \dfrac{2}{x-\sqrt{2-x^2}} = \dfrac{a}{x}.$

34. $\dfrac{1+x}{1+x+\sqrt{1+x^2}} = a - \dfrac{1-x}{1-x+\sqrt{1+x^2}}.$

ALGEBRA. — 15.

PROBLEMS.

265. 1. What number is that, to the square of which, if $\frac{5}{9}$ be added, the sum will be 1?

2. Find a number such that the square of $\frac{3}{4}$ of it will be 7 less than the square of it.

3. Find a number such that if 320 be divided by it, and the quotient added to the number itself, the sum will be equal to 6 times the number.

4. When a certain number is increased by 3 and also diminished by 3, the product of the sum and difference will be 55. What is the number?

5. Two numbers are to each other as 3 to 5, and the sum of their squares is 3400. What are the numbers?

6. A gentleman said that his son's age was $\frac{1}{4}$ of his own age, and that the difference of the squares of the numbers which represent their ages was 960. What were their ages?

7. A man lent a sum of money at 6% per annum, and found that, if he multiplied the principal by the number which expressed the number of dollars interest for 8 months, the product would be $900. What was the principal?

8. A gentleman has two square rooms whose sides are to each other as 2 to 3. He finds that it will require 20 square yards more of carpeting to cover the floor of the larger than of the smaller room. What is the length of one side of each room?

9. The sum of two numbers is 12, and their product is 27. What are the numbers?

SUGGESTION. — Let $6 + x =$ one number, and $6 - x =$ the other.

10. The sum of two numbers is 14, and their product is 48. What are the numbers?

11. The sum of two numbers is 13, and their product is 42. **What are the numbers?**

QUADRATIC EQUATIONS.

12. Divide 20 into two such parts that their product will be 96.

13. Divide 32 into two such parts that their product will be 240.

14. A merchant bought a piece of cloth for $24, paying $\frac{3}{8}$ as many dollars per yard as there were yards in the piece. How many yards were there?

15. A man purchased a rectangular field whose length was $1\frac{1}{9}$ times its breadth. It contained 9 acres. What was the length of each side?

16. Find two numbers which are to each other as 5 to 4, and the sum of whose squares is 164.

17. Find three numbers that shall be to each other as 3, 5, and 8, and the sum of whose squares shall be 392.

18. A man worked 10 times as many days as he received dollars a day, and earned $62.50. How many days did he work, and how much did he get a day?

19. The product of two numbers is 324, and the quotient of the greater divided by the less is 4. Find the numbers.

20. A man has two square lots containing 272 square rods. The side of the larger is as much greater than 10 rods as that of the other is less than 10 rods. What is the side of each?

21. The sum of the squares of two numbers is 170, and the difference of their squares is 72. What are the numbers?

AFFECTED QUADRATICS.

266. 1. How is a binomial squared? What is the square of $x+2$? Of $x+5$? Of $x+7$?

2. How may the first term of a binomial be found from its square?

3. Since the second term of the square of a binomial contains twice the product of both terms, how may the second term of the *binomial* be found from it, when the first term of the binomial is known?

4. What is it necessary to add to $x^2 + 4x$ to make a perfect square? How is the term found?

5. What must be added to $x^2 + 6x$ to make a perfect square? How is it found?

6. What is the square root of the completed square of which $x^2 + 4x$ are two terms?

7. What is the square root of the completed square of which $x^2 + 6x$ are two terms?

8. In the equation $x^2 + 4x + 4 = 9$, what is the square root of the first member? What is the square root of the second member?

9. Since, in the solution of the equation $x^2 + 4x + 4 = 9$, we obtain the result $x + 2 = \pm 3$, how many values has x?

267. An **Affected Quadratic Equation** is an equation which contains both the first and second powers of the unknown quantity.

Thus, $x^2 + 2x = 4$, $5x^2 + 6x = 8$, and $ax^2 + bx = c$, are affected quadratic equations.

268. An Affected Quadratic Equation is sometimes called a *Complete Quadratic Equation*.

Since affected quadratic equations contain both the second and the first powers of the unknown quantity, they may always be reduced to the general form of $ax^2 \pm bx = \pm c$, in which a and b represent respectively the coëfficients of x^2 and x, and c the other terms.

269. PRINCIPLE. — *Every affected quadratic equation has two roots, and only two.*

These roots are always numerically unequal, except when the second member of the equation reduces to 0 when the square is completed.

QUADRATIC EQUATIONS.

270. First method of completing the square.

EXAMPLES.

1. Given $x^2 + 4x = 96$, to find the values of x.

PROCESS.

$$x^2 + 4x = 96$$
$$x^2 + 4x + 4 = 96 + 4$$
$$x^2 + 4x + 4 = 100$$
$$x + 2 = \pm 10$$
$$x = 10 - 2 = 8$$
$$x = -10 - 2 = -12$$

EXPLANATION. — Completing the square in the first member by adding the square of one half the coefficient of x to each member, we have $x^2 + 4x + 4 = 96 + 4$. Extracting the square root of each member, we have $x + 2 = \pm 10$. Using, first, the positive value of 10, we obtain 8 for one value of x; using next the negative value of 10, we obtain -12 for the other value of x.

2. Given $x^2 - 5x = 24$, to find the values of x.

SOLUTION.

$$x^2 - 5x = 24$$

Completing the square, $\quad x^2 - 5x + \tfrac{25}{4} = 24 + \tfrac{25}{4}$ or $\tfrac{121}{4}$

Extracting the square root, $\quad x - \tfrac{5}{2} = \pm \tfrac{11}{2}$

Transposing, $\quad x = \tfrac{5}{2} + \tfrac{11}{2} = 8$

$$x = \tfrac{5}{2} - \tfrac{11}{2} = -3$$

$$\therefore x = 8 \text{ or } -3$$

3. Given $2x^2 - 7x = 30$, to find the values of x.

SOLUTION.

$$2x^2 - 7x = 30$$

Dividing by coëfficient of x^2, $\quad x^2 - \tfrac{7}{2}x = 15$

Completing the square, $\quad x^2 - \tfrac{7}{2}x + \tfrac{49}{16} = 15 + \tfrac{49}{16}$ or $\tfrac{289}{16}$

Extracting the square root, $\quad x - \tfrac{7}{4} = \pm \tfrac{17}{4}$

Transposing, $\quad x = \tfrac{7}{4} + \tfrac{17}{4} = 6$

$$x = \tfrac{7}{4} - \tfrac{17}{4} = -\tfrac{10}{4} = -2\tfrac{1}{2}$$

$$\therefore x = 6 \text{ or } -2\tfrac{1}{2}$$

RULE. — *Reduce the equation to the form $x^2 \pm bx = \pm c$, by dividing both members of the equation by the coëfficient of the highest power of the unknown quantity.*

Add the square of one half the coëfficient of the second term to each member of the equation, extract the square root of each member, and reduce the equation.

Inasmuch as it is impossible to extract the square root of a negative quantity, it is necessary that the term containing the second power of the unknown quantity should have the *positive* sign. If it should be negative, change the signs of all the terms in both members.

Find the values of x in the following equations:

4. $x^2 + 4x = 45$.
5. $x^2 + 6x = 27$.
6. $x^2 + 8x = 20$.
7. $x^2 + 10x = 11$.
8. $x^2 + 20x = 21$.
9. $x^2 + 18x = 19$.
10. $x^2 + 24x = 25$.
11. $x^2 - 12x = 45$.
12. $x^2 - 8x = 33$.
13. $x^2 - 14x = 51$.
14. $x^2 - 28x = 60$.
15. $x^2 - 30x = 64$.
16. $2x^2 + 3x = 14$.
17. $3x^2 + 4x = 39$.
18. $2x^2 + 7x = 39$.
19. $5x^2 + 15x = 50$.
20. $6x^2 - 21x = 12$.
21. $\dfrac{x-5}{10} = \dfrac{x+1}{x-6}$.
22. $\dfrac{x+2}{5} = \dfrac{8}{3x+4}$.
23. $3x - \dfrac{3x-3}{x-3} = \dfrac{3x-6}{2}$.

271. Other methods of completing the square.

By the previous method, when x^2 had a coëfficient, the equation was divided by that coëfficient so that the *term containing x^2 might always be a perfect square.* The same result may be secured in other ways.

Thus, if the term containing the highest power is $3x^2$, it may be made a perfect square by multiplying by 3; if $5x^2$, by multiplying by 5; if ax^2, by multiplying by a.

QUADRATIC EQUATIONS. 231

EXAMPLES.

1. Solve the equation $3x^2 + 6x = 24$.

PROCESS.

$$3x^2 + 6x = 24 \qquad (1)$$

(1) × 3, $\qquad 9x^2 + 18x = 72 \qquad (2)$

Completing square, $\quad 9x^2 + 18x + 9 = 81 \qquad (3)$

Extracting square root, $\quad 3x + 3 = \pm 9 \qquad (4)$

$$3x = 6 \text{ or } -12$$

$$x = 2 \text{ or } -4$$

EXPLANATION. — Since the coëfficient of x^2 is not a perfect square, it may be made a square by multiplying it by itself. If the coffiëcient of x^2 is multiplied by itself, both members of the equation must be multiplied by 3, the coëfficient of x^2, giving equation (2).

Since the second term of equation (2) is twice the product of the terms which are the square root of the completed square, if it is divided by twice the first term of the root the quotient will be the second term. The first term of the square root is $\sqrt{9x^2}$ or $3x$, and twice $3x$ is $6x$. $18x \div 6x = 3$, and the square of 3, which is 9, added to each member, completes the square.

In the complete square $a^2x^2 + 2abx + b^2$, it is evident that the third term, b^2, is the square of the quotient obtained by dividing the second term by twice the square root of the first.

2. Solve the equation $3x^2 + 4x = 39$.

SOLUTION.

$$3x^2 + 4x = 39 \qquad (1)$$

(1) × 3, $\qquad 9x^2 + 12x = 117 \qquad (2)$

$2\sqrt{9x^2} = 6x$, $\qquad 12x \div 6x = 2$, and 2^2 added to each member gives

$$9x^2 + 12x + 4 = 121 \qquad (3)$$

Extracting square root, $\quad 3x + 2 = \pm 11 \qquad (4)$

$$3x = 9 \text{ or } -13$$

$$x = 3 \text{ or } -4\tfrac{1}{3}$$

3. Solve the equation $5x^2 - 6x = 8$.

SOLUTION.

$$5x^2 - 6x = 8 \qquad (1)$$

$(1) \times 5,\qquad 25x^2 - 30x = 40 \qquad (2)$

$2\sqrt{25x^2} = 10x,\qquad 30x \div 10x = 3$, and 3^2 added to each member gives

$$25x^2 - 30x + 9 = 49 \qquad (3)$$

Extracting square root, $\qquad 5x - 3 = \pm 7 \qquad (4)$

$$5x = 10 \text{ or } -4$$
$$x = 2 \text{ or } -\tfrac{4}{5}$$

4. Solve the equation $3x^2 + 5x = 8$.

SOLUTION.

$$3x^2 + 5x = 8 \qquad (1)$$

$(1) \times 3,\qquad 9x^2 + 15x = 24 \qquad (2)$

$2\sqrt{9x^2} = 6x,\qquad 15x \div 6x = \tfrac{15}{6}$ or $\tfrac{5}{2}$, and $(\tfrac{5}{2})^2$ added to the members gives

$$9x^2 + 15x + \tfrac{25}{4} = 24 + \tfrac{25}{4} \qquad (3)$$

Clearing of fractions, $36x^2 + 60x + 25 = 96 + 25$

Reducing, $\qquad\qquad\qquad x = 1 \text{ or } -\tfrac{8}{3}$

Since the equation was first multiplied by the coëfficient of x^2 and afterwards by 4 to clear of fractions, it is evident that *fractions may be avoided by multiplying by four times the coëfficient* of the highest power of the unknown quantity.

GENERAL RULE. — *If the term containing the second power of the unknown quantity is not a perfect square, make it such by multiplying or dividing the members of the equation.*

Add to each member of the equation the square of the quotient obtained by dividing the term containing the first power of the unknown quantity by twice the square root of the term containing the second power.

Extract the root of each member and reduce the equation.

1. In multiplying the members of the equation by any quantity, multiply by the smallest one that will render the coëfficient a perfect square.

QUADRATIC EQUATIONS. 233

2. When the *coëfficient of the first power is an odd number*, fractions will be avoided by multiplying by four times the coëfficient of the highest power of the unknown quantity. The *third* term will then be the *square of the coëfficient of the first power* in the original equation.

This is often the most convenient method of solving quadratics.

Solve the following equations:

5. $3x^2 + 5x = 8.$

6. $2x^2 + 7x = 22.$

7. $4x^2 + 5x = 84.$

8. $5x^2 - 4x = 105.$

9. $3x^2 - 16x = 140.$

10. $4x^2 - 7x = 102.$

11. $9x^2 + 4x = 44.$

12. $8x^2 - 6x = 464.$

13. $5x^2 - 6x = 144.$

14. $3x^2 + 2x = 56.$

15. $x^2 - 6x - 14 = 2.$

16. $x^2 - 13x - 6 = 8.$

17. $x^2 + 17x - 18 = 0.$

18. $x^2 - 11x - 7 = 5.$

19. $2x^2 - 18x = -40.$

20. $2x^2 + 5x = 18.$

21. $3x^2 + 2x = 21.$

22. $2x^2 - 7x = 34.$

23. $5x^2 - 6x = 41.$

24. $\dfrac{4x}{x+3} - \dfrac{x-3}{2x+5} = 2.$

25. $3x - \dfrac{169}{x} = 26.$

26. $\dfrac{7}{4} - \dfrac{2x-5}{x+5} = \dfrac{3x-7}{2x}.$

27. $\dfrac{3x-5}{9x} - \dfrac{6x}{3x-25} = \dfrac{1}{3}.$

28. $\dfrac{1}{x-1} - \dfrac{2}{x+2} = \dfrac{1}{2}.$

29. $\dfrac{x}{x+60} = \dfrac{7}{3x-5}.$

30. $\dfrac{x+11}{x} = 7 - \dfrac{9+4x}{x^2}.$

31. $\dfrac{4x-10}{x+5} - \dfrac{7-3x}{x} = \dfrac{7}{2}.$

32. $\dfrac{3x^2}{x-7} - \dfrac{1-8x}{10} = \dfrac{x}{5}.$

33. $\dfrac{48}{x+3} = \dfrac{165}{x+10} - 5.$

34. $\dfrac{x^2}{2} - \dfrac{x}{3} = 2(x+2).$

35. $\dfrac{2}{x-1} = \dfrac{3}{x-2} + \dfrac{2}{x-4}.$

36. $\dfrac{x}{5-x} - \dfrac{5-x}{x} = \dfrac{15}{4}.$

37. $\dfrac{1}{7}(x-4) - \dfrac{2}{5}(x-2) = \dfrac{1}{x}(2x+3).$

38. $\sqrt{x+5} = \dfrac{12}{\sqrt{x+12}}.$

40. $\dfrac{2x-1}{x-1} + \dfrac{1}{6} = \dfrac{2x-3}{x-2}.$

39. $\dfrac{2x-1}{x} - \dfrac{3x}{3x-1} + \dfrac{1}{2} = 0.$

41. $x - \dfrac{x^2-3}{x+1} = \dfrac{14x-9}{8x-3}.$

42. $\dfrac{x+3}{x+2} + \dfrac{x-3}{x-2} = \dfrac{2x-3}{x-1}.$

43. $\dfrac{2x+3}{2(2x-1)} - \dfrac{7-x}{2(x+1)} = \dfrac{7-3x}{4-3x}.$

44. $\dfrac{1}{2(x-1)} + \dfrac{3}{x^2-1} = \dfrac{1}{4}.$

45. Solve the equation $ax^2 + bx = c.$

SOLUTION.

$$ax^2 + bx = c$$

Multiplying by $4a$ and adding the square of b (Note 2, Rule),

$$4a^2x^2 + 4abx + b^2 = 4ac + b^2$$

Extracting square root, $\quad 2ax + b = \pm\sqrt{4ac+b^2}$

$$\therefore x = \dfrac{-b \pm \sqrt{4ac+b^2}}{2a}$$

46. Solve the equation $x^2 - ax + bx = ab.$

SOLUTION.

$$x^2 - ax + bx = ab \tag{1}$$
$$x^2 - (a-b)x = ab \tag{2}$$

Multiplying by 4 and adding the square of $a-b$,

$$4x^2 - 4(a-b)x + (a-b)^2 = 4ab + (a-b)^2 \tag{3}$$
$$= a^2 + 2ab + b^2$$

Extracting square root, $\quad 2x - (a-b) = \pm(a+b)$

$$\therefore x = a \text{ or } -b$$

47. $x^2 + 2bx = b^2.$

52. $5cx - 2x^2 = 2c^2.$

48. $x^2 - 4bx = 12b^2.$

53. $12a^2x^2 - 5ax = 3.$

49. $x^2 = 4bx + 7b^2.$

54. $x^2 + 3ax = \dfrac{7a^2}{4}.$

50. $x^2 + 3ax = 10a^2.$

51. $x^2 + 5bx = 14b^2.$

55. $x^2 - (a+1)x = -a.$

QUADRATIC EQUATIONS.

56. $\dfrac{3cx^2}{4} + \dfrac{2cx}{3} = \dfrac{5c^2}{3c}.$

57. $x^2 + 2(a+8)x = -32a.$

58. $\dfrac{x(b-x)}{b+x} + \dfrac{x}{3} = 5.$

59. $x^2 + \dfrac{a^2-b^2}{ab}x = 1.$

60. $\dfrac{c^2}{(x+c)^2} = \dfrac{d^2}{(x-c)^2}.$

61. $\dfrac{2x(b-x)}{3b-2x} = \dfrac{b}{4}.$

62. $\dfrac{c}{3} + \dfrac{5x}{4} - \dfrac{x^2}{3c} = 0.$

63. $\dfrac{a^2 x^2}{b^2} + \dfrac{b^2}{c^2} = \dfrac{2ax}{c}.$

64. $\dfrac{x+1}{\sqrt{x}} = \dfrac{c+1}{\sqrt{c}}.$

65. $\dfrac{2c+x}{2c-x} + \dfrac{c-2x}{c+2x} = \dfrac{8}{3}.$

66. $\dfrac{x^2}{6b^2} - \dfrac{5x}{6ab} + \dfrac{1}{a^2} = 0.$

67. $(c^2 + 1)x = cx^2 + c.$

68. $\dfrac{1}{b-x} - \dfrac{1}{b+x} = \dfrac{3+x^2}{b^2-x^2}.$

69. $9x^2 - 3(b+2c)x + 2bc = 0.$

70. $x^2 - (b-a)c = ax - bx + cx.$

71. $\sqrt{(a+c)x - 4ac} = x - 2c.$

72. $(c+d)^2 x^2 - (c^2 - d^2)x = cd.$

73. $2\sqrt{x-n} + 3\sqrt{2x} = \dfrac{7n + 5x}{\sqrt{x-n}}.$

74. $\dfrac{x^2+1}{x} = \dfrac{a+b}{c} + \dfrac{c}{a+b}.$

75. $\dfrac{2x-3b}{x-2b} - \dfrac{3x}{x+2a} = \dfrac{a}{2(a-b)}.$

76. $(3c^2 + d^2)(x^2 - x + 1) = (c^2 + 3d^2)(x^2 + x + 1).$

EQUATIONS IN THE QUADRATIC FORM.

272. An equation which contains but two powers of an unknown quantity, the exponent of one power being twice that of the other power, is in the *Quadratic Form*.

These equations in the quadratic form can be reduced to the general form $ax^{2n} + bx^n = c$, in which n represents any number.

EXAMPLES.

1. Given $x^4 + 3x^2 = 28$, to find the values of x.

SOLUTION.

$$x^4 + 3x^2 = 28$$

Completing the square, $\quad x^4 + 3x^2 + \frac{9}{4} = \frac{121}{4}$

Extracting the square root, $\quad x^2 + \frac{3}{2} = \pm \frac{11}{2}$

$$x^2 = 4 \text{ or } -7$$

Extracting the square root, $\quad x = \pm 2 \text{ or } \pm\sqrt{-7}$

2. Given $x^{\frac{2}{3}} + 3x^{\frac{1}{3}} = 10$, to find the values of x.

FIRST SOLUTION.

$$x^{\frac{2}{3}} + 3x^{\frac{1}{3}} = 10$$

Completing the square, $\quad x^{\frac{2}{3}} + 3x^{\frac{1}{3}} + \frac{9}{4} = \frac{49}{4}$

Extracting the square root, $\quad x^{\frac{1}{3}} + \frac{3}{2} = \pm \frac{7}{2}$

$$x^{\frac{1}{3}} = 2 \text{ or } -5$$

Raising to third power, $\quad x = 8 \text{ or } -125$

SECOND SOLUTION.

Let $\quad\quad\quad\quad\quad\quad\quad\quad\quad x^{\frac{1}{3}} = p$

Then, $\quad\quad\quad\quad\quad\quad\quad\quad x^{\frac{2}{3}} = p^2$

Substituting in the given equation, $p^2 + 3p = 10$

Solving, $\quad\quad\quad\quad\quad\quad\quad p = 2 \text{ or } -5$

Hence, $\quad\quad\quad\quad\quad\quad\quad x^{\frac{1}{3}} = 2 \text{ or } -5$

Raising to third power, $\quad\quad\quad x = 8 \text{ or } -125$

QUADRATIC EQUATIONS.

3. Given $(x+2)^2 + (x+2) = 20$, to find the values of x.

FIRST PROCESS.

$$(x+2)^2 + (x+2) = 20$$

Completing the square, $(x+2)^2 + (x+2) + \tfrac{1}{4} = \tfrac{81}{4}$

Extracting the square root, $(x+2) + \tfrac{1}{2} = \pm \tfrac{9}{2}$

$$x + 2 = 4 \text{ or } -5$$
$$x = 2 \text{ or } -7$$

SECOND PROCESS.

Let $\qquad p = (x+2)$

Then, $\qquad p^2 = (x+2)^2$

Then, $\qquad p^2 + p = 20$

Solving, $\qquad p = 4 \text{ or } -5$

Hence, $\qquad x + 2 = 4 \text{ or } -5$

$$x = 2 \text{ or } -7$$

Solve the following equations:

4. $x^4 - 2x^2 = 8.$

5. $x^6 - 3x^3 = 40.$

6. $x^6 - 4x^3 = 32.$

7. $2x^4 - 4x^2 = 16.$

8. $4x^4 - x^2 = 3.$

9. $2x^6 - \dfrac{2x^3}{4} = 124.$

10. $x^4 + 4x^2 = 12.$

11. $x^{\tfrac{1}{2}} + 2x^{\tfrac{1}{4}} = 8.$

12. $x^3 + 3x^{\tfrac{3}{2}} = 88.$

13. $x^{\tfrac{2}{3}} + 3x^{\tfrac{1}{3}} = 4.$

14. $x^{\tfrac{1}{2}} - x^{\tfrac{1}{4}} = 20.$

15. $ax^{2n} + bx^n = c.$

16. $(x^2+4)^2 + (x^2+4) = 30.$

17. $(x-1)^2 + 5(x-1) = 14.$

18. $(x^2-9)^2 - 11(x^2-9) = 80.$

19. $(x^2+1)^2 + (x^2+1) = 30.$

20. $(x^2-x)^2 - (x^2-x) = 132.$

21. $x + 5 - \sqrt{x+5} = 6.$

22. $3x + 4 + 4\sqrt{3x+4} = 32.$

23. $\left(\dfrac{8}{x} + x\right)^2 + \left(\dfrac{8}{x} + x\right) = 42.$

24. $\left(\dfrac{6}{x} + x\right)^2 + \left(\dfrac{6}{x} + x\right) = 30.$

25. $(2x^2 - 4x + 1)^2 - (2x^2 - 4x + 1) = 42.$

26. $x^2 - 7x + 18 + \sqrt{x^2 - 7x + 18} = 42.$

27. $2x^2 + 3x + 9 - 5\sqrt{2x^2 + 3x + 9} = 6.$

28. $2(3x^2+1)^{\frac{1}{2}} + 3x^2 + 1 = 63.$

29. $\sqrt{x+12} + \sqrt[4]{x+12} = 6.$

30. $x + 2 + (x+2)^{\frac{1}{2}} = 20.$

31. $(x+4)^2 + x + 4 = 30.$

32. $\sqrt{x+31} - \frac{1}{3}\sqrt[4]{x+31} = 8.$

33. $4x^2 - 4\sqrt{4x^2+13} = 8.$

SUGGESTION. — Add 13 to each member of the equation.

34. $x^2 + b^2 + x - 6 = -2bx - b.$

35. $(x^2 + 4x + 4)^2 + 2(x^2 + 4x) = 7.$

36. $3x - \sqrt{9x^2 - 18x} = 5x - x^2 + 4.$

SUGGESTION. — The equation may be expressed thus:
$$x^2 - 2x - 3\sqrt{x^2 - 2x} = 4$$

37. $\sqrt{x+a} + 2b\sqrt[4]{x+a} = 3b^2.$

38. $x^2 - 10x - 2\sqrt{x^2 - 10x + 18} = -15.$

PROBLEMS.

273. 1. Find two numbers whose sum is 12, and whose product is 35.

SOLUTION.

Let $x =$ one

Then, $12 - x =$ the other.

$$12x - x^2 = 35$$
$$x^2 - 12x = -35$$
$$x^2 - 12x + 36 = 1$$
$$x - 6 = \pm 1$$
$$x = 7 \text{ or } 5$$
$$12 - x = 5 \text{ or } 7$$

2. The sum of two numbers is 10, and their product is 21. What are the numbers?

3. Divide 27 into two such parts that their product may be 140.

QUADRATIC EQUATIONS.

4. A rectangular field is 12 rods longer than it is wide, and contains 7 acres. What is the length of its sides?

5. A person purchased a flock of sheep for $100. If he had purchased 5 more for the same sum, they would have cost $1 less per head. How many did he buy?

6. An orchard containing 2000 trees had 10 rows more than it had trees in a row. How many rows were there? How many trees were there in each row?

7. The difference between two numbers is 2, and the sum of their squares is 244. What are the numbers?

8. One hundred and ten dollars was divided among a certain number of persons. If each person had received $1 more, he would have received as many dollars as there were persons. How many persons were there?

9. A man worked a certain number of days, receiving for his pay $18. If he had received $1 per day less than he did, he would have had to work 3 days longer to earn the same sum. How many days did he work?

10. Find the price of eggs, when 2 less for 12 cents raises the price 1 cent per dozen.

11. A person sold goods for $24, gaining a per cent. equal to the number of dollars which the goods cost him. What did they cost him?

SUGGESTION. — Let $x =$ the cost; then, $\frac{x}{100} =$ the gain per cent.

12. The expenses of a party of men amount to $10. If each man pays 30 cents more than there are persons, the bill will be settled. How many are there in the party?

13. A picture, which is 18 inches by 12, is to be surrounded with a frame of uniform width, whose area is equal to that of the glass. What is the width of the frame?

14. A man sold a quantity of goods for $39, and gained a per cent. equal to the number of dollars which the goods cost him. What did they cost him?

15. Two men dig a ditch 100 rods in length for $100, each receiving $50. A is to have 25 cents a rod more than B. How many rods does each dig? What is the price per rod?

16. A rectangular park, 60 rods long and 40 rods wide, is surrounded by a street of uniform width, containing 1344 square rods. How wide is the street?

17. A person purchased two pieces of cloth which together measured 36 yards. Each cost as many shillings per yard as there were yards in the piece. If one piece cost 4 times as much as the other, how many yards were there in each?

18. A person drew a quantity of pure wine from a vessel which was full, holding 81 gallons, and then filled the vessel up with water. He then drew from the mixture as much as he drew before of pure wine, when it was found that the vessel contained 64 gallons of pure wine. How much did he draw each time?

19. Two persons started at the same time and traveled toward a place 90 miles distant. A traveled one mile per hour faster than B, and reached the place one hour before him. At what rate did each travel?

20. A person found that he had in his purse, in silver and copper coins, just one dollar. Each copper coin was worth as many cents as there were silver coins, and each silver coin was worth as many cents as there were copper coins. There were in all 27 coins. How many were there of each?

SIMULTANEOUS QUADRATIC EQUATIONS.

274. A *Homogeneous Equation* is an equation in which the sum of the exponents of the unknown quantities in each term which contains unknown quantities, is the same.

Thus, $x^2 + 2y^2 = 16$ and $xy + y^2 = 12$ are homogeneous equations.

275. Simultaneous Quadratic Equations can usually be solved by the rules for quadratics, if they belong to one of the following classes:

QUADRATIC EQUATIONS.

1. *When one is simple and the other quadratic.*
2. *When the unknown quantities in each equation are combined in a similar manner.*
3. *When each equation is homogeneous and quadratic.*

276. The following solutions will illustrate the processes in many of the ordinary forms of simultaneous quadratics, but a little careful examination of the equations will sometimes enable the student to find the values of the unknown quantities more readily than they can be found by these general methods of solution.

(I.) **Simple and Quadratic.**

1. Given $\left\{ \begin{array}{l} x+y=5 \\ 2x^2+y^2=17 \end{array} \right\}$, to find the values of x and y.

SOLUTION.

$$x + y = 5 \quad (1)$$
$$2x^2 + y^2 = 17 \quad (2)$$
From (1), $\qquad x = 5 - y \quad (3)$
$$2x^2 = 50 - 20y + 2y^2 \quad (4)$$
Substituting in (2), $\quad 50 - 20y + 2y^2 + y^2 = 17 \quad (5)$
Collecting terms, etc., $\quad 3y^2 - 20y = -33 \quad (6)$
Solving, $\qquad y = 3 \text{ or } 3\frac{2}{3} \quad (7)$
Substituting in (1), $\quad x = 2 \text{ or } 1\frac{1}{3} \quad (8)$

(II.) **Unknown quantities similarly combined.**

2. Given $\left\{ \begin{array}{l} x+y=5 \\ xy=6 \end{array} \right\}$, to find the values of x and y.

SOLUTION.

$$x + y = 5 \quad (1)$$
$$xy = 6 \quad (2)$$
Squaring (1), $\qquad x^2 + 2xy + y^2 = 25 \quad (3)$
Multiplying (2) by 4, $\qquad 4xy = 24 \quad (4)$
Subtracting (4) from (3), $\quad x^2 - 2xy + y^2 = 1 \quad (5)$
Extracting square root of (5), $\quad x - y = \pm 1 \quad (6)$

ALGEBRA. — 16.

HIGH SCHOOL ALGEBRA.

Adding (1) and (6),	$2x = 6$ or 4	(7)
	$x = 3$ or 2	(8)
Subtracting (6) from (1),	$2y = 4$ or 6	(9)
	$y = 2$ or 3	(10)

3. Given $\left\{ \begin{array}{l} x + y = 5 \\ x^2y + xy^2 = 30 \end{array} \right\}$, to find the values of x and y.

Solution.

	$x + y = 5$	(1)
	$x^2y + xy^2 = 30$	(2)
Factoring (2),	$xy(x + y) = 30$	(3)
Dividing (3) by (1),	$xy = 6$	(4)
Squaring (1),	$x^2 + 2xy + y^2 = 25$	(5)
Multiplying (4) by 4,	$4xy = 24$	(6)
Subtracting (6) from (5),	$x^2 - 2xy + y^2 = 1$	(7)
Extracting square root of (7),	$x - y = \pm 1$	(8)
Adding (8) and (1),	$2x = 6$ or 4	(9)
	$x = 3$ or 2	(10)
Subtracting (8) from (1),	$2y = 4$ or 6	(11)
	$y = 2$ or 3	(12)

4. Given $\left\{ \begin{array}{l} x + y = 8 \\ x^3 + y^3 = 152 \end{array} \right\}$, to find the values of x and y.

Solution.

	$x + y = 8$	(1)
	$x^3 + y^3 = 152$	(2)
Dividing (2) by (1),	$x^2 - xy + y^2 = 19$	(3)
Squaring (1),	$x^2 + 2xy + y^2 = 64$	(4)
Subtracting (3) from (4),	$3xy = 45$	(5)
	$xy = 15$	(6)
Subtracting (6) from (3),	$x^2 - 2xy + y^2 = 4$	(7)
Extracting square root of (7),	$x - y = \pm 2$	(8)
Adding (8) and (1),	$2x = 10$ or 6	(9)
	$x = 5$ or 3	(10)
Subtracting (8) from (1),	$2y = 6$ or 10	(11)
	$y = 3$ or 5	(12)

QUADRATIC EQUATIONS. 243

5. Given $\begin{Bmatrix} x+y=5 \\ x^4+y^4=97 \end{Bmatrix}$, to find the values of x and y.

SOLUTION.

$$x + y = 5 \qquad (1)$$
$$x^4 + y^4 = 97 \qquad (2)$$

4th power of (1), $\quad x^4 + 4x^3y + 6x^2y^2 + 4xy^3 + y^4 = 625 \qquad (3)$

Subtracting (2) from (3), $\quad 4x^3y + 6x^2y^2 + 4xy^3 = 528 \qquad (4)$

Dividing by 2, $\quad 2x^3y + 3x^2y^2 + 2xy^3 = 264 \qquad (5)$

Squaring (1) and multiplying by $2xy$,

$$2x^3y + 4x^2y^2 + 2xy^3 = 50xy \qquad (6)$$

Subtracting (5) from (6), $\quad x^2y^2 = 50xy - 264 \qquad (7)$

Transposing, $\quad x^2y^2 - 50xy = -264 \qquad (8)$

Completing square, etc., $\quad xy = 6 \text{ or } 44 \qquad (9)$

Combining (1) and (9), two pairs of simultaneous equations are formed,

$$\begin{Bmatrix} x+y=5 \\ xy=6 \end{Bmatrix} \text{ and } \begin{Bmatrix} x+y=5 \\ xy=44 \end{Bmatrix}$$

Solving according to solution of Example 2, the values of x and y are found.

(III.) Homogeneous equations.

6. Given $\begin{Bmatrix} x^2 - xy = 15 \\ 2xy - y^2 = 16 \end{Bmatrix}$, to find the values of x and y.

SOLUTION.

$$x^2 - xy = 15 \qquad (1)$$
$$2xy - y^2 = 16 \qquad (2)$$

Assume $\quad x = vy \qquad (3)$

Substituting vy in (1), $\quad v^2y^2 - vy^2 = 15 \qquad (4)$

Substituting vy in (2), $\quad 2vy^2 - y^2 = 16 \qquad (5)$

From (4), $\quad y^2 = \dfrac{15}{v^2 - v} \qquad (6)$

From (5), $\quad y^2 = \dfrac{16}{2v - 1} \qquad (7)$

Equating (6) and (7), $\quad \dfrac{16}{2v-1} = \dfrac{15}{v^2 - v} \qquad (8)$

Clearing of fractions and reducing,

$$16v^2 - 46v = -15 \qquad (9)$$

Whence, $\quad v = \tfrac{5}{2} \text{ or } \tfrac{3}{8} \qquad (10)$

Substituting the value of v in (7), $y^2 = \dfrac{16}{5-1}$ or 4 \hfill (11)

And $\qquad\qquad\qquad\qquad y^2 = \dfrac{16}{\frac{3}{4} - 1}$ or -64 \hfill (12)

$$y = \pm 2 \text{ or } \pm 8\sqrt{-1}$$
$$x = \pm 5 \text{ or } \pm 3\sqrt{-1}$$

Find the values of the unknown quantities in the following:

7. $\begin{cases} x - y = 4. \\ x^2 - y^2 = 32. \end{cases}$

8. $\begin{cases} x + y = 5. \\ x^2 + y^2 = 13. \end{cases}$

9. $\begin{cases} xy = 15. \\ 2x + y = 13. \end{cases}$

10. $\begin{cases} xy = 6. \\ 2x - 3y = 9. \end{cases}$

11. $\begin{cases} x + 2y = 7. \\ 2x^2 - y^2 = 14. \end{cases}$

12. $\begin{cases} 2x + 3y = 22. \\ 2xy = 40. \end{cases}$

13. $\begin{cases} \dfrac{1}{x} + \dfrac{1}{y} = \dfrac{3}{10}. \\ \dfrac{1}{x^2} + \dfrac{1}{y^2} = \dfrac{1}{20}. \end{cases}$

14. $\begin{cases} xy = 24. \\ x^2 - 2y^2 = 4. \end{cases}$

15. $\begin{cases} xy = 10. \\ x - y = 3. \end{cases}$

16. $\begin{cases} xy = 12. \\ x + y = 7. \end{cases}$

17. $\begin{cases} x - y = 1. \\ x^3 - y^3 = 37. \end{cases}$

18. $\begin{cases} x + y = 4. \\ x^3 + y^3 = 28. \end{cases}$

19. $\begin{cases} x^3 + y^3 = 28. \\ x^2 y + xy^2 = 12. \end{cases}$

20. $\begin{cases} x^3 - y^3 = 26. \\ xy^2 - x^2 y = -6. \end{cases}$

21. $\begin{cases} x^2 - xy = 6. \\ x^2 + y^2 = 61. \end{cases}$

22. $\begin{cases} 3x^2 + xy = 18. \\ 4y^2 + 3xy = 54. \end{cases}$

23. $\begin{cases} x^2 + xy = 70. \\ xy - y^2 = 12. \end{cases}$

24. $\begin{cases} 2x + y = 22. \\ xy + 2y^2 = 120. \end{cases}$

25. $\begin{cases} x = 2y^2. \\ x - y = 15. \end{cases}$

26. $\begin{cases} x + 4y = 14. \\ y^2 - 2y + 4x = 11. \end{cases}$

27. $\begin{cases} 4xy + x^2 y^2 = 96. \\ x + y = 6. \end{cases}$

QUADRATIC EQUATIONS.

28. $\begin{cases} x^2 + y^2 = 52. \\ x + y + xy = 34. \end{cases}$

29. $\begin{cases} x^2 + y^2 = 13. \\ xy + y^2 = 15. \end{cases}$

30. $\begin{cases} \sqrt[3]{x} + \sqrt[3]{y} = 6. \\ x + y = 72. \end{cases}$

31. $\begin{cases} x^3 - y^3 = 56. \\ x - y = \dfrac{16}{xy}. \end{cases}$

32. $\begin{cases} x + y = 3. \\ x^4 + y^4 = 17. \end{cases}$

33. $\begin{cases} (x - y)(x^2 + y^2) = 13. \\ x^2 y - xy^2 = 6. \end{cases}$

34. $\begin{cases} x^2 + x + y = 18 - y^2. \\ xy = 6. \end{cases}$

35. $\begin{cases} x^2 - xy + y^2 = 21. \\ y^2 - 2xy + 15 = 0. \end{cases}$

36. $\begin{cases} x^2 + xy + y^2 = 39. \\ 2x^2 + 3xy + y^2 = 63. \end{cases}$

37. $\begin{cases} x^2 + xy + 2y^2 = 74. \\ 2x^2 + 2xy + y^2 = 73. \end{cases}$

38. $\begin{cases} x^2 + y^2 + xy = 49. \\ x^4 + y^4 + x^2 y^2 = 931. \end{cases}$

39. $\begin{cases} x^4 - x^2 + y^4 - y^2 = 84. \\ x^2 + x^2 y^2 + y^2 = 49. \end{cases}$

40. $\begin{cases} x + \sqrt{xy} + y = 19. \\ x^2 + xy + y^2 = 133. \end{cases}$

41. $\begin{cases} x^2 + 2y^2 = 44 - xy. \\ 2x^2 - 16 = 3xy - 2y^2. \end{cases}$

42. $\begin{cases} x - y = 1. \\ x^3 - y^3 = 61. \end{cases}$

43. $\begin{cases} x - y = 2. \\ x^5 - y^5 = 242. \end{cases}$

44. $\begin{cases} x^2 + y^2 + 4 = 3xy. \\ x^4 + y^4 = 272. \end{cases}$

45. $\begin{cases} x^2 + xy + 4y^2 = 6. \\ 3x^2 + 8y^2 = 14. \end{cases}$

46. $\begin{cases} 3x^2 + 7xy = 82. \\ x^2 + 5xy + 9y^2 = 279. \end{cases}$

47. $\begin{cases} x^2 + y^2 = 3 + xy. \\ x^4 + y^4 = 21 - x^2 y^2. \end{cases}$

48. $\begin{cases} x^2 + xy + 2y^2 = 44. \\ 2x^2 - xy + y^2 = 16. \end{cases}$

49. $\begin{cases} x^2 - xy + y^2 = 7. \\ x^4 + x^2 y^2 + y^4 = 133. \end{cases}$

50. $\begin{cases} x^2 + y^2 + x + y = 32. \\ xy = 12. \end{cases}$

51. $\begin{cases} x^2 + 4y^2 = 256 - 4xy. \\ 3y^2 - x^2 = 39. \end{cases}$

52. $\begin{cases} x + y + \sqrt{x + y} = 6. \\ x^2 + y^2 = 10. \end{cases}$

53. $\begin{cases} x^2 + y^2 - (x + y) = 78. \\ xy + (x + y) = 39. \end{cases}$

54. $\begin{cases} \dfrac{3(x + y)}{x - y} + \dfrac{3(x - y)}{x + y} = 10. \\ x^2 + y^2 = 45. \end{cases}$

55. $\begin{cases} x^2 - xy = a^2 + b^2. \\ xy - y^2 = 2ab. \end{cases}$ 57. $\begin{cases} x^2 + y^2 = 2(a^2 + b^2). \\ xy = a^2 - b^2. \end{cases}$

56. $\begin{cases} x + 2y = 3a + b. \\ xy + y^2 = 2a(a+b). \end{cases}$ 58. $\begin{cases} x + y + \sqrt{x+y} = 12. \\ x - y + \sqrt{x-y} = 2. \end{cases}$

59. $\begin{cases} x^2 + y^2 + 4\sqrt{x^2 + y^2} = 45. \\ x^4 + y^4 = 337. \end{cases}$

60. $\begin{cases} \sqrt{x} - \sqrt{y-x} = \sqrt{a-x}. \\ 2\sqrt{y-x} = 3\sqrt{a-x}. \end{cases}$

PROBLEMS.

277. 1. The sum of two numbers is 8, and their product is 12. What are the numbers?

2. The sum of two numbers is 12, and the sum of their squares is 104. What are the numbers?

3. Divide 13 into two such parts that the sum of their square roots is 5.

4. The product of two numbers is 99, and their sum is 20. What are the numbers?

5. The sum of two numbers is 100, and the difference of their square roots is 2. What are the numbers?

6. The difference of two numbers is 2, and the difference of their cubes is 56. What are the numbers?

7. Find two numbers whose sum multiplied by the second is 84, and whose difference multiplied by the first is 16.

8. The product of two numbers is 48, and the difference of their cubes is 37 times the cube of their difference. What are the numbers?

9. The sum of two numbers is a, and the sum of their squares is b. What are the numbers?

QUADRATIC EQUATIONS.

10. What two numbers are there such that their sum increased by their product is 34, and the sum of their squares diminished by their sum is 42?

11. There is a number expressed by two digits, such that the sum of the squares of the digits is equal to the number increased by the product of its digits, and if 36 be added to the number, the digits will be reversed. What is the number?

12. From two places, distant 720 miles, A and B set out to meet each other. A traveled 12 miles a day more than B, and the number of days before they met was equal to one half the number of miles B went per day. How many miles did each travel per day?

13. A merchant received $12 for a quantity of linen, and an equal sum, at 50 cents a yard less, for a quantity of cotton. The cotton exceeded the linen by 32 yards. How many yards did he sell of each?

14. A farmer has a field 18 rods long and 12 rods wide, which he wishes to enlarge, so that it may contain twice its former area, by making a uniform addition on all sides. What will be the sides of the field when it is enlarged?

15. A merchant bought a piece of cloth for $147, from which he cut off 12 yards which were damaged, and sold the remainder for $120.25, gaining 25 cents on each yard sold. How many yards did he buy? How much did it cost per yard?

16. A man gave $4 to be divided among some children, but 10 more joining the group, the share of each was reduced 2 cents. How many children were there, and what was the share of each?

17. A drover sold a lot of cows for $2400. He sold a second lot, containing 10 cows less, at $10 more a head, and received $100 more than he did for the first. How many were there in the first lot, and what was the price per head?

18. Mr. A has two fields, one a rectangle, and the other a square whose side is two thirds as long as the longer side of the rectangle. It takes 4 more rods of fence for the rectangular field than for the square one, and the square field contains 40 more square rods than the other. What is the length of each field?

19. A boy rode 10 miles on his bicycle, when it broke down, and he was compelled to return on foot. He found that it took 1 hour and 15 minutes longer to walk back than it did to ride out. How fast did he ride, if he walked 4 miles less per hour than he rode?

20. The square of John's number of marbles increased by the product of John's and Frank's is 525. The square of Frank's number increased by the product of John's and Frank's is 700. How many has each?

21. There is a number expressed by two figures, such that if the digits be squared their sum will be 18 more than the number itself. Find the number, if the sum of the digits is 13.

22. The product of two numbers is 16, and the sum of their fourth powers is 4112. What are the numbers?

23. A man loaned two sums of money at a rate equal to the number of hundreds represented by the first sum. His income from this was $96. If he had loaned them at a rate equal to the number of hundreds in the other sum, his income would have been increased $64. Find the sums loaned.

24. A regiment of 1104 men was marching in two columns, each having 10 more men in depth than in front. 380 men were taken from the rear of these two columns to make a detour, and after this it was found that each column was an exact square with the same number of men on each side as there was in front of that column at first. Find the number of men in each column at first.

25. The fore wheel of a carriage makes 6 revolutions more than the hind wheel, in going 360 feet. If the circumference

QUADRATIC EQUATIONS. 249

of each wheel had been 3 feet greater, the fore wheel would have made only 4 revolutions more than the hind wheel in going that distance. What is the circumference of each wheel?

26. Find two numbers whose sum, product, and difference of their squares are all equal.

27. The joint capital of A and B was $416. A's money was in trade 9 months, and B's 6 months. When they shared stock and gain, A received $228 and B $252. What was the capital of each?

28. A rectangular piece of ground has a perimeter of 100 rods, and its area is 589 square rods. What are its length and breadth?

29. Twenty persons sent together $48 to a benevolent society. One half the amount was contributed by women, and the other half by men; but each man gave a dollar more than each woman. How many women contributed? How many men? What was the contribution of each?

30. In a purse containing 9 coins, some are of gold, others of silver. Each gold coin is worth as many dollars as there are silver coins, and each silver coin is worth as many cents as there are gold coins, and the value of the whole is $20.20. How many are there of each?

31. The difference of two numbers is 15, and half their product equals the cube of the smaller. What are the numbers?

32. A and B set out from two places, C and D, at the same time. A started from C and traveled through D in the same direction in which B traveled. When A overtook B, it was found that they had together traveled 60 miles, that A had passed through D 5 hours before, and that it would have required 20 hours for B to return to C at the rate he had been traveling. What was the distance from C to D?

PROPERTIES OF QUADRATICS.

278. In the general equation of the second degree, $x^2 + px = q$, p and q may be either positive or negative. There are, therefore, four forms, to some one of which every equation of the second degree can be reduced.

(1) $\qquad x^2 + px = q \qquad\qquad x = -\dfrac{p}{2} \pm \sqrt{\dfrac{p^2}{4} + q} \qquad (1)$

(2) $\qquad x^2 - px = q \qquad\qquad x = \dfrac{p}{2} \pm \sqrt{\dfrac{p^2}{4} + q} \qquad (2)$

(3) $\qquad x^2 + px = -q \qquad\qquad x = -\dfrac{p}{2} \pm \sqrt{\dfrac{p^2}{4} - q} \qquad (3)$

(4) $\qquad x^2 - px = -q \qquad\qquad x = \dfrac{p}{2} \pm \sqrt{\dfrac{p^2}{4} - q} \qquad (4)$

279. Positive and negative roots.

1. How does the value of $\sqrt{\dfrac{p^2}{4} + q}$ compare with $\dfrac{p}{2}$? How does the value of $\sqrt{\dfrac{p^2}{4} - q}$ compare with $\dfrac{p}{2}$?

2. In form (1), if the positive sign of the radical is used, what sign, then, will the root have? What, if the negative sign is used? Which root is numerically the greater?

3. In form (2), if the positive sign of the radical is used, what sign will the root have? What, if the negative sign is used? Which root is numerically the greater?

4. In form (3), if the positive sign of the radical is used, what sign will the root have? What, if the negative sign is used?

5. In form (4), if the positive sign of the radical is used, what sign will the root have? What, if the negative is used?

QUADRATIC EQUATIONS.

280. Principles. — 1. *In form* (1) *one root is positive and the other negative, and the negative root is numerically the greater.*

2. *In form* (2) *one root is positive and the other negative, and the positive is numerically the greater.*

3. *In form* (3) *both roots are negative.*

4. *In form* (4) *both roots are positive.*

281. Equal and unequal roots.

1. How do the roots compare numerically in (1)? In (2)?

2. How does the second term of the root in (3) and (4), compare with the first term?

3. If $q = \dfrac{p^2}{4}$, to what does the radical quantity reduce? How, then, will the roots compare numerically? What signs will they have?

282. Principles. — 1. *In forms* (1) *and* (2) *the roots of each are unequal.*

2. *In forms* (3) *and* (4) *the roots of each are unequal except when* $\dfrac{p^2}{4} = q$.

283. Real and imaginary roots.

1. Since $\dfrac{p^2}{4}$ is the square of $\pm\dfrac{p}{2}$, what sign will it have whether p is positive or negative?

2. In forms (1) and (2), is the radical part, then, real or imaginary?

3. In forms (3) and (4) when will the radical part be real and when imaginary?

284. Principles. — 1. *In forms* (1) *and* (2) *both roots are real.*

2. *In forms* (3) *and* (4) *both roots are real except when q is numerically greater than* $\dfrac{p^2}{4}$. *They are then imaginary.*

EXAMPLES.

285. Determine the character of the roots in the following equations:

1. $x^2 - 3x = 4$.

SOLUTION. $p = 3$, $q = 4$, and $\pm\sqrt{\dfrac{p^2}{4}+q} = \pm\sqrt{\dfrac{25}{4}} = \pm\dfrac{5}{2}$

∴ The roots are one positive, one negative, unequal, the positive numerically greater, real.

2. $x^2 - 8x = 20$.
3. $x^2 + 4x = 12$.
4. $x^2 - 6x = 16$.
5. $x^2 - 7x = 8$.
6. $x^2 - x = 6$.
7. $x^2 - 12x = -11$.
8. $x^2 + 6x = -5$.
9. $x^2 + 20x = 21$.
10. $x^2 - 14x = 51$.
11. $x^2 - 30x = 64$.

286. Formation of Quadratic Equations.

1. Since the roots of the equation $x^2 + px = q$ are $-\dfrac{p}{2} + \sqrt{\dfrac{p^2}{4}+q}$ and $-\dfrac{p}{2} - \sqrt{\dfrac{p^2}{4}+q}$, what is the sum of the roots? How does it compare with the coëfficient of x?

2. What is the product of the two roots? How does it compare with the absolute term, or the second member of the equation?

287. When the unknown quantities are collected in the first member, and the known quantities united in the second member, the term of the second member is called the **Absolute Term.**

288. PRINCIPLES.—1. *The sum of the two roots of an affected quadratic equation having the form $x^2 \pm px = \pm q$, is equal to the coëfficient of the first power of the unknown quantity with its sign changed.*

2. *The product of the roots is equal to the absolute term, with its sign changed.*

QUADRATIC EQUATIONS.

Denoting the roots of the equation $x^2 + px = q$ by r_1 and r_2, their sum with the sign changed, $-(r_1 + r_2)$ may be substituted for p (Prin. 1) and their product with the sign changed, $-r_1r_2$, for q (Prin. 2).

Substituting, $\quad x^2 - (r_1 + r_2)x = -r_1r_2$
Transposing, etc., $x^2 - r_1x - r_2x + r_1r_2 = 0$
Factoring, $\quad x(x - r_1) - r_2(x - r_1) = 0$
Or, $\quad\quad\quad (x - r_2)(x - r_1) = 0$

Hence, to form a quadratic equation when the roots are given:

Subtract each root from x and place the product of the remainders equal to zero.

EXAMPLES.

1. Form an equation whose roots are 3 and -5.

Solution.

$$(x - 3)(x + 5) = 0$$
$$\therefore x^2 + 2x - 15 = 0 \text{ or } x^2 + 2x = 15$$

or,

The coëfficient of x is $+2$, the sum of the roots with the sign changed, and the absolute term is $+15$, the product of the roots with the sign changed. Hence the equation is $x^2 + 2x = 15$.

Form equations whose roots are:

2. $3, -2$.
3. $8, 6$.
4. $9, -8$.
5. $-7, -10$.
6. $15, -3$.
7. $1\frac{1}{2}, -2$.
8. $3, -\frac{3}{5}$.
9. $2\frac{1}{3}, 4$.
10. $\frac{2}{3}, \frac{3}{4}$.
11. $-\frac{5}{3}, -\frac{7}{2}$.
12. $3 + \sqrt{2}, 3 - \sqrt{2}$.
13. $2 + \sqrt{3}, 2 - \sqrt{3}$.
14. $1 + \sqrt{5}, 1 - \sqrt{5}$.
15. $a, a - b$.
16. $a - b, a + 2b$.
17. $a(1 + a), a(1 - a)$.
18. $\dfrac{a + \sqrt{b}}{2}, \dfrac{a - \sqrt{b}}{2}$.

289. To solve equations by factoring.

1. What are the factors of $x^2 + 4x - 5$?

2. Place the factors of $x^2 + 4x - 5$ for that expression in the equation $x^2 + 4x - 5 = 0$.

3. Since $(x+5)(x-1) = 0$, to what must one of the factors be equal to produce a product equal to 0?

4. Since when either $x+5 = 0$, or $x-1 = 0$ the equation is satisfied, what are the values of x?

5. Since any equation may be expressed with one member as 0, by transposing all the quantities to one member, how may the values of the unknown quantities be found by factoring?

It is evident that *only such equations* as can be resolved into factors *by inspection* can be solved in this way, but it is also evident that *very many* equations *cannot* be readily factored.

EXAMPLES.

1. Solve the equation $x^2 - 7x + 10 = 0$.

SOLUTION.

$$x^2 - 7x + 10 = 0$$
$$= (x-5)(x-2) = 0$$
$$\therefore x - 5 = 0, \text{ and } x = 5$$
also, $$x - 2 = 0, \text{ and } x = 2$$

2. Solve the equation $3x^2 - 10x + 3 = 0$.

SOLUTION.

$$3x^2 - 10x + 3 = 0$$
$$= (3x-1)(x-3) = 0$$
$$\therefore 3x - 1 = 0, \text{ and } x = \tfrac{1}{3}$$
also, $$x - 3 = 0, \text{ and } x = 3$$

QUADRATIC EQUATIONS.

Solve the following equations by factoring:

3. $x^2 + 3x + 2 = 0$.
4. $x^2 + 7x + 12 = 0$.
5. $x^2 - 4x - 21 = 0$.
6. $x^2 - 7x - 18 = 0$.
7. $x^2 + 6x + 8 = 0$.
8. $x^2 + 12x + 32 = 0$.
9. $x^2 - 10x - 39 = 0$.
10. $x^2 - 12x - 64 = 0$.
11. $4x^2 - 10x + 6 = 0$.
12. $9x^2 - 27x + 18 = 0$.
13. $4x^2 + 16ax + 12a^2 = 0$.
14. $9x^2 + 30bx + 24b^2 = 0$.

Equations of higher degrees can sometimes be solved by factoring.

15. Find the values of x in the equation $x^3 + 2x^2 - 3x - 6 = 0$.

SOLUTION.

$$x^3 + 2x^2 - 3x - 6 = 0$$

Factoring, $\quad x^2(x+2) - 3(x+2) = 0$

$$(x^2 - 3)(x + 2) = 0$$

$$\therefore x^2 - 3 = 0 \text{ and } x = \pm\sqrt{3}$$

also, $\quad x + 2 = 0$ and $x = -2$

16. Find the values of x in the equation $x^3 - 1 = 0$.

SOLUTION.

$$x^3 - 1 = 0$$

Factoring, $\quad (x - 1)(x^2 + x + 1) = 0$

$$\therefore x - 1 = 0 \text{ and } x = 1$$

also, $\quad x^2 + x + 1 = 0$

or, $\quad x^2 + x = -1$

Completing the square, etc., $\quad x = -\dfrac{1 \pm \sqrt{-3}}{2}$

Solve the following equations:

17. $x^3 - 5x^2 + 2x - 10 = 0$.
18. $2x^3 + 6x^2 - 3x - 9 = 0$.
19. $ax^3 + 2x^2 - 5ax - 10 = 0$.
20. $(x - 3)(x^2 + 3x + 2) = 0$.
21. $(x + 5)(x^2 - 4x - 21) = 0$.
22. $x^3 - 27 = 0$.

RATIO.

290. 1. What is the relation of $8x$ to $16x$? $5y$ to $10y$?

2. How does $8a$ compare with $2a$? What is the relation of $8a$ to $2a$?

3. How does $9xy$ compare with $3xy$? What is the relation of $9xy$ to $3xy$?

4. What is the relation of $2a$ to $4a$? What is the relation *between* $2a$ and $4a$?

5. What is the relation of $3x^2$ to $9x^2$? What is the relation *between* $3x^2$ and $9x^2$?

291. Ratio is the relation of one quantity to another of the same kind.

1. This relation is expressed either as the *quotient* of one quantity divided by the other, and is called *Geometrical Ratio* or simply *Ratio*, or as the *difference* between two quantities, and is called *Arithmetical Ratio*.

2. When it is required to determine *what is the relation of one quantity to another*, it is evident that the *first* is the *dividend* and the *second* the *divisor*. Thus, when the question is, "What is the relation of $5a$ to $10a$?" the answer is $\frac{1}{2}$.

3. When it is required to determine the relation *between* two quantities, *either* may be regarded as *dividend* or *divisor*. Thus, when the question is, "What is the relation *between* $5a$ and $10a$?" the answer is $\frac{1}{2}$, or 2.

292. The **Terms of a Ratio** are the quantities compared.

293. The **Sign** of ratio is a colon (:).

Thus, the ratio between $12a$ and $6a$ is expressed $12a : 6a$.

The colon is sometimes regarded as derived from the sign of division, by omitting the line.

RATIO.

294. The **Antecedent** is the first term of the ratio.

Thus, in the expression $5a : 3a$, $5a$ is the antecedent.

295. The **Consequent** is the second term of the ratio.

Thus, in the expression $5a : 3a$, $3a$ is the consequent.

296. A **Couplet** is the antecedent and consequent taken together.

297. The ratio of the squares of two quantities is called the **Duplicate** ratio of the quantities; the ratio of their cubes, their **Triplicate** ratio.

Thus, $a^2 : b^2$ and $a^3 : b^3$ are respectively the duplicate and triplicate ratios of a and b.

298. Since the ratio of two quantities, as the ratio of a to b, may be expressed by a fraction, as $\frac{a}{b}$, it follows that the changes which may be made upon a fraction without altering its value, may be made upon the terms of a ratio without changing the ratio of the terms, since the numerator is the antecedent and the denominator the consequent. Hence,

299. PRINCIPLE. — *Multiplying or dividing both terms of a ratio by the same quantity does not change the ratio of the terms.*

EXAMPLES.

1. What is the ratio of $3a$ to $6a$? $5a$ to $10a$?

2. What is the ratio of $7x$ to $35x$? $12ay$ to $13a$?

3. If the antecedent is $15a$, and the consequent $20a$, what is the ratio?

4. What is the ratio of $\frac{1}{2}$ to $\frac{1}{4}$? $\frac{1}{4}$ to $\frac{1}{3}$? $\frac{5}{6}$ to $\frac{2}{3}$?

When fractions are reduced to similar fractions, they have the ratio of their numerators.

5. When the antecedent is $2a$, and the ratio is $\frac{1}{2}$, what is the consequent?

ALGEBRA. — 17.

PROPORTION.

300. 1. What two numbers have the same relation to each other as 3 to 6? As 2 to 8? As 5 to 15?

2. What two quantities have the same relation to each other as $2a$ to $4a$? As $3b$ to $6b$? As $8b$ to $16b$?

3. What quantity has the same relation to $6a$ that $2b$ has to $4b$?

4. What quantity has the same relation to $10x$ that $3y$ has to $9y$?

5. What two quantities have the same ratio to each other that $5ay$ has to $10ay^2$?

6. What two quantities have the same ratio to each other that $8ax$ has to $4ax^2$?

7. How have the two ratios in each of the several examples given above compared in value?

301. A Proportion is an equality of ratios.

Thus, $5:6 = 10:12$, and $5xy:10xy = 4az:8az$, are proportions.

302. The **Sign** of proportion is a double colon ($::$).

This sign has been supposed to be the extremities of the lines which form the sign of equality. It is written between the ratios thus: $x:y::2a:2b$.

The sign of equality is frequently used instead of the double colon.

303. The **Antecedents** of a proportion are the antecedents of the ratios which form the proportion.

Thus, in the proportion $a:b::c:d$, a and c are the *antecedents*.

PROPORTION.

304. The **Consequents** of a proportion are the consequents of the ratios which form the proportion.

Thus, in the proportion $a:b::c:d$, b and d are the *consequents*.

305. The **Extremes** of a proportion are the first and fourth terms of the proportion.

Thus, in the proportion $a:b::c:d$, a and d are the *extremes*.

306. The **Means** of a proportion are the second and third terms of the proportion.

Thus, in the proportion $a:b::c:d$, b and c are the *means*.

307. A **Mean Proportional** is a quantity which serves as both means of a proportion.

Thus, in the proportion $a:b::b:c$, b is a *mean proportional*.

Since a proportion is an equality of ratios, and the ratio of one quantity to another is found by dividing the antecedent by the consequent, it follows that —

308. PRINCIPLE. — *A proportion may be expressed as an equation in which both members are fractions.*

Thus, the proportion $a:b::c:d$ may be expressed as $\dfrac{a}{b} = \dfrac{c}{d}$.

Since a proportion may be regarded as an equation in which both members are fractions, it follows that —

309. PRINCIPLE. — *The changes that may be made upon a proportion without destroying the equality of its ratios, are based upon the changes that may be made upon an equation without destroying the equality, and upon the terms of a fraction without altering the value of the fraction.*

PRINCIPLES OF PROPORTION.

310. 1. Let any four quantities form a proportion; as $a:b::c:d$.

2. Express the proportion as a fractional equation.

3. Clear the equation of fractions.

4. What does each member of the resulting equation contain?

5. How are the members of the equation produced from the terms of the proportion?

311. Principle 1. — *In any proportion the product of the extremes is equal to the product of the means.*

Thus, when $a:b::c:d$, $ad = bc$.

Since a *mean* proportional serves as both means of a proportion, as $a:b::b:c$,

The product of the extremes is equal to the square of the mean proportional.

DEMONSTRATION OF PRINCIPLE I.

Let $a:b::c:d$ represent any proportion.

Then (Art. 308), $\dfrac{a}{b} = \dfrac{c}{d}$.

Clearing of fractions, $ad = bc$.

Therefore, the product of the extremes is equal to the product of the means.

NUMERICAL ILLUSTRATIONS.

$$3:6::8:16$$
$$3 \times 16 = 8 \times 6$$
$$48 = 48$$

---

312. 1. Change the proportion $a:b::c:d$ into an equation, according to Principle 1.

2. Since $ad = bc$, how may the value of a be found? How the value of d? What parts of the proportion are a and d?

3. How, then, may either extreme of a proportion be found? How may either mean be found?

313. Principle 2. — *Either extreme is equal to the product of the means divided by the other extreme. Either mean is equal to the product of the extremes divided by the other mean.*

Thus, when $a:b::c:d$, $a = \dfrac{bc}{d}$, $d = \dfrac{bc}{a}$, $b = \dfrac{ad}{c}$, $c = \dfrac{ad}{b}$.

Demonstrate Prin. 2, and illustrate its truth with numbers.

PROPORTION.

314. 1. If $ad = bc$, what will be the resulting equation when both members are divided by bd?

2. Express the resulting equation as a proportion.

3. What does ad, the first member of the equation, form in the proportion? What does bc form?

315. PRINCIPLE 3. — *If the product of two quantities is equal to the product of two other quantities, the factors of one product may be made the extremes, and the factors of the other the means, of a proportion.*

Thus, when $ad = bc$, $a:b::c:d$.

Demonstrate Prin. 3, and illustrate its truth with numbers.

---

316. 1. Change the general proportion $a:b::c:d$ into an equation, according to Principle 1.

2. Divide the members of the equation by cd.

3. Express the result as a proportion.

4. What change has taken place in the order of antecedents and consequents, compared with the original proportion?

317. PRINCIPLE 4. — *If four quantities are in proportion, the antecedents will have the same ratio to each other as the consequents.*

Thus, when $a:b::c:d$, $a:c::b:d$.
When the antecedents are to each other as the consequents, the quantities are said to be in proportion by *Alternation.*

Demonstrate Prin. 4, and illustrate its truth with numbers.

---

318. 1. Change the general proportion $a:b::c:d$ into an equation, according to Principle 1.

2. Divide the members of the equation by ac.

3. Express the result as a proportion.

4. What change has taken place in the order of the terms in each couplet, compared with the original proportion?

262   HIGH SCHOOL ALGEBRA.

319. Principle 5. — *If four quantities are in proportion, the second will be to the first as the fourth to the third.*

Thus, when $a:b::c:d$, $b:a::d:c$.

When the terms of each ratio are written in the *inverse* order, the quantities are said to be in proportion by *Inversion*.

Demonstrate Prin. 5, and illustrate its truth with numbers.

---

320. 1. Express the proportion $a:b::c:d$ as a fractional equation.

2. Add 1 to each member of the equation.

3. Reduce each of the mixed quantities to the fractional form.

4. Express the result as a proportion.

5. How are the terms of this proportion formed from the terms of the original proportion?

6. Since, when $a:b::c:d$, $b:a::d:c$ (Prin. 5), if the changes just indicated are made in the second proportion, how may the terms of the resulting proportion be obtained from the terms of the original proportion?

321. Principle 6. — *If four quantities are in proportion, the sum of the terms of the first ratio is to either term of the first ratio as the sum of the terms of the second ratio is to the corresponding term of the second ratio.*

Thus, when $a:b::c:d$, $a+b:b::c+d:d$ and $a+b:a::c+d:c$.

When the sum of the terms of a ratio is to one of the terms as the sum of the terms of another ratio is to its corresponding term, the quantities are said to be in proportion by *Composition*.

Demonstrate Prin. 6, and illustrate its truth with numbers.

---

322. 1. Express the proportion $a:b::c:d$ as a fractional equation.

2. Subtract 1 from each member of the equation.

3. Reduce each of the mixed quantities to the fractional form.

PROPORTION. 263

4. Express the result as a proportion.

5. How are the terms of this proportion formed from the terms of the original proportion?

6. Since, when $a:b::c:d$, $b:a::d:c$ (Prin. 5), if the changes just indicated are made in the second proportion, how may the terms of the resulting proportion be obtained from the terms of the original proportion?

323. Principle 7. — *If four quantities are in proportion, the difference between the terms of the first ratio is to either term of the first ratio as the difference between the terms of the second ratio is to the corresponding term of the second ratio.*

Thus, when $a:b::c:d$, $a-b:b::c-d:d$ and $a-b:a::c-d:c$.

When the difference of the terms of a ratio is to one of the terms as the difference of the terms of another ratio is to its corresponding term, the quantities are said to be in proportion by *Division*.

Demonstrate Prin. 7, and illustrate its truth with numbers.

---

324. 1. Change the proportion $a:b::c:d$, according to Principle 6. Express the resulting proportion.

2. Change the same proportion according to Principle 7. Express the resulting proportion.

3. Change these proportions to fractional equations.

4. Divide the first equation by the second.

5. Express the result as a proportion.

6. How are the terms of this proportion formed from the terms of the original proportion?

325. Principle 8. — *If four quantities are in proportion, the sum of the quantities which form the first couplet is to their difference as the sum of the quantities which form the second couplet is to their difference.*

Thus, when $a:b::c:d$, $a+b:a-b::c+d:c-d$.

Demonstrate Prin. 8, and illustrate its truth with numbers.

326. 1. Express the proportion $a:b::c:d$ as a fractional equation.

2. Raise each member to the nth power.

3. Express the nth root of each member.

4. Express each of the equations as a proportion.

5. How may these proportions be formed from the original proportion?

327. PRINCIPLE 9. — *If four quantities are in proportion, the same powers of those quantities, or the same roots, will be in proportion.*

Thus, when $a:b::c:d$, $a^n:b^n::c^n:d^n$, and $a^{\frac{1}{n}}:b^{\frac{1}{n}}::c^{\frac{1}{n}}:d^{\frac{1}{n}}$.

Demonstrate Prin. 9, and illustrate its truth with numbers.

---

328. 1. Express the proportion $a:b::c:d$ as a fractional equation.

2. What may be done to a fraction without changing its value?

3. Multiply the terms of the first fraction by m, and the terms of the second by n.

4. Express the result as a proportion.

5. How are the terms of this proportion formed from the original proportion?

329. PRINCIPLE 10. — *If four quantities are in proportion, any equi-multiple of the terms of the first couplet will be proportional to any equi-multiple of the terms of the second couplet.*

Thus, when $a:b::c:d$, $ma:mb::nc:nd$.

Demonstrate Prin. 10, and illustrate its truth with numbers.

---

330. 1. Express the proportions $a:b::c:d$ and $x:y::z:w$ as fractional equations.

2. Multiply together the corresponding members of the equations.

PROPORTION.

3. Express the resulting equation as a proportion.

4. How are the terms of this proportion formed from the terms of the original proportions $a:b::c:d$ and $x:y::z:w$?

331. PRINCIPLE 11. — *The products of the corresponding terms of any number of proportions are in proportion.*

Thus, when $a:b::c:d$ and $x:y::z:w$, $ax:by::cz:dw$.

Demonstrate Prin. 11, and illustrate its truth with numbers.

Prove that the quotients will be in proportion if the proportions are *divided* term by term.

---

332. 1. Express the proportions $a:b::c:d$ and $a:b::e:f$ as fractional equations.

2. Since the first members of the equations are equal, what will the second members form?

3. Express the resulting equation as a proportion.

4. How are the terms of this proportion formed from the terms of the original proportions $a:b::c:d$ and $a:b::e:f$?

333. PRINCIPLE 12. — *If two proportions have a couplet in each the same, the other couplets will form a proportion.*

Thus, when $a:b::c:d$, and $a:b::e:f$, then $c:d::e:f$.

Demonstrate Prin. 12, and illustrate its truth with numbers.

EXAMPLES.

334. 1. In the proportion $5:8::4:x$, find the value of x.

SOLUTION.

$$5:8::4:x$$
Prin. 1, $$5x = 32$$
$$x = 6\tfrac{2}{5}$$

In solving examples like the following, the student should employ as many of the Principles of Proportion as are applicable.

Find the value of x in the following proportions:

2. $3 : x :: 4 : 6$.

3. $x : 5 :: 3 : 10$.

4. $4 : 6 :: x : 4$.

5. $3 : x :: x : 12$.

6. $x : 4 :: x^2 : 6$.

7. $x : 14 - x :: 4 : 3$.

8. $x : 12 :: x - 12 : 3$.

9. $x : 6 :: x + 6 : 10\frac{1}{2}$.

10. $x - 7 : x + 7 :: 2 : 9$.

11. $x^2 : 3 - x :: 8 :: 2$.

12. $x - 1 : x - 2 :: 2x + 1 : x + 2$.

13. Divide $40 between two men so that their shares will be in the proportion of 3 to 7.

14. There are two numbers in the ratio of 2 to 3, and if 3 be added to each, the ratio of the resulting numbers will be 5 to 7. What are the numbers?

15. There are two numbers which have to each other the ratio of 3 to 5, and if 4 be added to each, the results will have the ratio of 2 to 3. What are the numbers?

16. Mr. A's crop of wheat was to his crop of oats as 2 to 3. If he had raised 50 bushels more of each, the quantity of wheat would have been to the quantity of oats as 5 to 7. How many bushels of each kind of grain did he raise?

17. Find two numbers such that the greater is to the less as their sum is to 6, and the greater is to the less as their difference is to 2.

Solution.

Let $x =$ the greater; $y =$ the less.

By the conditions,	$x : y :: x + y : 6$	(1)
	$x : y :: x - y : 2$	(2)
By Prin. 12,	$x + y : 6 :: x - y : 2$	(3)
By Prin. 4,	$x + y : x - y :: 6 : 2$	(4)
By Prin. 8,	$2x : 2y :: 8 : 4$	(5)
By Prin., Art. 309,	$x : y :: 2 : 1$	(6)
From (1) and (6), Prin. 12,	$x + y : 6 :: 2 : 1$	(7)
From (2) and (6), Prin. 12,	$x - y : 2 :: 2 : 1$	(8)
From (7),	$x + y = 12$	(9)
From (8),	$x - y = 4$	(10)
	$\therefore x = 8, y = 4$	

PROPORTION.

18. The product of two numbers is 20, and the difference of their squares is to the square of their difference as 9 to 1.

Solution.

Let $x =$ the greater; $y =$ the less.

By the conditions, $\begin{cases} xy = 20 \\ x^2 - y^2 : (x-y)^2 :: 9 : 1 \end{cases}$ (1) (2)

Dividing first couplet by $(x-y)$, Prin., Art. 310,

$$x + y : x - y :: 9 : 1 \quad (3)$$

By Prin. 8, $2x : 2y :: 10 : 8$ (4)

By Prin., Art. 309, $x : y :: 5 : 4$ (5)

By Prin. 1, $4x = 5y$ (6)

$$x = \frac{5y}{4} \quad (7)$$

Substituting in (1), $\dfrac{5y^2}{4} = 20$ (8)

$$\therefore y = \pm 4$$
$$x = \pm 5$$

19. Find two numbers such that their sum is 8, and their product is to the sum of their squares as 15 to 34.

20. Find two numbers whose difference is 3, and whose product is to the sum of their squares as 10 is to 29.

21. What two numbers are those whose sum is to their difference as 7 to 1, and whose product is to the sum as 24 to 7?

22. The sum of two numbers is 12, and their product is to the sum of their squares as 2 to 5. What are the numbers?

23. The sum of two numbers is 6, and the sum of their squares is to the square of their sum as 5 to 9. What are the numbers?

24. What two numbers are those whose product is 12, and the difference of whose cubes is to the cube of their difference as 37 to 1?

FRACTIONAL EQUATIONS SOLVED BY THE PRINCIPLES OF PROPORTION.

335. Since a proportion is an equality of ratios, and the ratio of two quantities may be expressed as a fraction, it is evident that the Principles of Proportion are applicable to equations which have both members fractions.

Regarding the numerator of each fraction as an antecedent, and the denominator as a consequent, the terms of each fraction as a couplet, and the equation as a proportion, the Principles of Proportion may be readily applied.

1. Given $\dfrac{x - \sqrt{x+1}}{x + \sqrt{x+1}} = \dfrac{5}{11}$, to find x.

SOLUTION.

$$\dfrac{x - \sqrt{x+1}}{x + \sqrt{x+1}} = \dfrac{5}{11} \qquad (1)$$

By Prin. 8, the sum of the numerator and denominator of each member, divided by their difference, will form an equation.

Hence, $\qquad \dfrac{2x}{2\sqrt{x+1}} = \dfrac{16}{6} \qquad (2)$

By Prin., Art. 309, $\qquad \dfrac{x}{\sqrt{x+1}} = \dfrac{8}{3} \qquad (3)$

Squaring, $\qquad \dfrac{x^2}{x+1} = \dfrac{64}{9} \qquad (4)$

Clearing of fractions, etc., $9x^2 - 64x = 64 \qquad (5)$

Whence, $\qquad x = 8 \text{ or } -\tfrac{8}{9}$

2. Given $\dfrac{\sqrt{x+a} - \sqrt{x-a}}{\sqrt{x+a} + \sqrt{x-a}} = \dfrac{x}{2a}$, to find x.

SOLUTION.

$$\dfrac{\sqrt{x+a} - \sqrt{x-a}}{\sqrt{x+a} + \sqrt{x-a}} = \dfrac{x}{2a} \qquad (1)$$

By Prin. 8, $\qquad \dfrac{2\sqrt{x+a}}{2\sqrt{x-a}} = \dfrac{x + 2a}{2a - x} \qquad (2)$

PROPORTION.

By Prin., Art. 309,
$$\frac{\sqrt{x+a}}{\sqrt{x-a}} = \frac{x+2a}{2a-x} \quad (3)$$

Squaring,
$$\frac{x+a}{x-a} = \frac{x^2+4ax+4a^2}{x^2-4ax+4a^2} \quad (4)$$

By Prin. 8,
$$\frac{2x}{2a} = \frac{2x^2+8a^2}{8ax} \quad (5)$$

By Prin., Art. 309,
$$\frac{x}{a} = \frac{x^2+4a^2}{4ax} \quad (6)$$

Dividing denominators by a,
$$x = \frac{x^2+4a^2}{4x} \quad (7)$$

Whence,
$$x = \pm 2a\sqrt{\tfrac{1}{3}}$$

Solve the following by applying the Principles of Proportion when possible:

3. Given $\dfrac{\sqrt{6x}-2}{\sqrt{6x}+2} = \dfrac{4\sqrt{6x}-9}{4\sqrt{6x}+6}$, to find x.

4. Given $\dfrac{\sqrt{ax}-b}{\sqrt{ax}+b} = \dfrac{3\sqrt{ax}-2b}{3\sqrt{ax}+5b}$, to find x.

5. Given $\dfrac{\sqrt{4x+1}+\sqrt{4x}}{\sqrt{4x+1}-\sqrt{4x}} = 9$, to find x.

In the solution the second member may be written as $\tfrac{9}{1}$.

6. Given $\dfrac{\sqrt{a+x}+\sqrt{a-x}}{\sqrt{a+x}-\sqrt{a-x}} = b$, to find x.

7. Given $\dfrac{\sqrt{x^2+1}-\sqrt{x^2-1}}{\sqrt{x^2+1}+\sqrt{x^2-1}} = \dfrac{1}{2}$, to find x.

8. Given $\dfrac{a+x+\sqrt{2ax+x^2}}{a+x-\sqrt{2ax+x^2}} = b$, to find x.

9. Given $\dfrac{\sqrt{a^2+x^2}+x}{\sqrt{a^2+x^2}-x} = \dfrac{b}{c}$, to find x.

PROGRESSIONS.

336. 1. How does each of the numbers 2, 4, 6, 8, 10, 12, compare with the number that follows it?

2. How may each of the numbers 4, 6, 8, etc., be obtained from the one that precedes it in the series 2, 4, 6, 8, etc.?

3. How does each of the quantities $4x$, $8x$, $16x$, compare with the one that precedes it in the series $2x$, $4x$, $8x$, $16x$?

4. Write a series of six quantities, beginning with $2a$ and increasing by a constant addend $3a$.

5. Write a series of six quantities, beginning with a and increasing by a constant multiplier r.

337. A **Series** of quantities is a succession of quantities, each derived from the preceding according to some fixed law.

338. The first and last terms of a series are called the *extremes*, the intervening terms the *means*.

Thus, in the series, $a, a + d, a + 2d, a + 3d$, the quantities a and $a + 3d$ are *extremes*, and the others are *means*.

339. An **Ascending Series** is one in which the quantities *increase* regularly from the first term.

Thus, 2, 4, 8, 16, and $a, a + d, a + 2d$, etc., are ascending series.

340. A **Descending Series** is one in which the quantities *decrease* regularly from the first term.

Thus, 24, 12, 6, 3, and $a, a - d, a - 2d, a - 3d$, are descending series.

ARITHMETICAL PROGRESSION.

341. An Arithmetical Progression is a series of quantities each of which is derived from the preceding quantity by the addition of a constant quantity.

Thus, 4, 6, 8, 10, 12, 14, and a, $a - d$, $a - 2d$, $a - 3d$, are arithmetical progressions.

342. The constant quantity which is added is called the **Common Difference.**

The common difference may be either positive or negative. When it is positive, an ascending series is produced; when negative, a descending series.

343. To find the last term.

1. In the arithmetical progression, x, $x+2$, $x+4$, $x+6$, what is the common difference? How many times does it enter into the second term? How many times into the third term? How many times into the fourth term?

2. In the series a, $a+d$, $a+2d$, $a+3d$, how is the second term formed from the first term? The third term? The fourth term? The seventh term? The thirteenth term? Any term?

3. If the above series were descending, the first three terms would be a, $a-d$, $a-2d$. What would be the fifth term? The ninth term? The eleventh term? The twentieth term? The nth term?

344. When a represents the first term, d the common difference, l the last term, and n the number of terms,

$$\text{I. } l = a + (n-1)d. \text{ That is,}$$

345. PRINCIPLE. — *The last term of an arithmetical progression is equal to the first term increased by the common difference multiplied by the number of terms less 1.*

EXAMPLES.

1. Find the 15th term of the series 1, 3, 5, 7, etc.

PROCESS.

$l = a + (n-1)d$

$l = 1 + (15-1)2$

$l = 29$

EXPLANATION.—In this example $a = 1$, $d = 2$, and $n = 15$. Substituting these values in the formula, the value of l, or the last term, is 29.

2. Find the 18th term of the series 4, 7, 10, 13, etc.

3. Find the 12th term of the series 3, 7, 11, 15, etc.

4. Find the 10th term of the series $1\frac{1}{2}$, 2, $2\frac{1}{2}$, 3, etc.

5. Find the 12th term of the series 25, 23, 21, 19, etc.

SUGGESTION.—The common difference is -2.

6. Find the 20th term of the series 8, 4, 0, -4, etc.

7. Find the 30th term of the series a, $2a$, $3a$, $4a$, etc.

8. Find the 15th term of the series $\frac{7}{8}$, $\frac{13}{16}$, $\frac{3}{4}$, $\frac{11}{16}$, etc.

9. Find the nth term of the series 1, 3, 5, 7, etc.

10. A boy agreed to work for 50 days, at 25 cents for the first day, and an increase of 3 cents per day. What were his wages the last day?

11. A body falls $16\frac{1}{2}$ feet the first second, 3 times as far the second second, 5 times as far the third second. How far will it fall the seventh second?

346. To find the sum of the terms.

1. Express five terms of the series a, $a + d$, etc.

2. How is the term before the last obtained from the last? The second term from the last? The third term from the last?

3. Express four terms of the series before the last, when l is the last term, and d the common difference.

4. How will the sum of the last four terms and the first compare with the sum of the terms, in a series consisting of only five terms?

PROGRESSIONS.

347. Let a represent the first term, l the last term, d the common difference, n the number of terms, and s the sum of the terms.

Writing the sum of n terms in the usual order and then in the reverse order, and adding the terms, we have:

$$s = a + (a+d) + (a+2d) + (a+3d) + \cdots + l$$
$$s = l + (l-d) + (l-2d) + (l-3d) + \cdots a$$
$$2s = (a+l) + (a+l) + (a+l) + (a+l) + \cdots (a+l)$$
$$\therefore\ 2s = n(a+l)$$

II. $s = \dfrac{n}{2}(a+l)$ or $n\left(\dfrac{a+l}{2}\right).$ That is,

348. PRINCIPLE. — *The sum of any number of terms in arithmetical progression is equal to one half the sum of the extremes multiplied by the number of terms.*

EXAMPLES.

1. What is the sum of the series 2, 4, 6, 8, etc., containing 12 terms?

PROCESS.
$l = a + (n-1)d$
$l = 2 + (11 \times 2) = 24$
$s = n\left(\dfrac{a+l}{2}\right)$
$s = 12\left(\dfrac{2+24}{2}\right) = 156$

EXPLANATION. — Since the last term is not given, it is found by the previous case to be 24. Then, by the formula given for obtaining the sum, it is found to be 156.

2. What is the sum of 12 terms of the series 1, 3, 5, 7, etc.?

3. What is the sum of 9 terms of the series 4, 6, 8, 10, etc.?

4. What is the sum of 8 terms of the series 5, 8, 11, etc.?

5. What is the sum of 7 terms of the series 3, $4\frac{1}{2}$, 6, etc.?

6. What is the sum of 8 terms of the series $3a$, $5a$, $7a$, $9a$, etc.?

7. What is the sum of 9 terms of the series $a+b$, $a+b+c$, $a+b+2c$, etc.?

8. What is the sum of n terms of the series x, $3x$, $5x$, $7x$, etc.?

9. What is the sum of 8 terms of the series 2, 1, 0, -1, -2, etc.?

10. A man walked 15 miles the first day, and increased his rate 3 miles per day. How far did he walk in 11 days?

11. How many strokes does a clock strike in 12 hours?

12. A person received a gift of $100 per year from his birth until he was 21 years old. These sums were deposited in a bank, and drew simple interest at 6%. How much was due him when he became of age?

13. A debt can be discharged in 26 days by paying $1 the first day, $3 the second, $5 the third, and so on. What is the amount of the debt?

14. If a man travels 20 miles the first day of his journey and increases his rate 3 miles each day, how far will he travel in 12 days?

15. If a person travels 30 miles the first day, and each succeeding day a quarter of a mile less than he traveled the day before, how far will he travel in 45 days?

16. A sets out for a certain place, and travels 1 mile the first day, 2 the second, and so on. Five days afterward B starts from the same place to overtake A, and travels 12 miles a day. How far will A have traveled before he is overtaken?

17. Thirty flower pots are arranged in a straight line four feet apart. How far must a lady travel who, after watering each plant singly, returns to a well four feet from the first flower pot, and in a line with the plants?

18. A body falls $16\frac{1}{12}$ feet the first second, and in each succeeding second $32\frac{1}{6}$ feet more than in the next preceding one. How far will it fall in 20 seconds?

PROGRESSIONS.

349. From the fundamental formulas,

$$l = a + (n-1)d \text{ and } s = \frac{n}{2}(a+l),$$

when any three of the elements are given the other two may be found.

EXAMPLES.

1. How many terms are there in the series **3, 5, 7,** etc., if the sum is 168?

SOLUTION.

$l = a + (n-1)d$

$\therefore l = 3 + 2n - 2 = 1 + 2n$, to be substituted in next formula.

$s = \frac{n}{2}(a+l)$

$168 = \frac{n}{2}(3 + 1 + 2n) = \frac{n}{2}(4 + 2n) = 2n + n^2$

$n^2 + 2n = 168$

$n = 12 \text{ or } -14$

The negative value has no significance, therefore the number of terms is 12.

2. The last term of the series is 50, the common difference 4, and the sum of the terms is 338. How many terms are there?

SOLUTION.

$l = a + (n-1)d$
$50 = a + (n-1)4$
$50 = a + 4n - 4$
$n = \frac{54 - a}{4}$

$s = \frac{n}{2}(a+l)$

$338 = \frac{54 - a}{8}(a + 50)$

$338 = \frac{2700 + 4a - a^2}{8}$

$\therefore a = 2$

and $n = \frac{54 - 2}{4} = 13$

3. Given $d = 4$, $n = 21$, $s = 1197$, to find a.

4. Given $n = 10$, $l = 29$, $s = 155$, to find d.

5. Given $a = 3$, $d = 6$, $s = 507$, to find n.

6. Given $a = 25$, $d = 3$, $n = 12$, to find l.

7. Given $a = 6$, $l = 42$, $n = 9$, to find s.

8. Given $l = 57$, $d = 5$, $n = 21$, to find a.

9. Given $a = 6$, $d = 6$, $l = 1152$, to find n.

10. Given $a = 9$, $l = 41$, $s = 150$, to find n.

11. Given $a = 27$, $n = 9$, $s = 72$, to find l.

12. Given $l = \frac{7}{2}$, $d = \frac{1}{3}$, $s = 20$, to find a.

13. Given $a = -\frac{1}{3}$, $l = -\frac{4}{3}$, $s = -\frac{40}{3}$, to find d.

14. Given $a = \frac{3}{2}$, $l = -\frac{57}{2}$, $s = -\frac{351}{2}$, to find n.

15. Given $d = 21$, $l = 242$, $n = 12$, to find s.

350. To insert arithmetical means.

1. Insert 5 arithmetical means between 2 and 26.

SOLUTION.—Since there are 5 means there must be 7 terms; hence there are given the first term 2, the last term 26, and the number of terms 7, to find d, the common difference.

Substituting in $\qquad l = a + (n-1)d$

$$d = 4$$

∴ 2, 6, 10, 14, 18, 22, 26, is the series, and the means are 6, 10, 14, 18, 22.

2. Insert 4 arithmetical means between 3 and 18.

3. Insert 8 arithmetical means between 13 and 76.

4. Insert 4 arithmetical means between -4 and 17.

5. Insert 4 arithmetical means between 193 and 443.

6. Insert 8 arithmetical means between $\frac{1}{2}$ and $\frac{3}{4}$.

7. Insert 6 arithmetical means between -8 and -4.

8. Insert 9 arithmetical means between 1 and -1.

PROGRESSIONS. 277

SPECIAL APPLICATIONS.

351. In solving some of the problems in Arithmetical Progression there are several ways of representing the unknown terms of the series.

1. When x represents the first term of a series, and y the common difference, the series may be represented by

$$x,\ x+y,\ x+2y,\ x+3y,\ \text{etc.}$$

2. When there are *three* terms in the series, the middle term may be represented by x, and the common difference by y; as,

$$x-y,\ x,\ x+y.$$

3. When there are *five* terms in the series, the middle term may be represented by x, and the common difference by y; as,

$$x-2y,\ x-y,\ x,\ x+y,\ x+2y.$$

4. When there are *four* terms in the series, $x-3y$ may represent the first term, and $2y$ the common difference; as,

$$x-3y,\ x-y,\ x+y,\ x+3y.$$

It is obvious that by the notation adopted the *sum* of the quantities contains but one unknown quantity.

PROBLEMS.

1. There are three numbers in arithmetical progression, whose sum is 18 and the sum of whose squares is 116. What are the numbers?

SOLUTION.

Let
$\quad x - y =$ the first term,
$\quad x =$ the second term,
$\quad x + y =$ the third term,
$\quad y =$ the common difference.

By the conditions, $\quad \begin{cases} 3x = 18 \\ 3x^2 + 2y^2 = 116 \end{cases}$ (1) (2)

From (1), $\quad x = 6$ (3)
Substituting in (2), $\quad 108 + 2y^2 = 116$ (4)
Whence, $\quad y = 2$ (5)

$x - y = 4$, 1st term,
$x = 6$, 2d term,
$x + y = 8$, 3d term.

2. The first term of an arithmetical series is 5, the last term 92, and the sum of the terms 1455. What is the number of terms?

3. The first term of an arithmetical series is 2, the last term 30, and the sum of the terms 160. What is the number of terms?

4. The first term of an arithmetical series is 16, the common difference $-3\frac{1}{2}$, and the sum of the terms 30. What is the number of terms?

5. The sum of three numbers in arithmetical progression is 15, and the product of the second and third is 35. What are the numbers?

6. The sum of three numbers in arithmetical progression is 9, and their product is 15. What are the numbers?

7. The sum of three numbers in arithmetical progression is 18, and the sum of their squares is 126. What are the numbers?

8. There are three numbers in arithmetical progression such that the product of the first and third is 16, and the sum of the squares of the numbers is 93. What are the numbers?

9. There are three numbers in arithmetical progression such that the first is 3, and the product of the first and third is 21. What are the numbers?

10. The sum of four numbers in arithmetical progression is 10, and their product is 24. What are the numbers?

11. There are four numbers in arithmetical progression such that the product of the first and fourth is 27, and the product of the second and third is 35. What are the numbers?

12. There are four numbers in arithmetical progression such that the product of the fourth number by the common difference is 16, and the product of the second and third is 24. What are the numbers?

PROGRESSIONS. 279

13. There are five numbers in arithmetical progression such that their sum is 40, and the sum of their squares 410. What are the numbers?

14. The sum of five numbers in arithmetical progression is 25, and their product is 945. What are the numbers?

15. The product of four numbers in arithmetical progression is 280, and the sum of their squares is 166. What are the numbers?

16. A number is expressed by three digits which are in arithmetical progression. If the number is divided by the sum of the digits, the quotient will be 26, and if 198 be added to the number, the digits will be inverted. What is the number?

GEOMETRICAL PROGRESSION.

352. A **Geometrical Progression** is a series of quantities which increase or decrease by a constant multiplier.

Thus 2, 4, 8, 16, 32, and ab^3, ab^2, ab, a, are geometrical progressions.

In the first series the ratio is 2; in the second it is $\frac{1}{b}$.

353. The constant multiplier is called the **Ratio**.

The ratio may be integral or fractional, positive or negative.

354. To find the last term.

1. In the geometrical progression 2, 4, 8, 16, what is the ratio? How is the second term obtained from the first? How is the third term obtained from the first? How is the fourth term obtained from the first?

2. In the geometrical progression x, xy, xy^2, xy^3, xy^4, etc., what is the ratio? How many times does the ratio enter as a factor into the second term? How many times into the third term? How many times into the fourth term? How many times into the nth term?

HIGH SCHOOL ALGEBRA.

355. When a represents the first term, r the ratio, and l the last or nth term,

I. $l = ar^{n-1}$. That is,

356. PRINCIPLE. — *The last term of a geometrical progression is equal to the first term, multiplied by the ratio raised to a power whose index is 1 less than the number of terms.*

EXAMPLES.

1. Find the 8th term of the series 2, 4, 8, etc.

PROCESS.

$l = ar^{n-1}$

$l = 2 \times 2^7$

$l = 256$

EXPLANATION. — In this example $a = 2$, $r = 2$, and $n = 8$. Substituting these values in the formula, the value of l, or the last term, is 256.

2. Find the 6th term of the series 5, 10, 20, etc.

3. Find the 9th term of the series 2, 4, 8, etc.

4. Find the 7th term of the series 3, 9, 27, etc.

5. Find the 10th term of the series 1, 2, 4, 8, etc.

6. Find the 7th term of the series $2a$, $4a^2$, $8a^3$, etc.

7. Find the 9th term of the series 3, $6ax$, $12a^2x^2$, etc.

8. Find the nth term of the series 1, 2, 4, 8, etc.

9. Find the nth term of the series 3, 12, 48, etc.

10. Find the 8th term of the series 3, 1, $\frac{1}{3}$, etc.

11. If a person should be hired for 8 days at the rate of $1 the first day, $3 for the next day, $9 for the third day, and so on, what would his wages be for the last day?

12. If a man begins business with a capital of $1000, and doubles it every three years, how much will he have at the end of 15 years?

13. If a man should purchase 10 horses, giving $1 for the first, $2 for the second, $4 for the third, and so on, what would the last horse cost him?

PROGRESSIONS.

14. A man bought 9 bushels of wheat. If he had paid 3 cents for the first bushel, 9 cents for the second, 27 cents for the third, and so on, what would the last bushel have cost him?

15. A man worked 6 days on condition that he should receive for the 6 days as much as the last day's work would amount to at the rate of 1 cent for the first day, 5 cents for the second day, and so on, the wages of each day being 5 times the wages of the previous day. How much did he receive?

16. If a farmer plants a pint of corn and it produces 4 bushels, how much corn will he have at the expiration of 4 years, if he plants the entire harvest each succeeding spring?

17. A capital of $2000 is increased by $\frac{1}{10}$ of itself each year. What will it be at the beginning of the fifth year?

357. To find the sum of a series.

1. Express five terms of the series, a, ar, ar^2.

2. What is the nth or last term of the series?

358. Let a represent the first term, l the last term, r the ratio, n the number of terms, and s the sum of the terms. Then,

$$s = a + ar + ar^2 + ar^3 + \cdots + ar^{n-1} \qquad (1)$$

$(1) \times r \qquad rs = ar + ar^2 + ar^3 + \cdots ar^{n-1} + ar^n \qquad (2)$

$(2) - (1), \qquad rs - s = ar^n - a$

$$s(r-1) = ar^n - a$$

$$\text{II.} \quad s = \frac{ar^n - a}{r-1}.$$

Since $\qquad l = ar^{n-1}, \quad rl = ar^n$

Substituting rl for ar^n in the formula for the sum, we have

$$\text{III.} \quad s = \frac{rl - a}{r - 1}.$$

EXAMPLES.

1. Find the sum of 10 terms of the series 2, 4, 8, etc.

PROCESS.

$$s = \frac{ar^n - a}{r - 1}$$

$$s = \frac{2 \times 2^{10} - 2}{2 - 1} = 2046$$

EXPLANATION. — In this example, $a = 2$, $r = 2$, $n = 10$. Substituting in the first formula obtained for the sum, the sum is 2046.

2. Find the sum of 11 terms of the series 1, 2, 4, 8, etc.

3. Find the sum of 9 terms of the geometrical series 1, 3, 9, 27, etc.

4. Find the sum of 12 terms of the geometrical series 4, 8, 16, 32, 64, etc.

5. Find the sum of 11 terms of the geometrical series 3, 9, 27, 81, 243, etc.

6. Find the sum of 10 terms of the geometrical series $2a$, $4a$, $8a$, etc.

7. Find the sum of 10 terms of the geometrical series $2x^2$, $6x^2$, $18x^2$, etc.

8. Find the sum of n terms of the series 2, 4, 8, 16, etc.

9. Find the sum of 10 terms of the series 2, 1, $\frac{1}{2}$, $\frac{1}{4}$, etc.

10. Find the sum of 8 terms of the series 8, 2, $\frac{1}{2}$, $\frac{1}{8}$, etc.

11. The extremes of a geometrical progression are 4 and 1024, and the ratio is 4. What is the sum of the series?

12. The extremes of a geometrical progression are 2 and 512, and the ratio is 2. What is the sum of the series?

13. What is the sum of a series in which the first term is 2, the last term so small that it may be disregarded, and the ratio $\frac{1}{2}$; or what is the sum of the infinite series 2, 1, $\frac{1}{2}$, etc.?

14. What is the sum of the infinite series 6, 3, $1\frac{1}{2}$, etc.; or what is the sum of a series in which the first term is 6, and the ratio $\frac{1}{2}$?

PROGRESSIONS.

15. What is the sum of the infinite series 2, $\frac{2}{3}$, $\frac{2}{9}$, etc.?

16. What is the sum of the infinite series

$$1 + \frac{1}{x^2} + \frac{1}{x^4} + \frac{1}{x^6}, \text{ etc.?}$$

17. What is the sum of the infinite series

$$x - y + \frac{y^2}{x} - \frac{y^3}{x^2} + \frac{y^4}{x^3}, \text{ etc.?}$$

18. A man engaged to work for 8 months, upon condition that he should receive $2 for the first month, $4 for the second, $8 for the third, and so on. How much did he earn in the time?

19. A man rented a farm of 500 acres for 20 years, agreeing to pay $1 for the first year, $2 for the second year, $4 for the third year, and so on. What was the entire amount of rent paid for the farm?

20. A merchant made $500 the first year, and continued in business 5 years, making each year 3 times as much as the preceding year. How much did he make?

21. What sum of money can be paid by eight installments, the first of which is $1, the second $2, the third $4, and so on in a geometrical ratio?

22. If a body be put in motion by a force which would move it 10 feet the first second, 8 feet the second, 6.4 feet the third, and so on, how far would it move?

23. A man sold a horse upon the following terms: The purchaser was to pay 1 cent for the first mile the owner drove him, 2 cents for the second, 4 cents for the third, and so on for four hours. If he drove the horse 32 miles in 4 hours, how much did the horse cost?

359. To insert geometrical means between two terms.

EXAMPLES.

1. Insert 3 geometrical means between 2 and 32.

PROCESS.

$l = ar^{n-1}$

$32 = 2r^4$

$16 = r^4$

$2 = r$

EXPLANATION. — Since there are 3 means, the number of terms is 5. The first term is 2, the last term is 32. Substituting these values in the formula $l = ar^{n-1}$, the ratio is found to be 2, the series 2, 4, 8, 16, 32, and the means 4, 8, 16.

2. Insert 4 geometrical means between 1 and 32.

3. Insert 3 geometrical means between 31 and 496.

4. Insert 5 geometrical means between 1 and 729.

5. Insert 7 geometrical means between 2 and 13122.

6. Insert 4 geometrical means between $\frac{1}{8}$ and 128.

7. Insert 6 geometrical means between -2 and -4374.

8. Insert 4 geometrical means between 3 and $-\frac{729}{1024}$.

The geometrical mean between *two* quantities may also be found as follows:

Let a and b represent the quantities, and m the mean.

$\dfrac{m}{a}$ = the ratio, as also does $\dfrac{b}{m}$.

$\therefore \dfrac{m}{a} = \dfrac{b}{m}$, $m^2 = ab$, and $m = \sqrt{ab}$. That is,

360. PRINCIPLE. — *The geometrical mean between two quantities is equal to the square root of their product.*

9. Find the geometrical mean between 9 and 16.

10. Find the geometrical mean between 5 and 45.

11. Find the geometrical mean between 14 and 56.

12. Find the geometrical mean between -9 and -49.

PROGRESSIONS. 285

13. Find the geometrical mean between $13\frac{5}{7}$ and $4\frac{2}{3}$.

14. Find the geometrical mean between $\frac{16}{27}$ and $18\frac{3}{4}$.

15. Find the geometrical mean between $16x^3y$ and $25xy$.

361. From the formulas for the last term and the sum, when any three elements are given, the other two can be found.

1. Given l, r, and s, to find a.

2. The last term of a geometrical progression is 128, the ratio 2, and the sum 254. What is the first term?

3. The last term of a geometrical progression is -128, the ratio 2, and the sum -255. What is the first term?

4. Given a, n, and l, to find r.

5. The first term of a geometrical progression is 2, the number of terms 5, and the last term 512. What is the ratio?

6. The first term of a geometrical progression is $\frac{2}{3}$, the number of terms 4, and the last term 18. What is the ratio? Write the series.

7. Given r, n, and l, to find a.

8. The last term of a geometrical progression is 1215, the number of terms 6, and the ratio 3. What is the first term? Write the series.

9. Given a, l, and s, to find r.

10. The first term of a geometrical progression is 2, the last term $\frac{1}{32}$, and the sum $\frac{127}{32}$. What is the ratio? Write the series.

11. The first term of a geometrical progression is 3, the last term 6144, and the sum 12285. What is the ratio?

12. Given a, r, and s, to find l.

13. The first term of a geometrical progression is 5, the ratio 3, and the sum 1820. What is the last term?

SPECIAL APPLICATIONS.

362. In solving some problems in Geometrical Progression there are several ways of representing the unknown terms of the series.

1. When x represents the first term and y the ratio, the series may be represented as follows:

$$x,\ xy,\ xy^2,\ xy^3,\ xy^4,\ \text{etc.}$$

2. When there are *three* terms in the series, they may be represented as follows:

$$x^2,\ xy,\ y^2,\ \text{in which } \frac{y}{x} \text{ represents the ratio;}$$

or, by $\quad x,\ \sqrt{xy},\ y,\ $ in which $\sqrt{\dfrac{y}{x}}$ represents the ratio.

3. When there are *four* terms in the series, they may be represented as follows:

$$\frac{x^2}{y},\ x,\ y,\ \text{and } \frac{y^2}{x},\ \text{in which } \frac{y}{x} \text{ represents the ratio.}$$

PROBLEMS.

1. The sum of three numbers in geometrical progression is 7, and the sum of squares of these numbers is 21. What are the numbers?

Solution.

Let $x,\ \sqrt{xy},$ and y represent the numbers.

By the conditions, $\quad \begin{cases} x + \sqrt{xy} + y = 7 \\ x^2 + xy + y^2 = 21 \end{cases}$ \quad (1)
(2)

Dividing (2) by (1), $\quad x - \sqrt{xy} + y = 3$ \quad (3)

Adding (1) and (3), $\quad 2x + 2y = 10$ \quad (4)

$\quad\quad\quad\quad\quad\quad\quad\quad x + y = 5$ \quad (5)

Subtracting (5) from (1), $\quad \sqrt{xy} = 2$ \quad (6)

$\quad\quad\quad\quad\quad\quad\quad\quad xy = 4$ \quad (7)

Whence $\quad\quad\quad x = 1,\ \sqrt{xy} = 2,\ y = 4$

PROGRESSIONS.

2. The sum of a geometrical series containing 8 terms is 1785, and the ratio is 2. What is the first term?

3. The sum of a geometrical series containing 6 terms is 1365, and the ratio is 4. What is the first term?

4. The sum of a geometrical series is 1. The first term is $\frac{1}{3}$, and the last term 0. What is the ratio?

5. The first term of a geometrical series is 32, the last term 4000, and the number of terms 4. What is the ratio?

6. Find three terms in geometrical progression whose sum is 13, and the sum of whose squares is 91.

7. The product of three numbers in geometrical progression is 8, and the sum of their squares is 21. What are the numbers?

8. The sum of the first and third of four numbers in geometrical progression is 10, and the sum of the second and fourth is 30. What are the numbers?

SUGGESTION. — Represent the numbers by x, xy, xy^2, and xy^3.

9. The sum of four numbers in geometrical progression is 15, and the last term divided by the sum of the means is $\frac{4}{3}$. What are the numbers?

10. The sum of three numbers in geometrical progression is 14, and the sum of the extremes multiplied by the mean is 40. What are the numbers?

11. The sum of the first two of four numbers in geometrical progression is 10, and the sum of the last two is $22\frac{1}{2}$. What are the numbers?

12. Find three numbers in geometrical progression such that the sum of the first and last is 20, and the square of the mean 36.

13. The continued product of three numbers in geometrical progression is 216, and the sum of the squares of the extremes is 328. What are the numbers?

14. A man bought a farm for $5000, agreeing to pay principal and interest in five equal annual installments. What will be the annual payment, including interest at 6%?

SOLUTION.

Let P = any principal.

p = the annual payment.

r = the rate per cent.

$P(1 + r)$ = amount due at end of first year.

$P(1 + r) - p$ = sum due after first payment is made.

$P(1 + r)^2 - p(1 + r)$ = amount due at end of second year.

$P(1 + r)^2 - p(1 + r) - p$ = sum due after second payment is made.

$P(1 + r)^5 - p(1 + r)^4 - p(1 + r)^3 - p(1 + r)^2 - p(1 + r) - p$
= sum due after fifth payment is made.

Since the debt was discharged when the fifth payment was made,
$$P(1 + r)^5 - p(1 + r)^4 - p(1 + r)^3 - p(1 + r)^2 - p(1 + r) - p = 0$$
Whence,
$$p(1 + r)^4 + p(1 + r)^3 + p(1 + r)^2 + p(1 + r) + p = P(1 + r)^5$$
$$p = \frac{P(1 + r)^5}{(1 + r)^4 + (1 + r)^3 + (1 + r)^2 + (1 + r) + 1}$$

Or, since the denominator forms a geometrical series,
$$p = \frac{P(1 + r)^5}{\frac{(1 + r)^5 - 1}{r}} = \frac{Pr(1 + r)^5}{(1 + r)^5 - 1}$$

And, in general,
$$p = \frac{Pr(1 + r)^n}{(1 + r)^n - 1} = \frac{\$5000 \times .06 \times (1.06)^5}{(1.06)^5 - 1} = \$1186.98$$

15. If a man agrees to pay a debt of $3000, bearing interest at 7%, in 6 equal annual installments, what will be the annual payment?

EXAMPLES FOR REVIEW.

363. 1. Add $3x^{\frac{1}{2}}y - 4x\sqrt{y} + 5$, $\sqrt{xy} + 2xy^{\frac{1}{2}} + 4$, $6y\sqrt{x} - \sqrt{xy} - 7$, $4y\sqrt{x} - 3y^{\frac{1}{2}}x - 6$, and $2 + 5xy^{\frac{1}{2}} - 3yx^{\frac{1}{2}}$.

2. Add $2x^4y^2 + 2x^{-n}y^{\frac{3}{2}} - 3x^2$, $2b^2x^{-m}y^{\frac{3}{2}} - a + 6x^2$, $3x^4y^2 - 5cx^4y^2 - 2b^2x^{-m}y^{\frac{3}{2}} + 3a$, and $2x^3 + cy$.

REVIEW. 289

3. Add $4b - 2cy^{-\frac{2}{3}} + m$ to $7cy^{-\frac{2}{3}} + 8ax - 5b + 10ax - 2b + 8m - 3$, and subtract from the result the sum of $5ax - 4m + 3$, $5cy^{-\frac{2}{3}} - 3ax - 6$, and $3m - 10cy^{-\frac{2}{3}} - 2m$.

4. From $17xy^2 + 3xz + 10a$ subtract $7xy^2 + 4xz + 12x - 2bx$.

5. Multiply $a^4 + a^3b + a^2b^2 + ab^3 + b^4$ by $a - b$.

6. Multiply $3x^{m-1} - 2y^{n-2}$ by $2x - 3y^2$.

7. Multiply $3x^{-\frac{1}{2}} - 2y^{\frac{2}{3}}$ by $3x^{-\frac{1}{2}} + 2y^{\frac{2}{3}}$.

8. Multiply $2x^{\frac{m}{2n}} - 3y^{\frac{m-n}{2}}$ by $2x^{\frac{m}{2n}} + 3y^{\frac{m-n}{2}}$.

9. Expand $(x^{2n} + y^{2m})(x^{2n} + y^{2m})$.

10. Divide $x^4 - y^4$ by $x + y$.

11. Divide $x^7 + y^7$ by $x + y$.

12. Divide $x^n - y^n$ by $x - y$ to 6 terms.

13. Divide $\dfrac{x^3}{2} + x^2 + \dfrac{3x}{8} + \dfrac{3}{4}$ by $\dfrac{x}{2} + 1$.

14. Factor $4x^2 + 4xy + y^2$.

15. Factor $x^4 - y^4$.

16. Factor $x^2 - 2x - 35$.

17. Factor $x^2 - 6x - 27$.

18. Factor $x^8 - y^8$.

19. Find the highest common divisor of $x^2 - y^2$, $x^2 - 2xy + y^2$, and $xy - y^2$.

20. Find the highest common divisor of $6x^4 + 11x^2 + 3$ and $2x^4 - 5x^2 - 12$.

21. Find the highest common divisor of $4x^4 - 24x^3 + 34x^2 + 12x - 18$ and $4x^3 - 18x^2 + 19x - 3$.

22. Find the highest common divisor of $x^4 - 4x^3 - 16x^2 + 7x + 24$ and $2x^3 - 15x^2 + 9x + 40$.

23. Find the lowest common multiple of $2a^2x$, $3axy$, and $4a^2x^2y^3$.

ALGEBRA.— 19.

24. Find the lowest common multiple of $x^2 - y^2$, $x + y$, $x - y$, and $xy - y^2$.

25. Reduce $\dfrac{x^2 - 7x + 10}{2x^2 - x - 6}$ to its lowest terms.

26. Reduce $\dfrac{x^3 - 4x^2 + 9x - 10}{x^3 + 2x^2 - 3x + 20}$ to its lowest terms.

27. Reduce $\dfrac{a^2 - b^2 - 2bc - c^2}{a^2 + 2ab + b^2 - c^2}$ to its lowest terms.

28. Simplify $\dfrac{x}{x+1} - \dfrac{x}{1-x} + \dfrac{x^2}{x^2-1}$.

29. Simplify $\dfrac{3+2x}{2-x} - \dfrac{2-3x}{2+x} + \dfrac{16x-x^2}{x^2-4}$.

30. Simplify

$$\dfrac{1}{(a-b)(b-c)} + \dfrac{1}{(b-a)(a-c)} - \dfrac{1}{(c-a)(c-b)}.$$

31. Simplify $\dfrac{1+x}{1+x+x^2} + \dfrac{1-x}{1-x+x^2} - \dfrac{2}{1+x^2+x^4}$.

32. Simplify $\left(x + \dfrac{1}{x}\right)\left(x^2 + \dfrac{1}{x^2}\right)\left(x - \dfrac{1}{x}\right)$.

33. Simplify $\dfrac{a+b}{a-b} + \dfrac{a-b}{a+b} - \dfrac{2(a^2-b^2)}{a^2+b^2}$.

34. Simplify $\left(\dfrac{x}{1+\dfrac{1}{x}} + 1 - \dfrac{1}{x+1}\right) \div \left(\dfrac{x}{1-\dfrac{1}{x}} - x - \dfrac{1}{x-1}\right)$.

35. Divide $\left(\dfrac{x}{x-y} - \dfrac{y}{x+y}\right)$ by $\left(\dfrac{x^2}{x^2+y^2} + \dfrac{y^2}{x^2-y^2}\right)$.

REVIEW.

36. Multiply $\dfrac{x^2 - 2xy + y^2 - z^2}{x^2 + 2xy + y^2 - z^2}$ by $\dfrac{x+y-z}{x-y+z}$.

37. Simplify $\dfrac{2x - 3 + \dfrac{1}{x}}{\dfrac{2x-1}{x}}$.

38. Divide $2 - \dfrac{\sqrt{x}}{\sqrt{y}} - \dfrac{\sqrt{y}}{\sqrt{x}}$ by $\dfrac{\sqrt{x} - \sqrt{y}}{\sqrt{xy}}$.

39. Raise $x + y$ to the seventh power.

40. Expand $(2a + 3b)^5$.

41. Expand $(a + b)^{10}$.

42. Raise $\sqrt{x+y}$ to the sixth power.

43. Find the sum of $\sqrt[3]{x^7 y}$, $\sqrt[3]{8 x^4 y^4}$, $\sqrt[3]{xy^7}$.

44. From $\sqrt{3x^2 z + 6xyz + 3y^2 z}$ subtract $\sqrt{12 y^2 z}$.

45. From $6\sqrt[3]{32}$ subtract $6\sqrt[3]{\tfrac{4}{27}}$.

46. Multiply $\sqrt{x} - \sqrt{y}$ by $\sqrt{x} + \sqrt{y}$.

47. Multiply $\sqrt{a} + \sqrt{b}$ by $\sqrt{a} + \sqrt{b}$.

48. Solve $7 - [7 + 7 - (7 + x)] = 7$.

49. Solve $a - \dfrac{m+n}{x} = b - \dfrac{m-n}{x}$.

50. Solve $\dfrac{1}{a-b} + \dfrac{a-b}{x} = \dfrac{1}{a+b} + \dfrac{a+b}{x}$.

51. Solve $\dfrac{bx}{a} - \dfrac{d}{c} = \dfrac{a}{b} - \dfrac{cx}{d}$.

52. Solve $\dfrac{4x+3}{9} + \dfrac{7x-29}{5x-12} = \dfrac{8x+19}{18}$.

53. Solve $(x + \tfrac{5}{2})(x - \tfrac{3}{2}) - (x+5)(x-3) + \tfrac{3}{4} = 0$.

54. Solve $\dfrac{m(x+a)}{x+b} + \dfrac{n(x+b)}{x+a} = m + n$.

Solve:

55. $\dfrac{x^2-16}{x+4} = 6-x.$

56. $(1+x)^5 + (1-x)^5 = 242.$

57. $\sqrt{4+x} = \dfrac{3}{\sqrt{4-x}}.$

58. $\sqrt{x-32} = 16 - \sqrt{x}.$

59. $\sqrt{4x+21} = 2\sqrt{x}+1.$

60. $\dfrac{\sqrt{9x}-4}{\sqrt{x}+2} = \dfrac{15+3\sqrt{x}}{\sqrt{x}+40}.$

61. $\sqrt{2+x} + \sqrt{x} = \dfrac{4}{\sqrt{2+x}}.$

62. $\begin{cases} x+y=7. \\ x+z=8. \\ y+z=9. \end{cases}$

63. $\begin{cases} \tfrac{1}{2}x + \tfrac{1}{3}y + \tfrac{1}{4}z = 22. \\ \tfrac{1}{4}x + y + \tfrac{1}{2}z = 33. \\ x + \tfrac{1}{4}y + \tfrac{1}{8}z = 19. \end{cases}$

64. $\begin{cases} 7x - 2z + 3w = 17. \\ 4y - 2z + v = 11. \\ 5y - 3x - 2w = 8. \\ 4y - 3w + 2v = 9. \\ 3z + 8w = 33. \end{cases}$

65. $\begin{cases} cx + y + az = a + ac + c. \\ c^2 x + y + a^2 z = 3ac. \\ acx + 2y + acz = a^2 + 2ac + c^2. \end{cases}$

66. $\begin{cases} x^2 + xy = 12. \\ y^2 + xy = 24. \end{cases}$

67. $\begin{cases} 3x + 3y = 15. \\ xy = 6. \end{cases}$

68. $\begin{cases} x^2 y + xy^2 = 180. \\ x^3 + y^3 = 189. \end{cases}$

69. $\begin{cases} x - y = 8(\sqrt{x} - \sqrt{y}). \\ \sqrt{xy} = 15. \end{cases}$

70. $\begin{cases} x^3 - y^3 = 56. \\ x - y = \dfrac{16}{xy}. \end{cases}$

71. $\begin{cases} x^4 - y^4 = 369. \\ x^2 - y^2 = 9. \end{cases}$

72. $\begin{cases} x^2 + 2y^2 = 41. \\ x^2 + 2xy = 33. \end{cases}$

73. $\begin{cases} x^2 + x + y = 18 - y^2. \\ xy = 6. \end{cases}$

74. $\begin{cases} xy + xy^2 = 12. \\ x + xy^3 = 18. \end{cases}$

75. $\begin{cases} x^2 y^2 = 96 - 4xy. \\ x + y = 6. \end{cases}$

76. $\begin{cases} x + y + \sqrt{x+y} = 12. \\ x^3 + y^3 = 189. \end{cases}$

77. $\begin{cases} x^{\frac{3}{2}} + y^{\frac{2}{3}} = 3x. \\ x^{\frac{1}{2}} + y^{\frac{1}{3}} = x. \end{cases}$

78. Solve $\begin{cases} 2xy - 24(x+y) = -240. \\ x^2 + y^2 = 100. \end{cases}$

79. Solve $\begin{cases} (x-y)(x^2+y^2) = 13. \\ x^2y - xy^2 = 6. \end{cases}$

80. What two numbers, which are to each other as 3 to 4, have a product which is equal to twelve times their sum?

81. A person being asked the time of day, replied that the time past noon was equal to $\frac{4}{5}$ of the time to midnight. What was the time of day?

82. Find a number which being increased by 4 and the sum multiplied by 3, gives the same result as if half the number were multiplied by 8 and the product were diminished by 8.

83. Find two numbers in the proportion of 3 to 4 such that if 9 be added to each the sums will be as 6 to 7.

84. The sum of two numbers is 12, and the difference of their squares is 72. What are the numbers?

85. It is required to divide 99 into five such parts that the first may exceed the second by 3, may be less than the third by 10, greater than the fourth by 9, and less than the fifth by 16.

86. There are two numbers whose product is 6, and whose sum added to the sum of their squares is 18. What are the numbers?

87. What number is that to which if 12 be added, and from $\frac{1}{12}$ of the sum 12 be subtracted, the remainder will be 12?

88. A boy paid 20 cents for 200 apples and pears together, buying 25 apples for a cent and 25 pears for 3 cents. How many of each did he buy?

89. A steamboat, whose rate in still water is 10 miles per hour, descends a river whose velocity is 4 miles per hour, and returns. She was away for 10 hours. How far did she go?

90. Three years ago A's age was $\frac{1}{2}$ of B's, and 9 years hence it will be $\frac{5}{8}$ of it. What is the age of each?

91. There is a number whose three digits are the same; and when 4 times the sum of the digits is subtracted from the number, the remainder is 297. What is the number?

92. A woman being asked what she paid for her eggs, replied, "Six dozen cost as many cents as I can buy eggs for 32 cents." What was the price per dozen?

93. What fraction is that which will be doubled when the numerator is multiplied by 4 and 3 is added to the denominator; but will be halved when 2 is added to the numerator and the denominator is multiplied by 4?

94. The stones which paved a square court-yard would just cover a rectangular surface whose length was 6 yards longer and whose breadth was 4 yards shorter than the side of the square. What was the area of the court?

95. A gentleman had not room in his stables for 8 of his horses, so he built an additional stable one half the size of the other, and then had room for 8 horses more than he had. How many horses had he?

96. A gentleman purchased two square lots of ground for $300. Each of them cost as many cents per square rod as there were rods in a side of the other, and the sum of the perimeters of both was 200 rods. What was the cost of each?

97. There is a number consisting of three digits. The sum of the squares of the digits is 83, and the square of the middle digit exceeds the product of the other two by 4. If 396 be added to the number, the digits will be reversed. What is the number?

98. A and B hired a pasture, into which A put 4 horses, and B as many as cost him 18 shillings a week. Afterward B put in 2 additional horses, and found that he must pay 20 shillings per week. What was paid for the pasture per week?

99. The sum of two numbers is 40. If 3 times the smaller number be subtracted from 2 times the larger, the remainder will be 15. What are the numbers?

REVIEW.

100. A general, having lost a battle, found that he had only 3600 more than half his army left fit for action, 600 more than $\frac{1}{8}$ of his men being disabled by wounds, and the rest, which were $\frac{1}{5}$ of the whole army, being killed or taken prisoners. How many men had he in the army?

101. Four places are situated in the order of the letters, A, B, C, D. The distance from A to D is 34 miles; the distance from A to B is to the distance from C to D as 2 is to 3, and $\frac{1}{4}$ of the distance from A to B added to $\frac{1}{2}$ the distance from C to D is three times the distance from B to C. What are the respective distances?

102. Given $x + \sqrt{3} = \dfrac{2\sqrt{2}}{\sqrt{3} - x}$, to find the values of x.

103. Several persons incurred an expense of $12, which they were to share equally. If there had been 4 more in the company, the expense to each person would have been 50 cents less than it was. How many persons were there in the company?

104. It is between 11 and 12 o'clock, and the hour hand and minute hand make a straight line. What is the time?

105. A rectangular field, whose sides are to each other as 2 to 5, contains 4 acres. What are the length and breadth of the field?

106. Divide 18 into two such parts that the squares of those parts may be to each other as 25 to 16.

107. What will be the payment which will discharge a debt of $2000 in four years, paying principal and interest in equal annual installments, interest at 6 %?

108. A rectangular plat of ground has a walk 6 feet wide around the outside, which contains $\frac{1}{4}$ as much area as the plat itself. If the sides are to each other as 3 to 4, what are the length and breadth of the plat?

109. Given $\begin{cases} x + y = 24 \\ xy : x^2 + y^2 :: 3 : 10 \end{cases}$, to find x and y.

110. Given $\left\{ \begin{array}{l} xy = 320 \\ x^3 - y^3 : (x-y)^3 :: 61 : 1 \end{array} \right\}$, to find x and y.

111. There are four numbers in arithmetical progression such that the sum of the two least is 20, and the sum of the two greatest is 44. What are the numbers?

112. A farmer has two cubical granaries. The side of one is 3 yards longer than the side of the other, and the difference in their solid contents is 117 cubic yards. What is the side of each?

113. A merchant expended a sum of money in goods, which he sold for $56, and thereby gained a per cent. equal to the number of dollars which the goods cost him. How much did they cost him?

114. The sum of three numbers in geometrical progression is 13, and the sum of the extremes multiplied by the mean is 30. What are the numbers?

115. Given $\left\{ \begin{array}{l} x^2 + xy = 12 \\ xy - 2y^2 = 1 \end{array} \right\}$, to find x and y.

116. There are two rectangular boxes, one containing 20 cubic feet more than the other. Their bases are squares, the sides of each being equal to the depth of the other. If the capacities of the boxes are in the ratio of 4 to 5, what is the depth of each box?

117. What three numbers in geometrical progression are there whose sum is 14, and the sum of whose squares is 84?

118. What is the square root of $a^2x^4 + b^2x^2 + c^2 + 2abx^3 + 2acx^2 + 2bcx$?

119. A merchant has three pieces of cloth whose lengths are in geometrical progression. The aggregate length of the three pieces is 70 yards, and the longest piece is 30 yards longer than the shortest. What is the length of each?

120. A father divided $2100 among his three sons, so that the shares were in geometrical progression, and the second had $300 more than the third. What was the share of each?

121. A vintner has two casks of wine, from each of which he draws 6 gallons, and then finds that the quantities left are to each other as 4 to 7. He then puts into the smaller cask 3 gallons, and into the greater 4 gallons, and the quantities they contain are then to each other as 7 to 12. How many gallons were there in each at first?

122. Some smugglers discovered a cave which would exactly hold their cargo, which consisted of 13 bales of cotton and 33 casks of wine. While they were unloading, a revenue cutter hove in sight, when they sailed away with 9 casks and 5 bales, leaving the cave two thirds full. How many bales, or how many casks, would the cave hold?

123. A farmer sold a meadow at such a rate that the price per acre was to the number of acres as 2 to 3. If he had received $270 more for it, the price per acre would have been to the number of acres as 3 to 2. How many acres did he sell, and at what price per acre?

124. Given $\sqrt{x^5} - \dfrac{40}{\sqrt{x}} = 3x$, to find x.

125. The sum of two numbers is to their difference as 4 to 1, and the sum of their cubes is 152. What are the numbers?

126. A and B set out from two towns which were 204 miles apart, and traveled in a direct line until they met. A traveled 8 miles per hour; and the number of hours before they met was greater by 3 than the number of miles B traveled per hour. How far did each travel?

127. A merchant bought a number of pieces of cloth for $225, which he sold at $16 a piece, and gained by the sale as much as one piece cost him. How many pieces were there?

128. There are three numbers in arithmetical progression whose sum is 15. If 1, 4, and 19 be added to them respectively, they will be in geometrical progression. What are the numbers?

129. A and B agreed to reap a field of grain for 90 shillings, and they promised to complete the work in 5 days. A alone could have done the work in 9 days; but B, who did not work as quickly as he had expected, was obliged to call to his assistance C, who worked 2 days, in consequence of which B received 3s. 9d. less than would otherwise have been due him. In what time could B and C each reap the field alone?

130. Find two quantities such that their sum, their product, and the sum of their squares shall be equal to each other.

131. Find two quantities such that their product shall be equal to the difference of their squares, and the sum of their squares shall be equal to the difference of their cubes.

132. A sets out from London to York, and B, at the same time, from York to London, both traveling uniformly. A reaches York 25 hours and B reaches London 36 hours after they have met on the road. In what time did they each perform the journey?

133. From two towns, which were 102 miles apart, two persons, A and B, set out to meet each other. A traveled 3 miles the first day, 5 miles the second day, 7 miles the third day, and so on. B traveled 4 miles the first day, 6 the next, 8 the next, and so on. In how many days did they meet?

134. Given $x^4 - 2x^3 + x = 30$, to find x.

Resolve into factors by partially extracting the square root and factoring the remainder.

135. Given $x^3 - 6x^2 + 11x = 6$, to find x.

Multiply both members of the equation by x, and resolve into factors by extracting the square root partially and factoring the remainder.

136. Given $\begin{cases} x^2 + y^2 + 4\sqrt{x^2 + y^2} = 45 \\ x^4 + y^4 = 337 \end{cases}$, to find x and y.

137. Given $\begin{cases} x^4 + y^4 = 17 \\ x^3y + xy^3 = 10 \end{cases}$, to find x and y.

138. A railway train, after traveling 2 hours, is detained by an accident 1 hour. It then proceeds, for the rest of the distance, at $\frac{3}{5}$ of its former rate, and arrives $7\frac{2}{3}$ hours behind time. If the accident had occurred 50 miles further on, the train would have arrived $6\frac{1}{3}$ hours behind time. What was the whole distance traveled by the train?

IMAGINARY QUANTITIES.

364. 1. What is the square root of a^2? Of $-a^2$?

2. Into what factors may $\sqrt{-a^2}$ be separated so that one of them shall be a perfect square? $\sqrt{-4}$? $\sqrt{-9}$?

3. What is the square of any radical quantity of the second degree? What is the square of $\sqrt{5}$? Of $\sqrt{-5}$? Of $\sqrt{-3a}$? Of $\sqrt{-5a}$?

365. An **Imaginary Quantity** is an indicated even root of a negative quantity.

Thus, $\sqrt{-2a}$, $\sqrt[4]{-3x}$, $\sqrt[6]{-a}$ are imaginary quantities.

In contradistinction to imaginary quantities, all other quantities, whether rational or irrational, are called *real quantities*.

In accordance with the processes in radicals,

$$\sqrt{-4} = \sqrt{4} \times \sqrt{-1} = 2\sqrt{-1}$$
$$\sqrt{-16} = \sqrt{16} \times \sqrt{-1} = 4\sqrt{-1}$$
$$\sqrt{-5} = \sqrt{5} \times \sqrt{-1} = \sqrt{5}\sqrt{-1}$$
$$\sqrt{-a^2} = \sqrt{a^2} \times \sqrt{-1} = a\sqrt{-1}. \quad \text{Hence,}$$

366. PRINCIPLE. — *Every imaginary quantity may be reduced to the form of $a\sqrt[n]{-1}$, in which a is real and $\sqrt[n]{-1}$ is the imaginary factor.*

367. An expression which contains an imaginary quantity is called an *imaginary expression*.

Thus, $3 + \sqrt{-5}$ is an imaginary expression.

HIGH SCHOOL ALGEBRA.

368. To add or subtract imaginary quantities.

1. Add $\sqrt{-x^2}$ and $2\sqrt{-4x^2}$.

PROCESS.

$\sqrt{-x^2} = x\sqrt{-1}$
$2\sqrt{-4x^2} = 4x\sqrt{-1}$
$\overline{5x\sqrt{-1}}$

EXPLANATION. — Since the radical expressions are dissimilar, they must be reduced to similar radicals before adding. Reducing and adding coëfficients, the sum is $5x\sqrt{-1}$.

Simplify:

2. $\sqrt{-4} + \sqrt{-9}$.

3. $\sqrt{-8} + \sqrt{-18}$.

4. $6\sqrt{-4} + 2\sqrt{-16}$.

5. $\sqrt{-12} + \sqrt{-48}$.

6. $4\sqrt{-125} - 3\sqrt{-80}$.

7. $\sqrt{-a^2} + \sqrt{-4a^2}$.

8. $2\sqrt{-4b^2} + 2b\sqrt{-9}$.

9. $3\sqrt{-16a^2x^2} + a\sqrt{-25x^2}$.

10. $\sqrt{-9a^2} - a\sqrt{-16}$.

11. $\sqrt{-3m^2x^2} - \sqrt{-27m^2x^2}$

369. To multiply imaginary quantities.

1. Multiply $2\sqrt{-3}$ by $2\sqrt{-6}$.

$$2\sqrt{-3} = 2\sqrt{3} \times \sqrt{-1}$$
$$2\sqrt{-6} = 2\sqrt{6} \times \sqrt{-1}$$
$$(2\sqrt{3} \times \sqrt{-1}) \times (2\sqrt{6} \times \sqrt{-1})$$
$$= 4\sqrt{18} \times (\sqrt{-1})^2 = -4\sqrt{18} = -12\sqrt{2}$$

EXPLANATION. — In order to determine the sign of the product, we separate each of the imaginary quantities into two factors, one of which is $\sqrt{-1}$. We then multiply, as in radical quantities, observing that $(\sqrt{-1})^2 = -1$. $\sqrt{-1} \times \sqrt{-1}$ would, according to the ordinary rules for multiplication of radicals, give as a product $\sqrt{+1}$, which is equal to ± 1; but, inasmuch as we know that $\sqrt{+1}$ was derived from the product of two negative factors, viz., $\sqrt{-1}$ and $\sqrt{-1}$, the root of $\sqrt{+1}$, in this case, is -1, and not $+1$. Hence,

IMAGINARY QUANTITIES.

370. Principle. — *The product of two imaginary quantities of the second degree is real.*

Multiply:

2. $\sqrt{-4}$ by $\sqrt{-5}$.
3. $2\sqrt{-6}$ by $3\sqrt{-4}$.
4. $2\sqrt{-4}$ by $3\sqrt{-9}$.
5. $3\sqrt{-a^2}$ by $2\sqrt{-a^2x}$.
6. $1+\sqrt{-1}$ by $1-\sqrt{-1}$.
7. $1-\sqrt{-1}$ by $1-\sqrt{-1}$.
8. $3\sqrt{-3}$ by $2\sqrt{-2}$.
9. $-5\sqrt{-2}$ by $3\sqrt{-5}$.
10. $3\sqrt{-4}$ by $5\sqrt{-3}$.
11. $\sqrt{-2}$ by $\sqrt{-4}\times\sqrt{-8}$.
12. $7+3\sqrt{-9}$ by $6-2\sqrt{-9}$.
13. $\sqrt{x}+\sqrt{-y}$ by $\sqrt{y}+\sqrt{-x}$.

371. To divide imaginary quantities.

1. Divide $6\sqrt{-6}$ by $3\sqrt{-2}$.

    Solution.

    $$6\sqrt{-6} \div 3\sqrt{-2} = 2\sqrt{3}$$

2. Divide $4+\sqrt{-2}$ by $2-\sqrt{-2}$.

    Solution.

    $$(4+\sqrt{-2}) \div (2-\sqrt{-2}) = \frac{4+\sqrt{-2}}{2-\sqrt{-2}}$$

    Rationalizing the denominator, $= \dfrac{(4+\sqrt{-2})(2+\sqrt{-2})}{4-(-2)}$

    $$= \frac{6+6\sqrt{-2}}{6} = 1+\sqrt{-2}$$

Divide:

3. $3\sqrt{-8}$ by $\sqrt{-4}$.
4. $2\sqrt{-a^2}$ by $3\sqrt{-a^2}$.
5. $2\sqrt{-a^2x}$ by $3\sqrt{-x}$.
6. 2 by $1+\sqrt{-1}$.
7. $-2\sqrt{-1}$ by $1-\sqrt{-1}$.
8. $\sqrt{-6}$ by $2\sqrt{-2}$.

9. $4+\sqrt{4}$ by $2-\sqrt{-2}$.
10. $-6a^2\sqrt{x}$ by $3\sqrt{-a^2}$.
11. $-12\sqrt{6}$ by $2\sqrt{-6}$.
12. -36 by $2\sqrt{-4}$.
13. $8\sqrt{-x^2}$ by $4\sqrt{x}$.
14. $2+\sqrt{-4}$ by $1-\sqrt{-1}$.
15. $14+5\sqrt{-1}$ by $2-3\sqrt{-1}$.
16. $\sqrt{5}+\sqrt{-6}$ by $\sqrt{6}-\sqrt{-5}$.

ZERO AND INFINITY.

372. How much is 2 times 0? 3 times 0? 500 times 0? a times 0? Any number of times 0?

373. PRINCIPLE.—1. *When zero is multiplied by a finite quantity, the product is zero.*

How much is 0 divided by 2? 0 divided by 6? 0 divided by a? 0 divided by any number?

374. PRINCIPLE.—2. *When zero is divided by any finite quantity, the quotient is zero.*

1. What is the value of the fraction $\dfrac{n-nx^2}{1-x}$ when $x=\tfrac{1}{2}$? $\tfrac{3}{4}$? $\tfrac{7}{8}$? $\tfrac{9}{10}$?

2. What is the value of the fraction when $x=1$?

Ans. $\tfrac{0}{0}$ or $2n$.

3. Since $\tfrac{0}{0} = 2n$, and n may have any finite value, what value may the expression $\tfrac{0}{0}$ represent?

375. PRINCIPLE.—3. *The expression $\tfrac{0}{0}$ may equal any finite quantity, or it is a symbol of indetermination.*

1. What is the quotient of 2 divided by $\tfrac{1}{2}$? By $\tfrac{1}{4}$? By $\tfrac{1}{8}$? By $\tfrac{1}{16}$?

2. When the divisor is diminished while the dividend remains the same, what effect is produced upon the quotient?

3. If the divisor is made small enough, how large a quotient may be obtained?

ZERO AND INFINITY.

4. If a represents any finite quantity and 0 an exceedingly small quantity, what value will the expression $\frac{a}{0}$ represent?

376. Principle. — 4. *The expression $\frac{a}{0}$ may be regarded as a symbol of infinity.*

1. What is the quotient when 4 is divided by 2? By 4? By 8? By 16? By 32?

2. When the divisor is increased, the dividend remaining the same, what is the effect upon the quotient?

3. If the divisor is made large enough, how small a quotient may be obtained?

4. If a represents any finite quantity, and ∞ an exceedingly large quantity, what value will the expression $\frac{a}{\infty}$ represent?

377. Principle. — 5. *The expression $\frac{a}{\infty}$ may be regarded as a symbol for zero.*

378. It should be remembered that the expressions $\frac{0}{0}$, $\frac{a}{0}$, and $\frac{a}{\infty}$ are simply symbols, and that they are not to be understood in the strict arithmetical sense. For instance, the division of a by 0 is inconceivable, if 0 represents the absence of quantity. But if 0 represents a quantity, however small it may be, then the division of a by 0 is conceivable. Keeping this interpretation of the symbols in mind, the preceding principles may be expressed by algebraic formulas as follows:

Prin. 1, $0 \times a = 0$.   Prin. 3, $\frac{0}{0} =$ any finite quantity.

Prin. 2, $\frac{0}{a} = 0$.   Prin. 4, $\frac{a}{0} = \infty$.

Prin. 5, $\frac{a}{\infty} = 0$.

EXAMPLES.

379. 1. Find a number such that when 5 is added to 3 times the number, and the result is divided by the number increased by 2, the quotient will be 3.

The solution of this example shows the number to be ∞; that is, there is no finite number which will fulfill the conditions, and consequently the problem is *impossible*.

2. What number is there such that when $\frac{5}{6}$ of it is diminished by 4, the result is 3 less than $\frac{1}{2}$ of it plus $\frac{1}{3}$ of it?

3. I bought 400 sheep in two flocks, paying $1.50 per head for the first flock and $2 for the second. I lost 30 of the first flock and 56 of the second, and sold the rest of the first flock at $2 per head and the rest of the second for $2.50 per head without gain or loss. Required the number of sheep originally in each flock.

4. There is a number such that a third part of it exceeds a fourth part of it by as much as $\frac{29}{60}$ of it exceeds $\frac{2}{5}$ of it. What is the number?

5. A man being asked his age, replied that if from 3 times his age 10 years were subtracted, and to 4 times his age 8 years were added, the former result would be equal to $\frac{3}{4}$ of the latter. What was his age?

6. Two teachers, A and B, receive the same monthly salary. A is employed 10 months in the year, and his annual expenses are $600. B is employed 8 months in the year, and his annual expenses are $480. A saves as much money in 3 years as B does in 4 years. What is the monthly salary of each?

INTERPRETATION OF NEGATIVE RESULTS.

380. 1. A father who is 42 years of age has a son 12 years of age. In how many years will the son's age be $\frac{1}{4}$ of the father's?

SOLUTION.

Let $x =$ the number of years.

Then
$$\frac{42 + x}{4} = 12 + x$$
$$\therefore x = -2$$

INTERPRETATION OF NEGATIVE RESULTS.

Evidently the problem is absurd, for while -2 when substituted for x in the equation satisfies the equation, and thus shows that the result is correct *algebraically*, the problem must be interpreted *arithmetically*, and in that sense the result, -2 years, is an *absurdity*.

If $-x$ be substituted for x in the equation, the equation becomes

$$\frac{42-x}{4} = 12 - x$$

and $$x = 2$$

Thus, it is obvious that, if the conditions of the problem be changed to correspond with the changes in the equation, the problem will be true in an *arithmetical sense*.

The problem modified to correspond with the changes in the equation will be: A father who is 42 years of age has a son 12 years of age. *How many years ago* was the son's age $\frac{1}{4}$ of the father's?

2. There is a number such that when 7 is subtracted from the sum of $\frac{1}{3}$ of the number plus $\frac{1}{5}$ of it, the remainder is equal to the number. What is the number?

SOLUTION.

Let $x =$ the number.

Then $$\frac{x}{3} + \frac{x}{5} - 7 = x$$

$$\therefore x = -15$$

While the result -15 is *algebraically* correct, it is *arithmetically* absurd.
If $-x$ be substituted for x in the equation, the equation becomes

$$-\frac{x}{3} - \frac{x}{5} - 7 = -x$$

and $$x = 15$$

The problem when changed to express conditions which are *arithmetically* reasonable will be: There is a number such that when 7 is *added* to the sum of $\frac{1}{3}$ of the number plus $\frac{1}{5}$ of it, the result is equal to the number. What is the number?

381. From the discussion of the foregoing problems we may infer:

1. *When a negative value results from the solution of a problem producing an equation of the first degree, it indicates an impossibility in the conditions of the problem.*

2. *The impossibility consists in adding a quantity when it should be subtracted, or vice versa; or in applying a quantity in one direction when it should be applied in an opposite direction.*

3. *A possible problem analogous to the given problem may be formed by changing the absurd conditions to their opposites.*

EXAMPLES.

382. Modify the following problems so that they may be possible arithmetically.

1. What number is that whose third part exceeds half of it by 4?

2. A man was married when he was 30 years old and his wife 20. How soon will he be twice as old as his wife?

3. What fraction is that which becomes $\frac{3}{5}$ when its numerator is increased by 1 and $\frac{5}{7}$ when its denominator is increased by 1?

4. A man worked 7 days and had his son with him 3 days; he received for wages 22 shillings and the board of his son and himself. He afterward worked 5 days and had his son with him 1 day; he received 18 shillings and the board of himself and son. What were the daily wages of each?

5. The difference between two numbers is 24. If 3 times the greater be added to 5 times the less, the sum will be 48. What are the numbers?

INDETERMINATE EQUATIONS.

383 While Indeterminate Equations may have an infinite number of values for the unknown quantities (Art. 182), it is nevertheless true that by the introduction of a condition or conditions into a problem, the number of values may be limited and these values algebraically and accurately determined. **One common limitation in problems discussed under the pres**

INDETERMINATE EQUATIONS.

ent head is that the results shall be positive integers, or positive whole numbers.

In this entire discussion *whole number* is understood to mean *positive whole number*, and *integer*, *positive integer*.

EXAMPLES.

1. Find two integers whose sum is 3.

SOLUTION.

Let $x =$ one number,
and $y =$ the other.
Then, $x + y = 3$
and $x = 1$ or 2
$y = 2$ or 1

2. Find two integers whose sum is 5.

3. Find two integers whose sum is 7.

4. There are two integers such that one of them added to four times the other makes 17. Find all the possible numbers.

5. Given $3x + 4y = 29$. Find the values of x and y in whole numbers.

SOLUTION.

Since x and y are positive integers, it is evident that $3x$ must be equal to 3 or a multiple of 3, and $4y$ must be equal to 4 or a multiple of 4. Since the sum of these multiples is equal to 29, it is evident that, if we subtract from 29 successively the multiples of 3, some one of the remainders will be a multiple of 4, if the problem is possible.

By subtracting 3 and its multiples 6, 9, 12, etc., successively from 29, it is found that 9 and 21, or 3 times 3 and 7 times 3, are the only multiples of 3 which leave multiples of 4. Hence, $x = 3$ or 7.

Substituting the values of x in the equation, $y = 5$ or 2.

ANOTHER SOLUTION.

$$3x + 4y = 29 \qquad (1)$$

$$y = \frac{29 - 3x}{4} \qquad (2)$$

Expressing the value of y as a mixed quantity, $y = 7 - x + \dfrac{1+x}{4}$.

Since x and y are integers, $7 - x$ is integral, and $\dfrac{1+x}{4}$ is also integral; for if it were not, the value of y would contain a fraction.

$$\therefore \frac{1+x}{4} = w, \text{ a whole number or integer.}$$

Then
$$1 + x = 4w$$
$$x = 4w - 1$$

Making $\qquad w = 1, 2, 3,$ etc., successively,

$$x = 3, 7, 11, 15, \text{etc.}$$

The corresponding values of y are 5, 2, -1, -4, etc.

Since the values of x and y are positive, only the first two values fulfil the conditions, and

$$x = 3 \text{ and } y = 5, \text{ or } x = 7 \text{ and } y = 2$$

6. Separate 200 into two integers, one of which is a multiple of 7, and the other some other multiple of 13.

Solution.

Let $\qquad 7x =$ one number,

and $\qquad 13y =$ the other.

Then, $\qquad 7x + 13y = 200.$

Hence, $\qquad 7x = 161,$ and $13y = 39,$ the number sought.

7. Determine whether 500 can be separated into two parts, one of which shall be a multiple of 17, the other a multiple of 19.

Find the least values of x and y in the following equations:

8. $3x = 8y - 16.$ **11.** $7x - 9y = 29.$

9. $14x = 5y + 7.$ **12.** $19x - 5y = 119.$

10. $27x = 1600 - 16y.$ **13.** $17x - 49y = -8.$

14. A man has $500 which he wishes to spend for cows and sheep; cows cost $17 apiece, and sheep $5 apiece. How many can he buy of each?

15. A man buys 100 animals for $100, giving $10 each for calves, $2 each for sheep, and $½ each for geese. How many animals of each kind may he have purchased?

INDETERMINATE EQUATIONS. 309

SOLUTION.

Let $x =$ the number of calves,
$y =$ the number of sheep,
and $z =$ the number of geese.

Then $\quad x + y + z = 100 \quad (1)$

and $\quad 10x + 2y + \dfrac{z}{2} = 100 \quad (2)$

Eliminating z, $\quad 19x + 3y = 100$.

16. What is the least whole number which being divided by 19 will give a remainder 6, and being divided by 17 will give a remainder 12?

SUGGESTION. — $19x + 6 = 17y + 12$.

17. A boy being asked how many apples he had, replied that he had between two dozen and three dozen; that when he took them 4 at a time, he found there were 3 left, but when he took 5 at a time, there was but 1 left. How many apples had he?

18. Find two numbers either of which, upon being divided by 3, 4, and 5, gives the remainders 2, 3, and 4, respectively.

19. Find the least whole number which, being divided by 7, 8, and 9, respectively, leaves the remainders 5, 7, and 8.

20. Find the least whole number which, being divided by 39, leaves the remainder 16, and when divided by 56 leaves a remainder 27.

21. A person sold a number of sheep, calves, and lambs, 40 in all, for $48. How many may he have sold of each, if he received for each calf $1.75, for each sheep $1.25, and for each lamb $.75?

22. A farmer has oats worth 42 cents, barley worth 64 cents, and rye worth 87 cents per bushel. How many bushels of each may he take to make a mixture of 200 bushels, worth 75 cents per bushel?

23. A man bought calves, sheep, and lambs, 154 in all, for $154. He paid $3½ each for calves, $1⅓ each for sheep, and $½ each for lambs. How many may he have bought of each?

24. Four boys have a pile of marbles. A throws away 1, and takes $\frac{1}{4}$ of the remainder; B then throws away 1, and takes $\frac{1}{4}$ of the remainder; C then throws away 1, and takes $\frac{1}{4}$ of the remainder; D then throws away 1, and each boy takes $\frac{1}{4}$ of the remainder. At least, how many marbles must have been in the pile, and how many does each boy now have?

25. I bought a certain number of histories, at $2.50 apiece, and a number of lexicons at $1.75 apiece. The entire cost was $44.75. How many did I buy of each kind?

26. A man bought 30 animals for $130, paying $7 each for calves, $3 each for sheep, and $2 each for pigs. How many of each kind did he buy?

27. Divide 70 into three parts which shall give integral quotients when divided by 6, 7, 8, respectively, the sum of which quotients shall be 10.

28. A dealer in stock can buy 100 animals for $400, by paying $9 apiece for sheep, $2 apiece for pigs, and $1 apiece for lambs. How many of each kind may he buy?

29. A farmer bought 100 animals, consisting of sheep, turkeys, and geese, for $200. The sheep cost him $7 apiece, the turkeys 2\frac{2}{3}$ apiece, and the geese $1 apiece. How many of each kind did he buy?

INEQUALITIES.

384. An Inequality is an algebraic expression, indicating that one quantity is greater or less than another.

385. The Sign of inequality is $>$. It is placed with the opening of the angle toward the greater quantity.

Thus, $a > b$ is read, a is greater than b; $a < b$ is read, a is less than b.
Any negative quantity is regarded as less than 0, and the one having the greatest number of units is considered the least. Thus,

$$-2 > -3;\ 0 > -2.$$

386. The quantities on each side of the sign of inequality are called *members*.

INEQUALITIES.

387. When the first members of two inequalities are each greater or each less than the corresponding second members, the inequalities are said to *subsist in the same sense*.

When the first member is greater in one inequality and less in another, the inequalities are said to *subsist in a contrary sense*.

Thus, $a > b$ and $c > d$ subsist in the same sense, and $x > y$ and $v < z$ subsist in a contrary sense.

388. 1. In the inequality $10 > 6$, if 4 be added to each member, how will the inequalities subsist? If 4 be subtracted from each member, how will they subsist?

2. If to the equation $10 = 10$, the inequality $6 > 4$ be added member to member, how will the inequalities subsist? If the inequality be subtracted from the equation, member from member, how will the inequalities subsist?

3. If the members of the inequality $10 > 6$ be multiplied or divided by the same positive quantity, how will the inequalities subsist? If they be multiplied or divided by any negative quantity, how will they subsist?

4. If the corresponding members of two or more inequalities, subsisting in the same sense, be added, as $10 > 6$ and $4 > 3$, how will the inequalities subsist? If the corresponding members of the inequality $4 > 3$ be subtracted from $10 > 5$, how do the inequalities subsist? If $20 > 3$ be subtracted from $10 > 5$, how do they subsist?

5. If each member of the inequality $9 > 4$ be raised to the same power, how will the inequalities subsist? If the same root of each be taken, how will the inequalities subsist?

If each member of the inequality $-2 > -3$ be raised to the second power or any even power, how will the equalities subsist? If each member be raised to the third power or any odd power, how will they subsist?

If the same odd roots of each member of the inequality $-64 > -729$ be taken, how will the inequalities subsist?

6. If the members of the inequality $5 > 3$ be divided by -1, how will the inequalities subsist.

389. Principles. — 1. *If equal quantities be added to or subtracted from the members of an inequality, the inequalities will subsist in the same sense; if, however, the members of an inequality be subtracted from the corresponding members of an equation, the inequalities will subsist in a contrary sense.*

2. *If the members of an inequality be multiplied or divided by the same or equal positive quantities, the inequalities will subsist in the same sense; if, however, the multipliers or divisors be negative, the inequalities will subsist in a contrary sense.*

3. *If the corresponding members of two or more inequalities subsisting in the same sense be added, the inequalities will subsist in the same sense; if, however, the members of one inequality be subtracted from the corresponding members of another, the inequalities will not always subsist in the same sense.*

4. *If the same positive powers or roots of the members of an inequality be taken, the inequalities will subsist in the same sense, provided the members of the original inequality be positive, or if odd roots of negative quantities be taken; if, however, the members of an inequality be negative, and they be raised to the same even power, they will subsist in a contrary sense.*

5. *If the signs of all the terms of an inequality be changed, the resulting inequality must be expressed in a contrary sense.*

EXAMPLES.

1. Find the limit of x in the inequality $5x - 6 > 19$.

Solution.

$$5x - 6 > 19$$
(Prin. 1), $\qquad 5x > 25$
(Prin. 2), $\qquad x > 5$

∴ The inferior limit of x is 5.

2. Find the limits of x and y in $2x + 3y > 26$ and $2x + y = 16$.

INEQUALITIES.

SOLUTION.

$$2x + 3y > 26 \qquad (1)$$
$$2x + y = 16 \qquad (2)$$

(2)−(1) (Prin. 1), $\qquad 2y > 10 \qquad (3)$

(Prin. 2), $\qquad y > 5 \qquad (4)$

(2)−(4) (Prin. 1), $\qquad 2x < 11$

(Prin. 2), $\qquad x < 5\tfrac{1}{2}$

∴ The *inferior* limit of y is 5 and the *superior* limit of x is $5\tfrac{1}{2}$. Consequently, y may $= 6$ and x may $= 5$, and these values satisfy the equation.

Find the limit of x in the following: a and b being positive:

3. $4x - 7 > 21$.

4. $2x + \dfrac{3x}{2} - 17 > 11$.

5. $\dfrac{ax}{7} + bx - ab > \dfrac{a^2}{7}$.

6. $x + \dfrac{x}{2} + \dfrac{x}{3} + \dfrac{x}{6} > 12$.

7. $\dfrac{a^2 + b^2}{15} > \dfrac{a^2 + b^2}{5x}$.

8. $11x - 9 > 4x + 33$.

Find limits of x and y in the following, and, if possible, integral values of x and y which will satisfy the equations:

9. $3x + y = 9$.
$x + 2y < 12$.

10. $4x + 2y = 26$.
$3x + 4y > 38$.

11. $x + 6y = 13$.
$5x + 2y > 7$.

12. $\dfrac{x}{2} + \dfrac{y}{3} = 3$.
$\dfrac{x}{5} + \dfrac{y}{2} > 2$.

13. $\dfrac{5x}{6} + \dfrac{2y}{5} = 14$.
$\dfrac{3x}{4} - \dfrac{2y}{5} > 4$.

14. $\dfrac{3x}{5} + \dfrac{7y}{4} = 10$.
$\dfrac{2x}{7} - \dfrac{y}{5} < \dfrac{2}{3}$.

15. If a and b are unequal, prove that $a^2 + b^2 > 2ab$.

16. If a and b are unequal and positive, find the inferior limit of $\dfrac{a}{b} + \dfrac{b}{a}$.

17. When a and b are unequal and positive, which is greater, $\dfrac{a^2 + b^2}{a + b}$ or $\dfrac{a^3 + b^3}{a^2 + b^2}$?

LOGARITHMS.

390. 1. What power of 3 is 9? 27? 81? 243?

2. What power of 4 is 4? 16? 64? 256?

3. What power of 10 is 10? 100? 1000? 1?

391. The **Logarithm** of a number is the index of the power to which a constant number must be raised to produce the given number.

Thus, when 4 is the constant number, 2 is the logarithm of 16, for $4^2 = 16$.

392. The constant number which must be raised to some power in order to produce the given numbers is called the **Base** of the system of logarithms.

393. Logarithms may be computed with any positive number except unity as a base, but the base of the **Common System of Logarithms** is 10.

Since $10^0 = 1$, the logarithm of 1 is 0.
Since $10^1 = 10$, the logarithm of 10 is 1.
Since $10^2 = 100$, the logarithm of 100 is 2.
Since $10^3 = 1000$, the logarithm of 1000 is 3.
Since $10^{-1} = \frac{1}{10}$, the logarithm of .1 is -1.
Since $10^{-2} = \frac{1}{100}$, the logarithm of .01 is -2.
Since $10^{-3} = \frac{1}{1000}$, the logarithm of .001 is -3.

394. It is evident, therefore, that the logarithm of any number between 1 and 10 is less than 1 and greater than 0; between 10 and 100, 1 plus a fraction; between 100 and 1000, 2 plus a fraction, etc.

LOGARITHMS.

395. The integral part of a logarithm is called the **Characteristic**; the fractional part, the **Mantissa**.

Thus, in log. 3.16857, the characteristic is 3 and the mantissa .16857.

From the examples given in Art. 390, the following principles may be deduced.

396. Principles. — 1. *The characteristic of the logarithm of an integral number is positive, and numerically 1 less than the number of figures in the given number.*

2. *The characteristic of the logarithm of a decimal fraction is negative, and numerically 1 greater than the number of zeros immediately following the decimal point.*

Thus, the characteristic of 42 is 1; of 423 is 2; of 4234 is 3; of .01 is -2; of .42 is -1; of .324 is -1; of .00325 is -3.

397. The following examples will illustrate the *characteristic* and *mantissa*, and their significance:

Log. of $231.4 = 2.364363$, or $231.4 = 10^{2.364363}$.

Log. of $23.14 = 1.364363$, or $23.14 = 10^{1.364363}$.

Log. of $2.314 = 0.364363$, or $2.314 = 10^{.364363}$.

Log. of $.2314 = \bar{1}.364363$, or $.2314 = 10^{\bar{1}.364363}$.

Log. of $.02314 = \bar{2}.364363$, or $.02314 = 10^{\bar{2}.364363}$.

From an examination of the examples given it is seen that as the number decreases in a tenfold ratio, the logarithm is diminished by unity. The logarithm of .2314 is then $0.364363 - 1$ or -0.635637, and that of $.02314 - 1.635637$. But it is found convenient to write it $\bar{1}.364363$, the understanding being that the *characteristic only is negative*. The mantissa is always positive. It is evident also that in the logarithms of numbers expressed by the same figures, the decimal part, or *mantissa*, is the same, and the logarithms differ only in the *characteristic*. Hence, tables of logarithms of numbers contain only the *mantissas*.

TABLES OF LOGARITHMS.

398. The tables of logarithms on the next two pages give the *decimal part,* or *mantissa,* of the common logarithms of all numbers from 1 to 999 extended to five decimal places.

The logarithms given in the tables begin with the mantissa of 10, but since the mantissas of 10, 20, 30, 40, etc., are the same as the mantissas of 1, 2, 3, 4, etc., the table may be said to give the logarithms of numbers from 1 to 1000.

Explanation of Tables.

The left-hand column of each page of the table is a column of numbers. It is designated by **N**.

The mantissas of the logarithms of these numbers are opposite them in the next column.

At the top of each page and extending across the top are found the figures **0, 1, 2, 3, 4, 5, 6, 7, 8, 9,** each standing over a column of figures. These figures are the right-hand figures of numbers whose left-hand figures are given in the left-hand column, and the figures under them are the corresponding mantissas of the numbers.

It will be seen that the first column of mantissas contains five figures, while the others contain only four. This difference is due to the fact that the left-hand figure in the mantissas, which is *usually* the same for a whole horizontal column, is omitted except in the first column. When, however, the first figure of the mantissa in any column after the first is 0, and the corresponding figure in the previous column is 9, the left-hand figure for this mantissa and all others following it in the same horizontal line is one greater than the first figure of the left-hand column of mantissas in this line.

By subtracting these mantissas, each from the one next succeeding, it is found that those in the same horizontal line have nearly the same difference.

This *Average Difference* is found in the column marked **D**.

It is evident that any tables of logarithms, however extensive, cannot be *absolutely* accurate. They will approach accuracy, however, in proportion as the number of decimal places in the mantissa is increased.

The tables given here are designed to exhibit the methods of computation with logarithms, but fuller ones are required in actual practice.

LOGARITHMS.

Table of Common Logarithms.

N.	0	1	2	3	4	5	6	7	8	9	D.
10	00000	0432	0860	1284	1703	2119	2531	2938	3342	3743	414
11	04139	4532	4922	5308	5690	6070	6446	6819	7188	7555	378
12	07918	8279	8636	8991	9342	9691	0037	0380	0721	1059	348
13	11394	1727	2057	2385	2710	3033	3354	3672	3988	4301	322
14	14613	4922	5229	5534	5836	6137	6435	6732	7026	7319	300
15	17609	7898	8184	8469	8752	9033	9312	9590	9866	0140	280
16	20412	0683	0952	1219	1484	1748	2011	2272	2531	2789	263
17	23045	3300	3553	3805	4055	4304	4551	4797	5042	5285	248
18	25527	5768	6007	6245	6482	6717	6951	7184	7416	7646	235
19	27875	8103	8330	8556	8780	9003	9226	9447	9667	9885	223
20	30103	0320	0535	0750	0963	1175	1387	1597	1806	2015	212
21	32222	2428	2634	2838	3041	3244	3445	3646	3846	4044	202
22	34242	4439	4635	4830	5025	5218	5411	5603	5793	5984	193
23	36173	6361	6549	6736	6922	7107	7291	7475	7658	7840	185
24	38021	8202	8382	8561	8739	8917	9094	9270	9445	9620	177
25	39794	9967	0140	0312	0483	0654	0824	0993	1162	1330	170
26	41497	1664	1830	1996	2160	2325	2488	2651	2813	2975	164
27	43136	3297	3457	3616	3775	3933	4091	4248	4404	4560	158
28	44716	4871	5025	5179	5332	5484	5637	5788	5939	6090	152
29	46240	6389	6538	6687	6835	6982	7129	7276	7422	7567	147
30	47712	7857	8001	8144	8287	8430	8572	8714	8855	8996	142
31	49136	9276	9415	9554	9693	9831	9969	0106	0243	0379	138
32	50515	0651	0786	0920	1055	1188	1322	1455	1587	1720	134
33	51851	1983	2114	2244	2375	2504	2634	2763	2892	3020	130
34	53148	3275	3403	3529	3656	3782	3908	4033	4158	4283	126
35	54407	4531	4654	4777	4900	5023	5145	5267	5388	5509	122
36	55630	5751	5871	5991	6110	6229	6348	6467	6585	6703	119
37	56820	6937	7054	7171	7287	7403	7519	7634	7749	7864	116
38	57978	8092	8206	8320	8433	8546	8659	8771	8883	8995	113
39	59106	9218	9329	9439	9550	9660	9770	9879	9988	0097	110
40	60206	0314	0423	0531	0639	0746	0853	0959	1066	1172	107
41	61278	1384	1490	1595	1700	1805	1909	2014	2118	2221	105
42	62325	2428	2531	2634	2737	2839	2941	3043	3144	3246	102
43	63347	3448	3548	3649	3749	3849	3949	4048	4147	4246	100
44	64345	4444	4542	4640	4738	4836	4933	5031	5128	5225	98
45	65321	5418	5514	5610	5706	5801	5896	5992	6087	6181	95
46	66276	6370	6464	6558	6652	6745	6839	6932	7025	7117	93
47	67210	7302	7394	7486	7578	7669	7761	7852	7943	8034	91
48	68124	8215	8305	8395	8485	8574	8664	8753	8842	8931	90
49	69020	9108	9197	9285	9373	9461	9548	9636	9723	9810	88
50	69897	9984	0070	0157	0243	0329	0415	0501	0586	0672	86
51	70757	0842	0927	1012	1096	1181	1265	1349	1433	1517	84
52	71600	1684	1767	1850	1933	2016	2099	2181	2263	2346	83
53	72428	2509	2591	2673	2754	2835	2916	2997	3078	3159	81
54	73239	3320	3400	3480	3560	3640	3719	3799	3878	3957	80

HIGH SCHOOL ALGEBRA.

TABLE OF COMMON LOGARITHMS.

N.	0	1	2	3	4	5	6	7	8	9	D.
55	74036	4115	4194	4273	4351	4429	4507	4586	4663	4741	78
56	74819	4896	4974	5051	5128	5205	5282	5358	5435	5511	77
57	75587	5664	5740	5815	5891	5967	6042	6118	6193	6268	76
58	76343	6418	6492	6567	6641	6716	6790	6864	6938	7012	74
59	77085	7159	7232	7305	7379	7452	7525	7597	7670	7743	73
60	77815	7887	7960	8032	8104	8176	8247	8319	8390	8462	72
61	78533	8604	8675	8746	8817	8888	8958	9029	9099	9169	71
62	79239	9309	9379	9449	9518	9588	9657	9727	9796	9865	69
63	79934	0003	0072	0140	0209	0277	0346	0414	0482	0550	68
64	80618	0686	0754	0821	0889	0956	1023	1090	1158	1224	67
65	81291	1358	1425	1491	1558	1624	1690	1757	1823	1889	66
66	81954	2020	2086	2151	2217	2282	2347	2413	2478	2543	65
67	82607	2672	2737	2802	2866	2930	2995	3059	3123	3187	64
68	83251	3315	3378	3442	3506	3569	3632	3696	3759	3822	63
69	83885	3948	4011	4073	4136	4198	4261	4323	4386	4448	62
70	84510	4572	4634	4696	4757	4819	4880	4942	5003	5065	62
71	85126	5187	5248	5309	5370	5431	5491	5552	5612	5673	61
72	85733	5794	5854	5914	5974	6034	6094	6153	6213	6273	60
73	86332	6392	6451	6510	6570	6629	6688	6747	6806	6864	59
74	86923	6982	7040	7099	7157	7216	7274	7332	7390	7448	58
75	87506	7564	7622	7679	7737	7795	7852	7910	7967	8024	58
76	88081	8138	8195	8252	8309	8366	8423	8480	8536	8593	57
77	88649	8705	8762	8818	8874	8930	8986	9042	9098	9154	56
78	89209	9265	9321	9376	9432	9487	9542	9597	9653	9708	55
79	89763	9818	9873	9927	9982	0037	0091	0146	0200	0255	55
80	90309	0363	0417	0472	0526	0580	0634	0687	0741	0795	54
81	90849	0902	0956	1009	1062	1116	1169	1222	1275	1328	53
82	91381	1434	1487	1540	1593	1645	1698	1751	1803	1855	53
83	91908	1960	2012	2065	2117	2169	2221	2273	2324	2376	52
84	92428	2480	2531	2583	2634	2686	2737	2788	2840	2891	51
85	92942	2993	3044	3095	3146	3197	3247	3298	3349	3399	51
86	93450	3500	3551	3601	3651	3702	3752	3802	3852	3902	50
87	93952	4002	4052	4101	4151	4201	4250	4300	4349	4399	50
88	94448	4498	4547	4596	4645	4694	4743	4792	4841	4890	49
89	94939	4988	5036	5085	5134	5182	5231	5279	5328	5376	49
90	95424	5472	5521	5569	5617	5665	5713	5761	5809	5856	48
91	95904	5952	5999	6047	6095	6142	6190	6237	6284	6332	47
92	96379	6426	6473	6520	6567	6614	6661	6708	6755	6802	47
93	96848	6895	6942	6988	7035	7081	7128	7174	7220	7267	46
94	97313	7359	7405	7451	7497	7543	7589	7635	7681	7727	46
95	97772	7818	7864	7909	7955	8000	8046	8091	8137	8182	45
96	98227	8272	8318	8363	8408	8453	8498	8543	8588	8632	45
97	98677	8722	8767	8811	8856	8900	8945	8989	9034	9078	45
98	99123	9167	9211	9255	9300	9344	9388	9432	9476	9520	44
99	99564	9607	9651	9695	9739	9782	9826	9870	9913	9957	44

LOGARITHMS.

399. To find the logarithm of a number.

1. Find the logarithm of 3824.

EXPLANATION. — Since the tables on the preceding pages contain the mantissas of no numbers expressed by more than three figures, the mantissa of 382 is first found, which is the same as the mantissa of 3820. It is found to be .58206.

Since the mantissa of the next larger number, 383 or 3830, is 114 hundred-thousandths greater than the mantissa of 3820, it may be assumed that every unit added to 3820 will add .1 of 114 hundred-thousandths to the mantissa, and 4 will add .4 of 114 hundred-thousandths, or 46 hundred-thousandths. This added to .58206 gives .58252, the mantissa of 382, which is sufficiently accurate for practical purposes.

Since the number is expressed by 4 figures, the characteristic is 3.
Therefore, the logarithm of 3824 is 3.58252.

Find the logarithms of:

2. 318.	16. 2105.	30. 8966.	44. 540.6.
3. 285.	17. 1508.	31. 7847.	45. 78.99.
4. 486.	18. 3054.	32. 3521.	46. 8.754.
5. 335.	19. 2247.	33. 4712.	47. 6278.
6. 33.6.	20. 3156.	34. 8363.	48. .0111.
7. 2.68.	21. 2984.	35. 9274.	49. .0001.
8. .384.	22. 4138.	36. 4995.	50. .3001.
9. 4831.	23. 3645.	37. 6726.	51. .0142.
10. 3846.	24. 5039.	38. 5.787.	52. .0201.
11. 2785.	25. 3755.	39. 62.88.	53. 2451.
12. 3169.	26. 4167.	40. 423.9.	54. 4367.
13. 1875.	27. 3483.	41. 7620.	55. 8590.
14. 2.345.	28. 5.718.	42. 83.43.	56. .0078.
15. 1.684.	29. 62.45.	43. 8.562.	57. .0142.

400. To find a number whose logarithm is given.

1. Find the number whose logarithm is 3.95323.

PROCESS.

Given log.,	3.95323
Log. next less,	3.95279
Difference of logs.,	44
Tabular difference,	49

$$44 \div 49 = .89+$$

Number corresponding to mantissa, .95279 is 897.

Annexing to 897 the rest of number, .89+, the whole number is 8978.9+, since it is expressed by four figures.

EXPLANATION. — The logarithm next less than the given logarithm is 3.95279. This, subtracted from the given logarithm, gives 44 as a remainder. Since the difference between the logarithms is .00044, and the average difference between the logarithms on this line as given in the table is .00049 for each unit of increase in the numbers, it will be sufficiently accurate to assume that an addition of $\frac{44}{49}$ of a unit must be made to the number whose logarithm is next less, so that the number corresponding to the given logarithm may be found. The addition is .89. The number corresponding to the logarithm 3.95279 consists of 4 integral figures, the first three of which are found from the table to be 897. Annexing the part found by dividing the difference of the logarithms by the average difference, the number is 8978.9+.

Find the numbers corresponding to the following:

2. 2.38257.
3. 2.18625.
4. 0.23146.
5. $\overline{1}$.28643.
6. $\overline{2}$.98465.
7. 3.18425.
8. 2.86435.
9. $\overline{3}$.24685.
10. 2.98456.
11. 2.58643.
12. 3.94875.
13. 0.21654.
14. 2.72586.
15. $\overline{1}$.21436.
16. 2.51326.
17. 3.14215.
18. $\overline{2}$.64823.
19. 4.80167.

401. Multiplication by logarithms.

Since logarithms are the exponents of the powers to which a constant quantity is to be raised, how may the product of quantities be found when their logarithms are known?

LOGARITHMS.

1. Multiply 32.4 by 26.

PROCESS.

Log. of 32.4 = 1.51055
Log. of 26 = 1.41497
───────────────────
Sum of logs. = 2.92552

2.92552 is log. of 842.4.

∴ 32.4 × 26 = 842.4.

EXPLANATION. — We find the logarithm of each of the given numbers, and, inasmuch as the logarithms are exponents of a constant quantity, the product of these numbers will be the constant quantity, with an exponent equal to the sum of the exponents of this constant quantity. The sum of these exponents or logarithms is 2.92552. The number corresponding to this logarithm is 842.4, the product of the numbers.

NOTE. — The student should bear in mind that logarithms cannot be relied upon to give accurate results with large numbers. The logarithms are themselves but approximations toward accuracy, and they should therefore be used only when approximate results will answer the purpose, or where the ordinary processes of arithmetic cannot be applied.

Multiply:

2. 2.3 by 3.7.
3. 25 by 3.5.
4. 216 by 3.5.
5. 312 by .24.
6. 123 by 3.4.
7. 2.24 by 2.6.
8. .0023 by .26.
9. .0015 by .015.
10. .034 by 26.
11. .057 by 5.7.
12. .068 by 7.2.
13. .0018 by .25.
14. .0432 by .082.
15. 856.7 by 1.38.
16. .0796 by .152.
17. 4876 by 37.5.
18. 3.04 by .003.
19. 468 by 34.5.
20. .002 by .008.
21. 3456 by .035.
22. 2874 by .836.
23. 41.03 by .507.
24. .0009 by .708.
25. 8976 by 2.04.

402. Division by logarithms.

Since in multiplication we add the logarithms, or the exponents, of the constant quantity, how may division be performed?

1. Divide .05475 by 15.

PROCESS.

Log. of .05475 is $\bar{2}$.73839
Log. of 15 is 1.17609
───────────────────
Difference of logs. is $\bar{3}$.56230

$\bar{3}$.56230 is log. of .00365.

∴ .05475 ÷ 15 = .00365.

EXPLANATION. — We find the logarithm of each number, and then subtract the logarithm of the divisor from that of the dividend. The number corresponding to this difference between the logarithms is the quotient.

322   HIGH SCHOOL ALGEBRA.

Divide:

2. 2.45 by 9.8.	9. 34.43 by .011.	16. 3.485 by .205.
3. 18.312 by 24.	10. 259.2 by .012.	17. 7.865 by .013.
4. 105.7 by 3.5.	11. 87.36 by 2.1.	18. 8925 by 42.5.
5. 135.05 by .037.	12. 97.24 by .022.	19. 3250 by .52.
6. .04905 by .327.	13. 13.696 by 32.	20. .21084 by .042.
7. 69854 by 217.	14. 216.83 by 2.3.	21. 2938 by 14.2.
8. 823.68 by 307.	15. 38142 by 3.7.	22. 3685 by .273.

403. Involution by logarithms.

Since logarithms are exponents, how may quantities, whose logarithms are known, be raised to any power?

1. What is the second power of 25?

PROCESS.

Log. of 25 is 1.39794
2
———
Log. of the power is 2.79588

2.79588 is log of 625.
$\therefore (25)^2 = 625.$

EXPLANATION. — Since in involution we multiply the exponent of the quantity by the exponent of the power to which it is to be raised, in involution by logarithms we may find the logarithm of the given quantity, and multiply it by the exponent of the power to which it is to be raised; the number corresponding to the resulting logarithm will be the power sought.

Find by logarithms the values of the following:

2. 65^2.	9. 91^3.	16. 6.75^3.	23. 3.040^3.
3. 84^2.	10. 102^3.	17. 723^4.	24. $.0065^3$.
4. 75^3.	11. 107^4.	18. 799^3.	25. $.0862^2$.
5. 64^4.	12. 146^3.	19. 8.76^2.	26. 768.4^2.
6. 87^3.	13. 475^2.	20. $.960^3$.	27. 4.132^2.
7. 37^3.	14. 612^2.	21. 3.54^3.	28. 5.184^2.
8. 49^4.	15. 584^3.	22. 29.3^2.	29. 37.25^3.

LOGARITHMS.

404. Evolution by logarithms.

Since in involution the logarithms, or exponents, are multiplied by the index of the power to produce the power, what must be done when roots of numbers are to be extracted?

1. What is the square root of 625?

PROCESS.

Log. of 625 is 2.79588
Dividing by 2, 1.39794
1.39794 is the log. of 25.
$\therefore (625)^{\frac{1}{2}} = 25.$

EXPLANATION. — Since in evolution we divide the exponent of the quantity by the number corresponding to the root to be extracted, in evolution by logarithms we find the logarithm of the given number, and divide it by the index of the required root; the number corresponding to the resulting logarithm will be the root sought.

Find by logarithms the indicated roots of the following:

2. $196^{\frac{1}{2}}.$ **6.** $1681^{\frac{1}{2}}.$ **10.** $13824^{\frac{1}{3}}.$

3. $256^{\frac{1}{2}}.$ **7.** $3969^{\frac{1}{2}}.$ **11.** $15.625^{\frac{1}{3}}.$

4. $576^{\frac{1}{2}}.$ **8.** $4096^{\frac{1}{3}}.$ **12.** $74088^{\frac{1}{3}}.$

5. $1156^{\frac{1}{2}}.$ **9.** $5184^{\frac{1}{2}}.$ **13.** $91125^{\frac{1}{3}}.$

Find the values of the following to three decimal places:

14. $\sqrt[3]{29}.$ **18.** $\sqrt[7]{15}.$ **22.** $\sqrt[15]{973}.$

15. $\sqrt[3]{65}.$ **19.** $\sqrt[8]{67}.$ **23.** $\sqrt[3]{3.27}.$

16. $\sqrt[4]{87}.$ **20.** $\sqrt[9]{105}.$ **24.** $\sqrt[5]{9.181}.$

17. $\sqrt[5]{96}.$ **21.** $\sqrt[10]{111}.$ **25.** $\sqrt[11]{62.73}.$

26. $\dfrac{\sqrt{3} + \sqrt[3]{2}}{\sqrt{2} + \sqrt[3]{3}}.$ **28.** $\dfrac{\sqrt{7} + \sqrt[5]{9}}{\sqrt[3]{12} - \sqrt[5]{17}}.$

27. $\dfrac{\sqrt[3]{5} - \sqrt[4]{8}}{\sqrt{7} + \sqrt[3]{2}}.$ **29.** $\dfrac{\sqrt{6} + \sqrt[3]{9}}{\sqrt[3]{30} - \sqrt[4]{49}}.$

405. Logarithms applied to the solution of problems in compound interest and annuities.

1. What is the amount of $1 for 1 year at 6%? By what must the amount for 1 year be multiplied to find the amount for 2 years at 6% compound interest?

2. By what must the amount for 2 years be multiplied to obtain the amount for 3 years compound interest? By what must the amount for 3 years be multiplied to obtain the amount for 4 years compound interest?

3. Since the amount of $1 at 6% compound interest for 1 year is 1.06 times the principal; for 2 years, 1.06×1.06 times the principal; for 3 years, $1.06 \times 1.06 \times 1.06$ or $(1.06)^3$ times the principal, what will be the amount (A) of any principal (P) for n years at any rate per cent (r)?

FORMULA. $\qquad A = P(1 + r)^n.$

Expressing the formula by logarithms:

$$\log A = \log P + n \times \log(1 + r) \qquad (1)$$

$$\therefore \log P = \log A - n \times \log(1 + r) \qquad (2)$$

also, $\qquad \log(1 + r) = \dfrac{\log A - \log P}{n} \qquad (3)$

and $\qquad n = \dfrac{\log A - \log P}{\log(1 + r)} \qquad (4)$

EXAMPLES.

406. 1. What will be the amount of $650 loaned for 5 years at 7% compound interest?

SOLUTION. — To find the amount formula (1) is used:

$$\log \text{ of } (1 + r) \text{ or } 1.07 = 0.02938$$
$$n = 5$$
$$n \times \log(1 + r) = \overline{0.14690}$$
$$\log \text{ of } P \text{ or } 650 = 2.81291$$
$$\therefore \log \text{ of } A = \overline{2.95981}$$

The number corresponding to log 2.95981 is 911.62.
∴ The amount is $911.62.

LOGARITHMS.

2. What will be the amount of $800 loaned for 4 years at 6% compound interest?

3. How much will $580 amount to in 5 years at 5% compound interest?

4. Find the amount of $475 for 8 years at 5% compound interest.

5. What principal will amount to $500 in 7 years at 6% compound interest?

6. What principal at 5% compound interest will amount to $800 in 8 years?

7. What must be the sum of money loaned that will amount to $1000 at 4% compound interest in 10 years?

8. What is the rate per cent when $500 loaned at compound interest amounts to $680.30 in 7 years?

9. A man agreed to loan $1000 at 6% compound interest for a time sufficiently long for the principal to double itself. How long was the money on interest?

407. An **Annuity** is a sum of money to be paid annually for a given number of years, during the life of a person, or forever.

Payment upon annuities is often deferred for a number of years, and interest is allowed upon the sums due and unpaid.

408. To find the amount of an annuity left unpaid for a number of years, compound interest being allowed.

1. Suppose an annuity of a dollars is left unpaid, how much is due at the end of the first year?

2. Upon what sum will compound interest be computed at the end of the *second* year? What will be the amount of that sum when the rate is r? What will be the whole sum due at the end of the *second* year?   *Ans.* $a + a(1+r)$.

3. Upon what sum will compound interest be computed at

the end of the *third* year? What will be the amount of that sum at the given rate? *Ans.* $a(1+r)+a(1+r)^2$.

What will be the whole sum due at the end of the third year? *Ans.* $a+a(1+r)+a(1+r)^2$.

4. What will be the whole sum due at the end of the *fourth* year? At the end of the nth year?

Let a represent the annuity, n the number of years, r the rate, and A the amount. Then,

$$A = a + a(1+r) + a(1+r)^2 + a(1+r)^3 + \cdots + a(1+r)^{n-1}$$
$$= a\{1 + (1+r) + (1+r)^2 + (1+r)^3 + \cdots + (1+r)^{n-1}\}.$$

These terms form a geometrical progression in which $1+r$ is the ratio. By Art. 358 the sum of the series is obtained, and

$$A = a \cdot \frac{(1+r)^n - 1}{r}$$

Sometimes annuities, drawing interest, are not payable until after a certain number of years. It is often necessary, therefore, to find the present value of such annuities.

409. The Present Worth of the annuity is a sum which, if it were put at interest at the given rate for the given time, would amount to the value of the annuity at the end of the given time.

1. If P denotes the present worth of an annuity due in n years, allowing $r\%$ compound interest, to what sum will P be equal in that time at the given rate? *Ans.* $P(1+r)^n$.

2. Since the amount of the present worth put at interest for the given time and rate is equal to the amount of the annuity for the same time and rate, equate the two sums and find the value of P.

$$P(1+r)^n = a \cdot \frac{(1+r)^n - 1}{r}$$

$$\therefore P = a \cdot \frac{(1+r)^n - 1}{r(1+r)^n}$$

$$= \frac{a}{r} \cdot \frac{(1+r)^n - 1}{(1+r)^n}$$

LOGARITHMS.

EXAMPLES.

1. What will be the amount of an annuity of $650 remaining unpaid for 10 years at 5 % compound interest?

Solution.　　By logarithms $(1.05)^{10} = 1.62889$

$$(1.05)^{10} - 1 = .62889$$

log of $650 = 2.81291$

log of $.62889 = \overline{1}.79858$

sum of logs $= 2.61149$

log of $.05 = \overline{2}.69897$

dif. of logs $= 3.91252$

The number corresponding to the log 3.91252 is 8175.62. Therefore, the amount of the annuity is $8175.62.

2. Find the present worth of an annuity of $650 to continue 10 years, allowing compound interest at 5 %.

Solution.　　By logarithms $(1.05)^{10} = 1.62889$

$$(1.05)^{10} - 1 = .62889$$

log of $.62889 = \overline{1}.79858$

log of $(1.05)^{10} = 0.02119$

dif. of logs $= \overline{1}.77739$

$\dfrac{a}{r} = \dfrac{650}{05} = 13000$, whose log $= 4.11394$

sum of logs $= 3.89133$

The number corresponding to the log 3.89133 is 7786.25. Therefore, the present worth of the annuity is $7786.25.

3. What will be the amount of an annuity of $700 remaining unpaid for 12 years at 4 % compound interest?

4. To what sum will an annuity of $144 amount in 20 years at 6 % compound interest?

5. What is the present worth of an annuity of $1000 for 5 years at 4 % compound interest?

6. What will be the amount of an annuity of $1200 remaining unpaid for 6 years at 6 % compound interest?

7. What is the present worth of an annuity of $800 to continue 8 years, allowing compound interest at 4 %?

8. What is the present worth of an annuity of $1000 to continue 20 years, allowing compound interest at 5 %?

PERMUTATIONS AND COMBINATIONS.

410. Permutations of things are the different orders in which things can be arranged by placing them in every possible order.

In forming the permutations a *part* or the *whole* of the things may be taken at a time.

411. To find the number of permutations.

1. Find the permutations of the letters a, b, c, taking 2 at a time.

2. Since they are ab, ac, ba, bc, ca, cb, in forming them what letters are written after a? after b? after c?

3. Since each of the other letters is written after a, b, and c, respectively, with how many letters is each letter arranged when the letters are taken 2 at a time? How may the number of permutations be found when the letters are taken 2 at a time?

4. What will be the number of permutations of 4 letters taken 2 at a time? Of 5 letters taken 2 at a time? Of 6 letters taken 2 at a time? Of n letters taken 2 at a time?

Ans. $n(n-1)$.

5. Find the permutations of the letters a, b, c, taking 3 at a time.

6. Since they are abc, acb, bac, bca, cab, cba, how may they be formed from the permutations of the same letters taken 2 at a time? How many letters are remaining to annex to each of the permutations taken 2 at a time?

7. Write the permutations of 4 letters taken 2 at a time. How many other letters are left to annex to each of the permutations when 2 are taken at a time to form permutations taken 3 at a time? How many letters are left to combine with each of the permutations of 2 letters at a time to form permutations taken 3 at a time when there are 5 letters?

PERMUTATIONS AND COMBINATIONS.

When there are 6 letters taken 3 at a time? When there are n letters taken 3 at a time?   *Ans.* $n-2$.

8. Since with each of the permutations of n letters taken 2 at a time, $n-2$ other letters are annexed to form the permutations of the letters taken 3 at a time, how many permutations are there of n letters taken 3 at a time?

Ans. $n(n-1)(n-2)$.

9. How many permutations are there when n things are arranged 4 in a set, or taken 4 at a time?

Ans. $n(n-1)(n-2)(n-3)$.

10. How many permutations are there when n things are arranged r in a set?

Ans. $n(n-1)(n-2)(n-3)\cdots n-(r-1)$.

412. PRINCIPLE 1. — *The number of permutations of n things taken r at a time is equal to the continued product of the natural numbers from n to $n-(r-1)$ inclusive.*

FORMULA. $P_r = n(n-1)(n-2)\cdots n-(r-1)$.

When all the letters are taken together, then r is equal to n, and the last factor becomes 1. Therefore

413. PRINCIPLE 2. — *The number of permutations of n things taken n at a time is equal to the continued product of the natural numbers from n to 1 inclusive.*

FORMULA. $P_n = n(n-1)(n-2)\cdots 1$.

414. To find the number of combinations.

415. Combinations are the different ways in which a selection of a certain number of things can be made from a given number of things without regard to the order in which the things are placed.

Thus, a and b admit of *two* permutations when taken 2 at a time; viz., ab and ba, but only *one* combination ab; a, b, c admit, taken 3 at a time, of 6 permutations, abc, acb, bac, bca, cab, cba, but *one* combination abc.

1. How many permutations are there of 4 things taken 2 at a time? How many combinations are there?

2. How many permutations are there of 4 things taken 3 at a time? How many combinations are there?

3. How many permutations are there of 4 things taken 4 at a time? How many combinations are there?

4. Since the number of permutations of *n* things taken 2 at a time is twice the number of combinations, the number of permutations of *n* things taken 3 at a time 6 times the number of combinations, etc., how may the number of *combinations* of *n* things taken *r* in a set be obtained from the number of *permutations* of *n* things taken *r* in a set?

416. PRINCIPLE. — *The number of combinations of n things taken r at a time is equal to the number of permutations of n things taken r at a time divided by the number of permutations of r things taken altogether.*

FORMULA. $$C_r = \frac{n(n-1)(n-2)\cdots n-(r-1)}{r(r-1)(r-2)\cdots 1}.$$

EXAMPLES.

1. How many different permutations may be formed of 9 different letters taken 4 at a time?

2. How many different permutations may be formed of 8 things taken 5 at a time?

3. How many different combinations may be formed of 8 things taken 4 at a time?

4. How many changes may be rung with 6 bells out of 8?

5. How many different integral numbers may be expressed by writing the first five significant digits in succession, each figure to be taken once, and only once, in each number?

6. The colors of the rainbow are violet, indigo, blue, green, yellow, orange, and red. In how many different orders may these colors be arranged?

COMBINATIONS AND PERMUTATIONS.

7. From 12 books, how many ways can a selection of 4 books be made?

8. How many different sums of money can be paid with a cent, a three cent piece, a half dime, a dime, and a twenty-five cent piece?

9. From a company of 18 persons, how many different parties of 6 each can be formed?

10. A certain family consists of father, mother, 3 sons, and 5 daughters. In how many different orders may they arrange themselves around the dinner table?

11. If a single combination of 6 things is formed from 10 different things, how many things are left?

12. Since, whenever a combination of 6 things is formed from 10 different things, a combination of 4 things can be formed from those that are left, how will the number of combinations of 10 different things taken 6 at a time compare with the number of combinations of 10 different things taken (10 − 6) or 4 at a time?

13. How, then, will the number of combinations of n things taken r at a time compare with the number of combinations of n things taken $n - r$ at a time?

14. Express the conclusion in the form of a principle.

The application of this principle will sometimes abridge the process employed in finding the number of combinations.

Thus, if it is required to find the number of combinations of 18 things, taken 16 at a time, the solution by Art. 416 will be

$$\frac{18 \cdot 17 \cdot 16 \cdot 15 \cdot 14 \cdot 13 \cdot 12 \cdot 11 \cdot 10 \cdot 9 \cdot 8 \cdot 7 \cdot 6 \cdot 5 \cdot 4 \cdot 3}{16 \cdot 15 \cdot 14 \cdot 13 \cdot 12 \cdot 11 \cdot 10 \cdot 9 \cdot 8 \cdot 7 \cdot 6 \cdot 5 \cdot 4 \cdot 3 \cdot 2 \cdot 1} = 153.$$

But by the principle just established, the number of combinations of 18 things taken 16 at a time is equal to the number of combinations of 18 things taken 2 at a time, consequently the solution will be

$$\frac{18 \cdot 17}{1 \cdot 2} = 153.$$

THE BINOMIAL THEOREM.

417. The Binomial Theorem derives a formula by means of which any binomial, whether its exponent be positive or negative, integral or fractional, can be expanded without employing the ordinary process of involution.

POSITIVE INTEGRAL EXPONENTS.

418. From the binomials whose powers are given in Art. 202, certain Principles were derived.

Assuming that those principles or laws are applicable to the expansion of any binomial, as $(a + x)$, having a positive integral exponent, as n, we have

$$(a+x)^n = a^n + na^{n-1}x + \frac{n(n-1)}{1 \cdot 2}a^{n-2}x^2$$
$$+ \frac{n(n-1)(n-2)}{1 \cdot 2 \cdot 3}a^{n-3}x^3 + \cdots \qquad (1)$$

This formula is called the **Binomial Formula**.

419. We know that the laws are applicable to the expansion of binomials to the third, fourth, and fifth powers; we shall now show that they are true for any power.

It is plain that if they hold true for the $(n+1)$th power, they hold true for the sixth power, since we know that they hold true for the fifth; and if they are true for the sixth power, they are true for the seventh, and so on for any power with a positive integral exponent, inasmuch as the $(n+1)$th power is a power whose exponent is one greater than any assumed power.

THE BINOMIAL THEOREM.

420. Assuming formula (1) to be the expansion of $(a+x)^n$, if each member is multiplied by $(a+x)$, we shall have the expansion of $(a+x)^{n+1}$. Multiplying, we have

$$(a+x)^{n+1} = a^{n+1} + na^n x + \frac{n(n-1)}{1 \cdot 2} a^{n-1} x^2 + \frac{n(n-1)(n-2)}{1 \cdot 2 \cdot 3} a^{n-2} x^3$$

$$+ \cdots + a^n x + na^{n-1} x^2 + \frac{n(n-1)}{1 \cdot 2} a^{n-2} x^3 + \cdots \quad (2)$$

Uniting similar terms,

$$= a^{n+1} + (n+1)a^n x + \left(\frac{n(n-1)}{1 \cdot 2} + n\right) a^{n-1} x^2$$

$$+ \left(\frac{n(n-1)(n-2)}{1 \cdot 2 \cdot 3} + \frac{n(n-1)}{1 \cdot 2}\right) a^{n-2} x^3$$

$$= a^{n+1} + (n+1)a^n x + \frac{n(n-1+2)}{1 \cdot 2} a^{n-1} x^2$$

$$+ n(n-1)\left(\frac{n-2+3}{1 \cdot 2 \cdot 3}\right) + \cdots$$

$$= a^{n+1} + (n+1)a^n x + n \frac{(n+1)}{1 \cdot 2} a^{n-1} x^2$$

$$+ \frac{(n+1)n(n-1)}{1 \cdot 2 \cdot 3} a^{n-2} x^3 + \cdots$$

By examining the formula as it is given in its last reduction, it is seen that it has the same form as the expansion of $(a+x)^n$, $n+1$ simply taking the place of n. Substituting $n+1$ for n in equation (1), it reduces to the form of equation (2) as expressed in its last reduction. Therefore, if the formula is true for the nth power, it holds true for the $(n+1)$th power.

Since the formula (1) is true for the fifth power, it must be true for $(n+1)$th or sixth power, and for the power next higher or the seventh power, and so on, and, consequently, for any power, whose exponent is a positive integer.

421. When $(a-x)$ is raised to the nth power, it is evident that the terms containing the even powers of x will be positive and those containing the odd powers will be negative; that is, the second term will be negative because it contains the first power of x, the third term will be positive because it contains the second power of x, etc. Therefore,

$$(a-x)^n = a^n - na^{n-1}x + \frac{n(n-1)}{1\cdot 2}a^{n-2}x^2$$
$$-\frac{n(n-1)(n-2)}{1\cdot 2\cdot 3}a^{n-3}x^3 + \cdots$$

422. In the binomial $(a+x)$, if x and a be interchanged and the binomial be expanded, the coëfficients of the terms will be the same as in formula (1). Hence, *the coëfficients of the latter half of a power of $(a+x)$ are the same as those of the first half written in the reverse order.*

EXAMPLES.

Expand the following:

1. $(a+x)^5$.
2. $(a-x)^7$.
3. $(a^2+b^3)^4$.
4. $(a^{-1}+c^{-2})^4$.
5. $(c^{\frac{1}{2}}+d^{-2})^3$.
6. $(a^{\frac{2}{3}}-x^{\frac{3}{4}})^4$.
7. $(x^{-\frac{1}{2}}+y^3)^5$.
8. $\left(\dfrac{2x}{y}-\dfrac{3y}{x}\right)^3$.
9. $(x+\sqrt{y})^4$.
10. $(\sqrt{a^3}+2\sqrt[3]{x})^6$.
11. $(3x^2-\frac{1}{3}\sqrt{x})^5$.
12. $\left(\dfrac{2\sqrt{x}}{\sqrt[3]{y^2}}-\dfrac{\sqrt{y}}{\sqrt{x}}\right)^4$.

423. To find any term.

EXPONENTS.— From formula (1) it is apparent that the exponent of x in the second term is 1, in the third term 2, in the rth term $r-1$.

Since the sum of the exponents in any term is n, the exponent of a in any term will be $n-r+1$.

COËFFICIENTS.— Since the coëfficient in the third term is $\dfrac{n(n-1)}{1\cdot 2}$, in the fourth term $\dfrac{n(n-1)(n-2)}{1\cdot 2\cdot 3}$, the coëfficient in the rth term is $\dfrac{n(n-1)(n-2)(n-3)\cdots(n-r+2)}{1\cdot 2\cdot 3\cdot 4\cdots(r-1)}$.

THE BINOMIAL THEOREM.

EXAMPLES.

1. Find the sixth term of $(a+b)^9$.

2. Find the fifth term of $(a-x)^{10}$.

3. Find the seventh term of $(a+c)^{11}$.

4. Find the ninth term of $(a+b^2)^{12}$.

5. Find the fourth term of $(3-x^3)^7$.

6. Find the fourth term of $\left(a-\dfrac{x}{2}\right)^9$.

7. Find the fifth term of $\left(\dfrac{a}{x}+\dfrac{x}{a}\right)^7$.

8. Find the fourth term of $(a^4-\sqrt{a})^6$.

9. Find the fourth term of $(\tfrac{1}{3}x - y\sqrt{y})^7$.

10. Find the sixth term of $\left(a^{-2}+\dfrac{ax}{2}\right)^8$.

11. Find the fifth term of $(\sqrt{a}-\sqrt{x})^8$.

12. Find the eighth term of $(\sqrt[3]{a^2}-\sqrt{c})^9$.

13. Find the seventh term of $(ax^2+2\sqrt{x})^6$.

14. Find the fourth term of $\left(\sqrt{\dfrac{a}{2}}-\sqrt{\dfrac{2}{3}b}\right)^6$.

424. A polynomial may be expanded by the binomial formula by grouping its terms in the form of a binomial. Thus, $(a+b+c)^n = [(a+b)+c]^n$ and $(a+b+c+d)^n = [(a+b)+(c+d)]^n$.

Expand:

1. $(x+y+z)^3$.
2. $(1+x-x^2)^4$.
3. $(x^2-x+2)^4$.
4. $(a^2+ax+x^2)^3$.
5. $(a+x^2+z+1)^3$.
6. $(1+2x-x^2+2)^4$.

UNDETERMINED COËFFICIENTS.

425. 1. Expand the fraction $\dfrac{1}{1-x}$ into a series, by dividing the numerator by the denominator. How many terms are there in the quotient? What name may, therefore, be given to the series?

2. If $x = \frac{1}{2}$, what is the value of the fraction? What does the series become when $\frac{1}{2}$ is substituted for x? What is the sum of the first 2 terms? Of the first 3 terms? Of the first 4 terms? How will the sum of the largest conceivable number of terms compare with 2, the value of the fraction?

3. If $x = 1$, what is the value of the fraction? What does the series become when 1 is substituted for x? What is the sum of the largest conceivable number of terms?

4. If $x = -1$, what is the value of the fraction? What does the series become when -1 is substituted for x? What is the sum of the first 2 terms? Of the first 3 terms? Of the first 4 terms? An even number of terms? An odd number of terms? How does the sum of the terms correspond with the value of the fraction?

5. If $x = 2$, what is the value of the fraction? What does the series become when 2 is substituted for x? What is the value of 3 terms? Of n terms? What relation exists between the sum of the terms and the value of the fraction?

426. A **Convergent Series** is an infinite series, in which the sum of the terms, however many may be taken, can never exceed a certain finite value. This finite value is the sum of the series as determined from the value of the fraction.

Thus, the fraction $\dfrac{1}{1-x}$, expanded into a series by division, becomes $1 + x + x^2 + x^3 + \cdots$. Whether the series is convergent or not depends upon the value of x. If $x = \frac{1}{2}$, the series becomes $1 + \frac{1}{2} + \frac{1}{4} + \frac{1}{8} + \cdots$, and the sum of any number of terms, however great, is less than 2, the value of the fraction $\dfrac{1}{1-x}$. Hence, under that supposition, the series is *convergent*.

UNDETERMINED COEFFICIENTS.

1. It is evident that as the number of terms is increased, their sum approaches 2, the sum of the series.

2. It is evident, also, that when $x = 1$, the series becomes $1 + 1 + 1 + 1 + \cdots$, and hence is not convergent, because the sum may exceed any finite value. The same is true when x is equal to any positive quantity greater than 1.

427. A **Divergent Series** is an infinite series in which the sum of the terms is greater than any definite finite quantity, if enough terms are taken.

Thus, the sum of the terms of the series $1 + x + x^2 + x^3 + \cdots$ may be made larger than any definite finite quantity when $x = 2$, or 3, or any value not lying between $+1$ and -1, if enough terms are taken.

Since the sum of the terms of a divergent series has no limit, and since the sum of the terms is not equal to the value of the fraction from which it was expanded, it is evident that it cannot be used in discussions of series. Hence, in all discussions concerning series, the series must be understood to be *convergent*.

428. An **Identical Equation** is an equation in which the members are identical, or may be reduced to identity.

Thus, $ax + b = ax + b$ is an identical equation.

Also, $\dfrac{a^2 - x^2}{a - x} = a + x$ is an identical equation.

429. **Undetermined Coëfficients** are unknown coëfficients assumed in connection with quantities.

Thus, $(1 + x)^3$ may be assumed equal to $A + Bx + Cx^2 + Dx^3$ in which A, B, C, and D are undetermined coëfficients of x^0, x^1, x^2, and x^3.

430. Coëfficients in identical equations.

Assume $A + Bx + Cx^2 + Dx^3 + \text{etc.} = A' + B'x + C'x^2 + D'x^3 + \text{etc.}$, in which both series are convergent, and in which A, B, A', B', etc., are *constant* quantities, independent of x, and x a quantity to which various values may be assigned, or a *variable* quantity.

Transposing $A - A' + (B - B') x + (C - C') x^2 + (D - D') x^3 + \text{etc.} = 0$.

If $A - A'$ is not equal to 0, let their difference $= p$.
Then $(B - B') x + (C - C') x^2 + (D - D') x^3 + \text{etc.} = -p$.

ALGEBRA. — 22.

Now since A and A' are constant quantities, their difference p must be a constant quantity. But the value of $-p$ in the equation $(B-B')x+(C-C')x^2+(D-D')x^3+$ etc., is variable since it depends upon the values of x, which is a variable quantity. Consequently the value of x is both fixed and variable, which is absurd. Hence, it is impossible that there is a difference between A and A'.

$$\therefore A - A' = 0 \text{ and } A = A'$$

Hence, $(B-B')x+(C-C')x^2+(D-D')x^3=0$

Dividing by x, $B-B'+(C-C')x+(D-D')x^2=0$

Reasoning as before, $B=B'$; $C=C''$; $D=D'$, etc.

431. Principles. — 1. *The coëfficients of the same powers of x in two equal and identical convergent series are equal.*

432. Expanding fractions into series.

EXAMPLES.

1. Expand $\dfrac{1-x-x^2}{1+x+x^2}$ into a series.

Solution.

Assume $\dfrac{1-x-x^2}{1+x+x^2} = A + Bx + Cx^2 + Dx^3 + Ex^4 \cdots$

Clearing of fractions and collecting terms.

$$\begin{array}{r|c|c|c|c}
1-x-x^2 = A + B & x + C & x^2 + D & x^3 + E & x^4 \\
+ A & x + B & x^2 + C & x^3 + D & x^4 \\
 & + A & x^2 + B & x^3 + C & x^4
\end{array}$$

Equating coëfficients of like powers of x (Prin. 1) and remembering that the coëfficient of any power that is wanting is 0, we have

$$A = 1$$
$$B + A = -1 \quad \therefore B = -2$$
$$C + B + A = -1 \quad \therefore C = 0$$
$$D + C + B = 0 \quad \therefore D = 2$$
$$E + D + C = 0 \quad \therefore E = -2$$
$$\therefore \frac{1-x-x^2}{1+x+x^2} = 1 - 2x + 0x^2 + 2x^3 - 2x^4$$

The same result may be obtained by division.

PARTIAL FRACTIONS.

2. Expand $\dfrac{a}{a + bx}$ into a series.

SOLUTION.

Assume $\dfrac{a}{a + bx} = A + Bx + Cx^2 + Dx^3 + \cdots$

Clearing of fractions, $a = Aa + \begin{vmatrix} Ba \\ + Ab \end{vmatrix} x + \begin{vmatrix} Ca \\ + Bb \end{vmatrix} x^2 + \begin{vmatrix} Da \\ + Cb \end{vmatrix} x^3 + \cdots$

Equating coëfficients of like powers of x (Prin. 1),

$$Aa = a \quad \therefore A = 1$$

$$Ba + Ab = 0 \quad \therefore B = -\dfrac{b}{a}$$

$$Ca + Bb = 0 \quad \therefore C = +\dfrac{b^2}{a^2}$$

$$Da + Cb = 0 \quad \therefore D = -\dfrac{b^3}{a^3}, \text{ etc.}$$

$$\therefore \dfrac{a}{a + bx} = 1 - \dfrac{b}{a} + \dfrac{b^2}{a^2} - \dfrac{b^3}{a^3} + \dfrac{b^4}{a^4}, \text{ etc.}$$

The same result may be obtained by division.

Expand the following to five terms:

3. $\dfrac{1-x}{1+x}$.

4. $\dfrac{1-2x}{1-3x}$.

5. $\dfrac{1+x}{2+3x}$.

6. $\dfrac{1+2x}{1-x-x^2}$.

7. $\dfrac{1-x}{1-3x-2x^2}$.

8. $\dfrac{2-3x}{1+x+x^2}$.

9. $\dfrac{1}{1-3x+2x^2}$.

10. $\dfrac{3-2x}{1-x+x^2}$.

11. $\dfrac{x}{1+x+x^2}$.

12. $\dfrac{2-x}{1-2x+x^2}$.

13. $\dfrac{1}{1+2x^2+3x^4}$.

14. $\dfrac{x+5x^2}{1-4x+4x^2}$.

433. To resolve a fraction into its partial fractions.

To resolve a fraction into partial fractions is to separate it into fractions whose sum is equal to the given fraction.

It is evident that the denominators of the partial fractions must be factors of the denominator of the given fraction.

HIGH SCHOOL ALGEBRA.

EXAMPLES.

1. Resolve $\dfrac{5x-14}{x^2-6x+8}$ into partial fractions.

SOLUTION.

The factors of the denominator are $(x-2)$ and $(x-4)$; consequently the denominators of the partial fractions are $(x-2)$ and $(x-4)$ respectively.

Assume $\dfrac{5x-14}{x^2-6x+8} = \dfrac{A}{x-2} + \dfrac{B}{x-4}$, an identical equation.

Clearing of fractions, $\quad 5x - 14 = Ax - 4A + Bx - 2B$
$$= (A+B)x - 4A - 2B$$

By Prin. Art. 431, $\quad A + B = 5$
$$-4A - 2B = -14$$

Solving the equations, $\quad A = 2$ and $B = 3$

$$\therefore \dfrac{5x-14}{x^2-6x+8} = \dfrac{2}{x-2} + \dfrac{3}{x-4}$$

2. Resolve $\dfrac{3x^2-1}{x^3-x}$ into partial fractions.

SOLUTION.

The factors of the denominator are x, $(x-1)$, and $(x+1)$.

Assume $\dfrac{3x^2-1}{x^3-x} = \dfrac{A}{x} + \dfrac{B}{x-1} + \dfrac{C}{x+1}$.

Clearing of fractions,

$$3x^2 \pm 0x - 1 = Ax^2 - A + Bx^2 + Bx + Cx^2 - Cx$$
$$= -A + (B-C)x + (A+B+C)x^2$$

By Prin. Art. 431, $\quad -A = -1 \quad \therefore A = 1$
$$B - C = 0 \quad \therefore B = C = 1$$
$$A + B + C = 3 \quad \therefore C = 1$$

$$\therefore \dfrac{3x^2-1}{x^3-x} = \dfrac{1}{x} + \dfrac{1}{x-1} + \dfrac{1}{x+1}$$

3. Resolve $\dfrac{7x^2+x}{2x^2+x-1}$ into partial fractions.

SOLUTION.

Assume $\quad\dfrac{7x^2+x}{2x^2+x-1} = \dfrac{A}{x+1} + \dfrac{B}{2x-1}$

PARTIAL FRACTIONS.

Clearing of fractions, $\quad 7x^2 + x = (2A + B)x + B - A$

$$= 0x^2 + (2A + B)x + B - A$$

$\therefore 7 = 0$, an absurdity.

Therefore the numerators of the fractions cannot be A and B.
Assume the numerators to be Ax and Bx. Then the fractions are readily found.

4. Resolve $\dfrac{4x^2 - 9x + 8}{(x+1)^3}$ into partial fractions.

Solution.

Here the factors of the denominator are equal.

Assume $\quad \dfrac{4x^2 - 9x + 8}{(x+1)^3} = \dfrac{A}{(x+1)^3} + \dfrac{B}{(x+1)^2} + \dfrac{C}{x+1}$

Clearing of fractions,

$$4x^2 - 9x + 8 = A + B(x+1) + C(x^2 + 2x + 1)$$
$$4x^2 - 9x + 8 = A + B + C + (B + 2C)x + Cx^2$$

By Prin. Art. 431, $\quad 8 = A + B + C$

$$-9 = B + 2C$$
$$4 = C \quad \therefore -17 = B \quad \therefore 21 = A$$

$\therefore \dfrac{4x^2 - 9x + 8}{(x+1)^3} = \dfrac{21}{(x+1)^3} - \dfrac{17}{(x+1)^2} + \dfrac{4}{(x+1)}$

Resolve into partial fractions:

5. $\dfrac{3x - 3}{x^2 - x - 2}.$

6. $\dfrac{5x + 7}{x^2 + 3x + 2}.$

7. $\dfrac{2x - 5}{x^2 - 5x + 6}.$

8. $\dfrac{7x + 4}{x^2 + 2x - 8}.$

9. $\dfrac{9x - 35}{x^2 - 8x + 15}.$

10. $\dfrac{11x + 13}{2x^2 - 7x + 3}.$

11. $\dfrac{5x^2 - 2}{x^3 - x^2 - 2x}.$

12. $\dfrac{3x^2 - 6x + 2}{x^3 - 3x^2 + 2x}.$

13. $\dfrac{3x + 1}{(x-2)^2}.$

14. $\dfrac{x + 5x^2}{(x-1)^3}.$

REVERSION OF SERIES.

434. To Revert a Series is to express the value of the unknown quantity in it by means of another series involving some other unknown quantity.

Let it be required to revert the series in the equation

$$y = ax + bx^2 + cx^3 + dx^4 + \cdots \qquad (1)$$

That is, to find the value of x in a series involving y.

Assume $\quad x = Ay + By^2 + Cy^3 + Dy^4 + \cdots \qquad (2)$

in which the coëfficients of y are undetermined.

Substituting this value of x in (1), and disregarding all terms of the series involving a power higher than the 4th, we have

$$y = aAy + \begin{vmatrix} aB \\ +bA^2 \end{vmatrix} y^2 + \begin{vmatrix} aC \\ +2bAB \\ +cA^3 \end{vmatrix} y^3 + \begin{vmatrix} aD \\ +bB^2 \\ +2bAC \\ +3cA^2B \\ +dA^4 \end{vmatrix} y^4 + \cdots$$

Equating the coëfficients of like powers of y (Art. 431):

$aA = 1.$ $\hspace{4cm} \therefore A = \dfrac{1}{a}.$

$aB + bA^2 = 0.$ $\hspace{2cm} \therefore B = -\dfrac{bA^2}{a} = -\dfrac{b \cdot \dfrac{1}{a^2}}{a} = -\dfrac{b}{a^3}.$

$aC + 2bAB + cA^3 = 0.$ $\hspace{0.5cm} \therefore C = -\dfrac{2bAB - cA^3}{a} = \dfrac{2b^2 - ac}{a^5}.$

$aD + bB^2 + 2bAC + 3cA^2B + dA^4 = 0.$ $\therefore D = -\dfrac{a^2d - 5abc + 5b^3}{a^7}.$

Hence $x = \dfrac{1}{a}y - \dfrac{b}{a^3}y^2 + \dfrac{2b^2 - ac}{a^5}y^3 - \dfrac{a^2d - 5abc + 5b^3}{a^7}y^4 + \cdots$

REVERSION OF SERIES. 343

EXAMPLES.

1. Revert the series in the equation

$$y = x + 2x^2 + 4x^3 + 8x^4 + \cdots$$

SOLUTION. — In this series $a = 1$, $b = 2$, $c = 4$, $d = 8$.

Substituting these values for the values of A, B, C, D, etc., as determined above, the coëfficients of the reverted series are found, and the equation is

$$x = y - 2y^2 + 4y^3 - 8y^4 + \cdots$$

Revert the series in the following equations:

2. $y = x + x^2 + x^3 + x^4 + \cdots$

3. $y = x + 3x^2 + 5x^3 + 7x^4 + \cdots$

4. $y = x - \dfrac{x^2}{2} + \dfrac{x^3}{3} - \dfrac{x^4}{4} + \cdots$

5. $y = 2x + 3x^2 + 4x^3 + 5x^4 + \cdots$

6. $y = 2x + 4x^2 + 6x^3 + 8x^4 + \cdots$

7. Revert the series in the equation

$$\tfrac{1}{4} = 2x - \frac{4x^2}{3} + \frac{6x^3}{5} - \frac{8x^4}{7} + \cdots$$

and find the approximate value of x.

SOLUTION. — Put y for $\tfrac{1}{4}$.

Then
$$y = 2x - \frac{4x^2}{3} + \frac{6x^2}{5} - \frac{8x^4}{7} + \cdots$$

Reverting,
$$x = \frac{y}{2} + \frac{y^2}{6} + \frac{13y^3}{360} + \frac{5y^4}{1512} + \cdots$$

Restoring the value of y,

$$x = \frac{\tfrac{1}{4}}{2} + \frac{(\tfrac{1}{4})^2}{6} + \frac{13(\tfrac{1}{4})^3}{360} + \frac{5(\tfrac{1}{4})^4}{1512} + \cdots$$

$$= .125 + .010416 + .000564 + .000013 + \cdots$$

$$= .135993 +$$

Find the approximate value of x in the following equations:

8. $\frac{1}{2} = x + \dfrac{x^2}{6} + \dfrac{x^3}{12} + \dfrac{x^4}{20} + \cdots$

9. $\frac{1}{4} = x - \dfrac{3x^2}{8} + \dfrac{5x^3}{24} - \dfrac{7x^4}{48} + \cdots$

RECURRING SERIES.

435. A **Recurring Series** is a series in which every term sustains a fixed relation to one or more of the preceding terms.

Thus, $1 + 2x + 11x^2 + 50x^3 + 233x^4 + \cdots$ is a recurring series in which every term is formed by multiplying the first of the two preceding terms by $3x^2$ and adding to that product the term immediately preceding multiplied by $4x$.

436. The **Scale of Relation** is the expression by means of which any term of a series may be formed from the preceding terms.

Thus, in the series $1 + 2x + 11x^2 + 50x^3 + 233x^4 + \cdots$, $3x^2, 4x$ is the scale of relation.

437. A recurring series in which each term depends upon the one preceding is termed a series of the *first order;* when it depends upon the *two* preceding terms, it is of the *second order;* when upon the *three* preceding terms, it is of the *third order,* etc.

438. To find the scale of relation of a recurring series.

First. When the series is of the *first order*, and each term depends upon the one next preceding, we have simply a geometrical progression. If the series be $a + b + c + d + e + \cdots$ and m represents the scale of relation,

$$m = \frac{b}{a}$$

RECURRING SERIES.

Second. When the series is of the second order, let m and n represent the scale of relation. Then, if the series be $a + b + c + d + e + \cdots$,

$$c = ma + nb$$
$$d = mb + nc$$

Finding the values of m and n

$$m = \frac{c^2 - bd}{ac - b^2} \text{ and } n = \frac{ad - bc}{ac - b^2}$$

Third. When the series is of the third order, let m, n, r represent the scale of relation. Then, if the series be $a + b + c + d + e + f + \cdots$,

$$d = ma + nb + rc$$
$$e = mb + nc + rd$$
$$f = mc + nd + re$$

From these equations, m, n, and r may be found, and in a similar manner the scales of relation for series of the 4th, 5th, 6th, and nth orders.

EXAMPLES.

Find the scale of relation in the following series:

1. $1 + 6x + 12x^2 + 48x^3 + 120x^4 + \cdots$

SOLUTION. — Assume m, n to be the scale of relation.

Then, $\quad\quad\quad 12x^2 = m + 6nx$
$\quad\quad\quad\quad\quad 48x^3 = 6mx + 12nx^2$

Solving, $\quad\quad\quad m = 6x^2$ and $n = x$

$\therefore 6x, x$ is the scale of relation.

2. $1 + 2x + 3x^2 + 5x^3 + 8x^4 + 13x^5 + \cdots$
3. $1 + 3x + 5x^2 + 11x^3 + 21x^4 + 43x^5 + \cdots$
4. $1 + 7x + 17x^2 + 55x^3 + 161x^4 + 487x^5 + \cdots$
5. $1 + 2x + 8x^2 + 28x^3 + 100x^4 + 356x^5 + \cdots$

439. To find the sum of an infinite recurring series.

First. Assume that the series is an infinite series of the *first order*, and that it is convergent.

Let $a + b + c + d + e + \cdots$ be the series.
Let m represent the scale of relation.

Then, $$b = ma$$
$$c = mb$$
$$d = mc$$
$$e = md, \text{ etc., } ad \; infinitum.$$

Adding, $b + c + d + e + \cdots = m(a + b + c + d + e + \cdots)$

Representing the sum of the series by s, the first member of the above equation becomes $s - a$, and the second, ms. That is,

$$s - a = ms$$

Whence $$s = \frac{a}{1 - m} \qquad (1)$$

Since a recurring series of the first order is simply a geometrical progression, the *scale of relation* is the *ratio*, and the sum of the recurring series the same as the sum of an infinite geometrical progression.

Second. Assume that the series is an infinite series of the *second order*, and that it is convergent.

Let $a + b + c + d + e + \cdots$ be the series.
Let m, n represent the scale of relation.

Then, $$c = ma + nb$$
$$d = mb + nc$$
$$e = mc + nd$$
$$f = md + ne, \text{ etc., } ad \; infinitum.$$

BINOMIAL THEOREM — ANY EXPONENT.

Adding,

$$c + d + e + f + \cdots = m(a + b + c + d + e + f + \cdots)$$
$$+ n(b + c + d + e + f + \cdots)$$

Representing the sum by s, the equation becomes

$$s - (a + b) = ms + n(s - a)$$

Whence $$s = \frac{a + b - an}{1 - m - n} \qquad (2)$$

1. The sum of a series of any higher order may be found by a similar process.

2. Every infinite recurring series may be assumed to arise from the development of a rational fraction; hence the *sum* of the series will be the *generating fraction*.

EXAMPLES.

Find the sum of each of the following series:

1. $1 + 2x + 3x^2 + 5x^3 + 8x^4 + 13x^5 + \cdots$

SOLUTION. — The scale of relation is found to be x^2, x, and the series is therefore of the second order.

Substituting 1 for a, $2x$ for b, x^2 for m, and x for n in formula (2) for the sum

$$s = \frac{1 + 2x - x}{1 - x^2 - x} = \frac{1 + x}{1 - x + x^2}$$

2. $1 + 3x + 4x^2 + 7x^3 + 11x^4 + 18x^5 + \cdots$
3. $1 + 3x + 5x^2 + 11x^3 + 21x^4 + 43x^5 + \cdots$
4. $1 + 2x + 8x^2 + 28x^3 + 100x^4 + 356x^5 + \cdots$
5. $1 + 3x + 5x^2 + 7x^3 + 9x^4 + 11x^5 + \cdots$

BINOMIAL THEOREM — ANY EXPONENT.

440. It has been shown (Art. 418) that when n is a positive integer,

$$(1 + x)^n = 1 + nx + \frac{n(n-1)}{1 \cdot 2} x^2 + \frac{n(n-1)(n-2)}{1 \cdot 2 \cdot 3} x^3 + \cdots,$$

since any power of 1 is 1.

It is yet to be shown that this formula is true when the exponent n is a positive fraction, a negative integer, or a negative fraction.

441. *First.* Let n be a positive fraction $\dfrac{p}{q}$.

Let $(1+x)^{\frac{p}{q}} = 1 + ax + \cdots$ and let Ax represent all the terms after the first.

$$\therefore (1+x)^{\frac{p}{q}} = 1 + Ax$$

also $$(1+x)^p = (1+Ax)^q$$

Expanding, $$1 + px + \cdots = 1 + qAx + \cdots$$
$$= 1 + qax + \cdots$$

Equating coëfficients, $\quad p = qa$

$$\therefore a = \frac{p}{q}$$

and $$(1+x)^{\frac{p}{q}} = 1 + \frac{p}{q}x + \cdots$$

That is, the formula is true for *two* terms when n is a *positive fraction*.

442. *Second.* Let n be negative, and either integral or fractional.

$$(1+x)^{-n} = \frac{1}{(1+x)^n} \qquad \text{(Art. 226)}$$

$$= \frac{1}{1 + nx + \cdots} \qquad \text{(Art. 418)}$$

Dividing the numerator by the denominator,

$$(1+x)^{-n} = 1 - nx + \cdots$$

That is, the formula is true for *two* terms when n is negative and either integral or fractional.

Therefore the formula is true for *two* terms whatever be the exponent, and the coëfficient of the second term is n.

BINOMIAL THEOREM — ANY EXPONENT. 349

443. The general law of coëfficients is yet to be determined.

Let $\quad (1+x)^n = 1 + nx + Ax^2 + Bx^3 + Cx^4 + \cdots \quad (A)$

in which A, B, and C, etc., are the undetermined coëfficients of x^2, x^3, x^4, etc.

To find the value of the undetermined coëfficients, we involve them in an identical equation, so that the coëfficients of like powers of x may be equated (Art. 429).

To do this, put $x = x + z$. The expression $(1 + x)^n$ then becomes $(1 + x + z)^n$, in which $(1 + x)$ or $(x + z)$ may be regarded as one term.

$$[(1+x)+z]^n = (1+x)^n + n(1+x)^{n-1}z$$
$$+ A(1+x)^{n-2}z^2 + B(1+x)^{n-3}z^3 + \cdots$$

$$[1+(x+z)]^n = 1 + n(x+z) + A(x+z)^2 + B(x+z)^3 + \cdots$$
$$= 1 + nx + Ax^2 + Bx^3 + \cdots$$
$$+ (n + 2Ax + 3Bx^2 + \cdots)z + \cdots$$

Equating the two right-hand members, we have an identical equation, when both members of the equation are convergent.

Therefore,

$$(1+x)^n + n(1+x)^{n-1}z + A(1+x)^{n-2}z^2 + \cdots$$
$$= 1 + nx + Ax^2 + Bx^3 + \cdots + (n + 2Ax + 3Bx^2 + \cdots)z$$

Equating coëfficients of z,

$$n(1+x)^{n-1} = n + 2Ax + 3Bx^2 + \cdots$$

Multiplying each member by $1 + x$,

$$n(1+x)^n = n + (2A+n)x + (3B+2A)x^2 + \cdots$$

Eq. $(A) \times n$, $n(1+x)^n = n + n^2x + nAx^2 + nBx^3 + \cdots$

$\therefore\ n + n^2x + nAx^2 + nBx^3 + \cdots = n + (2A+n)x$
$$+ (3B+2A)x^2 + \cdots$$

Equating the coëfficients of like powers of x (Art. 429),

$$2A + n = n^2 \quad \therefore\ A = \frac{n(n-1)}{1 \cdot 2}$$

$$3B + 2A = nA \quad \therefore\ 3B = nA - 2A = A(n-2)$$

$$\therefore\ B = \frac{n(n-1)(n-2)}{1 \cdot 2 \cdot 3}$$

Substituting these values in equation (A),

$$(1+x)^n = 1 + nx + \frac{n(n-1)}{1 \cdot 2} x^2 + \frac{n(n-1)(n-2)}{1 \cdot 2 \cdot 3} x^3 + \cdots \quad (B)$$

444. The expansion of $(1+x)^n$ is not the expansion of the most general form of a binomial, since the first term is 1; consequently the equation must be changed so as to express the expansion of $(a+x)^n$ for the general form.

Putting $\dfrac{x}{a}$ for x in (B), then

$$\left(1+\frac{x}{a}\right)^n = 1 + n\frac{x}{a} + \frac{n(n-1)}{1 \cdot 2}\frac{x^2}{a^2} + \frac{n(n-1)(n-2)}{1 \cdot 2 \cdot 3}\frac{x^3}{a^3} + \cdots$$

Multiplying both numbers by a^n, observing that,

$$a^n\left(1+\frac{x}{a}\right)^n = a^n\left(\frac{a+x}{a}\right)^n = \frac{a^n(a+x)^n}{a^n} = (a+x)^n$$

$$(a+x)^n = a^n + na^{n-1}x + \frac{n(n-1)}{1 \cdot 2}a^{n-2}x^2$$

$$+ \frac{n(n-1)(n-2)}{1 \cdot 2 \cdot 3}a^{n-3}x^3$$

Therefore, the general formula has been shown to be applicable for the expansion of any binomial with any exponent, provided the second member is convergent when it forms an infinite series.

BINOMIAL THEOREM — ANY EXPONENT.

EXAMPLES.

1. Expand $(a+b)^{\frac{3}{4}}$ to four terms.

SOLUTION.

The coëfficient of the first term is 1

The coëfficient of the second term is $\frac{3}{4}$

The coëfficient of the third term is $\dfrac{\frac{3}{4}(\frac{3}{4}-1)}{1\cdot 2}$ or $-\dfrac{3}{32}$

The coëfficient of the fourth term is $-\dfrac{3}{32}\dfrac{(\frac{3}{4}-2)}{3}$ or $\dfrac{15}{384}$

$\therefore (a+b)^{\frac{3}{4}} = a^{\frac{3}{4}} + \frac{3}{4}a^{-\frac{1}{4}}b - \frac{3}{32}a^{-\frac{5}{4}}b^2 + \frac{15}{384}a^{-\frac{9}{4}}b^3 + \cdots$

Expand to four terms:

2. $(a+b)^{\frac{1}{2}}$.   **6.** $(a-b)^{-3}$.   **10.** $(1+3x)^{\frac{5}{3}}$.

3. $(a+b)^{\frac{2}{3}}$.   **7.** $(a-b)^{\frac{3}{4}}$.   **11.** $(2a+3b)^{\frac{3}{4}}$.

4. $(a+b)^{-2}$.   **8.** $(a+b)^{-\frac{1}{4}}$.   **12.** $\sqrt[4]{1-2x}$.

5. $(a+b)^{-\frac{1}{2}}$.   **9.** $(a+b)^{-\frac{3}{2}}$.   **13.** $\dfrac{1}{\sqrt[5]{(a-b)^3}}$.

14. Find the approximate square root of 11.

SOLUTION.

$$\sqrt{11} = \sqrt{9+2} = \sqrt{3^2+2} = (3^2+2)^{\frac{1}{2}}$$

Expanding, $(3^2+2)^{\frac{1}{2}} = (3^2)^{\frac{1}{2}} + \frac{1}{2}(3^2)^{-\frac{1}{2}}\cdot 2 - \frac{1}{8}(3^2)^{-\frac{3}{2}}\cdot 2^2 + \cdots$

$= 3 + \frac{1}{2}\cdot 3^{-1}\cdot 2 - \frac{1}{8}\cdot 3^{-3}\cdot 2^2 + \cdots$

$= 3 + \dfrac{1}{2\cdot 3}\cdot 2 - \dfrac{1}{8\cdot 3^3}\cdot 4 + \cdots$

$= 3 + \frac{1}{3} - \frac{1}{54} + \cdots$

$= 3 + .3333 - .0185 +$

$= 3.3148 +$

Find the approximate values of the following:

15. $\sqrt{7}$.   **17.** $\sqrt{26}$.   **19.** $\sqrt[4]{85}$.

16. $\sqrt{18}$.   **18.** $\sqrt[3]{29}$.   **20.** $\sqrt[5]{245}$.

THEORY OF EQUATIONS.

445. The object of this discussion is to discover methods of solving equations of a higher degree than the second, and the processes of approximating to the roots of numerical equations.

It is impossible to solve general equations of a higher degree than the fourth; but if the equations are numerical, approximate roots may be found.

446. Equations containing one unknown quantity may be reduced to the general form,

$$x^n + px^{n-1} + qx^{n-2} + \cdots + sx^2 + tx + u = 0 \qquad (1)$$

In equation (1) p, q, etc., are positive or negative, integral or fractional, real or imaginary, and n is a positive integer.

It is sometimes called the *reduced equation*. The reduction is effected by transposing, collecting the terms of the same degrees, and by dividing by the coëfficient of the highest power of the unknown quantity.

447. A **Root** of an equation is any expression which, upon being substituted for the unknown quantity, satisfies the equation.

In the reduced equation, a root of the equation reduces the first member to 0, when it is substituted for x.

448. Divisibility of the members of equations.

Suppose a is a root of the reduced equation,

$$x^n + px^{n-1} + qx^{n-2} + \cdots sx^2 + tx + u = 0$$

Then, $x = a$, and $x - a = 0$

Divide the members of the equation by $x - a$, representing the first member of the reduced equation by X, the quotient by Q, the remainder, if any, by R; then,

$$X = (x - a)Q + R$$

Inasmuch as a is a root of the equation, it may be substituted for x, and then by the definition of a *root*, X, the first member of the reduced equation, equals 0 and $(x - a) = 0$.

THEORY OF EQUATIONS.

Since $(x-a)$ equals 0, R equals 0. That is, there is no remainder, and $x-a$ is an exact divisor of the first member of equation (1). Hence,

449. Principle 1. — *If a is a root of the reduced equation, the first member of the equation is exactly divisible by $x-a$.*

450. Suppose the first member of the reduced equation to be exactly divisible by $x-a$. Then,

$$X = Q(x-a)$$

Now, whatever value of x reduces X to 0 is a root of the equation, and since, when $x=a$, X is equal to 0, a is a root of the equation.

451. Principle 2. — *If the first member of a reduced equation is exactly divisible by $x-a$, a is a root of the equation.*

EXAMPLES.

1. Show that 2 is a root of the equation $x^3 - 7x + 6 = 0$.

2. Show that 3 is a root of the equation $x^3 - 6x^2 + 11x - 6 = 0$.

3. Show that 4 is a root of the equation $x^3 - 5x^2 - 8x + 48 = 0$.

4. Show that 5 is a root of the equation $x^3 - 15x^2 + 66x - 80 = 0$.

452. Number of roots of an equation.

Suppose a to be a root of the reduced equation,

$$x^n + px^{n-1} + qx^{n-2} + \cdots + sx^2 + tx + u = 0$$

then by Prin. 1 the first member of the equation is divisible by $x-a$, and the quotient may be expressed as follows, if p', q', etc., denote the coëfficients of the powers of x in the quotient,

$$x^{n-1} + p'x^{n-2} + \cdots t'x + u' = 0$$

Suppose b to be a root of this last equation. Then it may be factored as follows:

$$(x-b)(x^{n-2}+p''x^{n-3}+\cdots t''x+u'')=0$$

It is evident that this process may be continued until the highest power of x in the second factor is x^{n-n}, or until x disappears from the second factor. The process will then have been performed n times, each time giving one root. Hence the equation has n factors, and,

$$(x-a)(x-b)(x-c)\cdots(x-l)=0$$

Consequently the equation has n roots.

Since the first member of the equation has only the factors $x-a$, $x-b$, $x-c$, $\cdots x-l$, it can have no other roots.

453. PRINCIPLE. — *Every equation of the nth degree containing but one unknown quantity has n roots, and no more.*

EXAMPLES.

1. One root of the equation $x^3-2x^2+7x-30=0$ is 3. Find the other roots.

SOLUTION.

$$x^3-2x^2+7x-30=0$$

Art. 446, Prin. 1, $\quad (x-3)(x^2+x+10)=0$

$$\therefore x^2+x+10=0$$

Solving, $\quad x=-\frac{1}{2}+\frac{1}{2}\sqrt{-39}$ and $-\frac{1}{2}-\frac{1}{2}\sqrt{-39}$, the other roots.

2. One root of the equation $x^3-4x^2+x+6=0$ is 2. Find the other roots.

3. One root of the equation $x^3-2x^2-x+2=0$ is -1. Find the other roots.

4. One root of the equation $x^3-3x^2-10x+24=0$ is 4. What are the other roots?

5. One root of the equation $x^3-15x^2+66x-80=0$ is 5. What are the other roots?

6. Two roots of the equation $x^4+3x^2-16x-60$ are -2 and 3. Find the other roots.

THEORY OF EQUATIONS. 355

454. Coëfficients of the terms of an equation.

Suppose $\qquad x = a,\ b,\ c,\ d$

Then, $x - a = 0;\ x - b = 0;\ x - c = 0;\ x - d = 0$

$\qquad \therefore (x-a)(x-b)(x-c)(x-d) = 0$

$$\begin{array}{l|l|l|l}
x^4 - a & x^3 + ab & x^2 - abc & x + abcd = 0 \\
- b & + ac & - abd & \\
- c & + ad & - acd & \\
- d & + bc & - bcd & \\
 & + bd & & \\
 & + cd & &
\end{array}$$

From this equation the following principles may be deduced:

455. Principles. — 1. *The coëfficient of the second term of an equation of the nth degree in its reduced form is the sum of all its roots with their signs changed.*

2. *The coëfficient of the third term is the sum of their products taken two and two.*

3. *The coëfficient of the fourth term is equal to the sum of their products taken three and three with their signs changed, etc.*

4. *The last term is the product of all the roots but with its sign changed, if the degree of the equation is odd.*

From the preceding equation and principles, it is evident that:

1. If the roots are all positive, the signs are alternately positive and negative.

2. If the coëfficient of the second term is wanting, the sum of the roots is 0; for otherwise the sum of the roots would be some quantity.

3. If there is no absolute term, *one* root must be 0; for otherwise the term would not be wanting.

4. Every rational root is a divisor of the last term; for the last term is the product of all the roots

EXAMPLES.

1. Form the equation whose roots are 2, 3, and 4.

2. Form the equation whose roots are 3, 1, and 5.

3. Form the equation whose roots are -1, -3, and 2.

4. Form the equation whose roots are 1, 2, 3, and -4.

5. Form the equation whose roots are 1, 3, and $\frac{1}{2}$.

6. Form the equation whose roots are 2, $-\frac{1}{2}$, -3, and $\frac{1}{3}$.

7. Two roots of the equation $x^4 - 10x^3 + 35x^2 - 50x + 24 = 0$ are 1 and 2. Find all the roots.

8. Find all the roots of the equation $y^4 - 5y^3 - 2y^2 + 12y + 8 = 0$, if two of the roots are 2 and -1.

456. Positive and negative roots of equations.

457. A **Variation of Sign** is the change of sign in two successive terms.

Thus, in $x - y$, there is one variation of sign; in $x - y + z$, there are two.

458. A **Permanence of Sign** is the continuation of the same sign with two successive terms.

Thus, in $x + y$, there is one permanence of sign; in $x + y - z - v$, there are two.

459. Assume the signs of the terms in a complete equation to be $+ + - + - + + +$, and suppose a new factor, $x - a = 0$, corresponding to a new positive root, to be introduced.

The signs of the product may be found as follows:

$$
\begin{array}{r}
+ + - + - + + + \\
+ - - \\
\hline
+ + - + - + + + \\
- - + - + - - - \\
\hline
+ \pm - + - + \pm \pm -
\end{array}
$$

THEORY OF EQUATIONS. 357

In the original equation there were 4 variations, and in the product there are 5, whether the ambiguous signs are considered positive or negative. Hence, the introduction of a positive root has caused at least one *additional* variation of sign.

Since this is true for any positive root, there must be as many *variations* of signs as there are *positive roots*.

By introducing the factor $x + a = 0$, in which a is a negative root, it may be shown in a similar manner that there are as many *permanences* of sign as there are *negative roots*.

460. PRINCIPLES. — 1. *An equation cannot have more positive roots than it has variations of sign.*

2. *An equation cannot have more negative roots than it has permanences of sign.*

3. *A complete equation whose terms are all positive can have no positive roots, and one whose terms are alternately positive and negative can have no negative roots.*

461. Since the roots enter into the coëfficients, two imaginary roots, by multiplication, give a real quantity. If, however, there is an odd number of imaginary roots, some one of them will appear in the coëfficients. Hence,

462. PRINCIPLE. — *When the coëfficients of the reduced equation are all real, if there be imaginary roots, they exist in pairs.*

EXAMPLES.

463. The roots of the following equations are all real. Determine their signs:

1. $x^3 - 3x^2 - 4x + 12 = 0$.
2. $x^3 - 6x^2 + 12x - 18 = 0$.
3. $x^3 - x^2 - 10x + 24 = 0$.
4. $x^3 + x^2 - 4 = 0$.
5. $x^3 - 3x - 2 = 0$.
6. $x^4 - x^3 - 7x^2 + x + 6 = 0$.

TRANSFORMATION OF EQUATIONS.

464. To transform an equation into another in which the roots are equal to the roots of the given equation but have opposite signs.

Let the given equation be

$$x^n + px^{n-1} + qx^{n-2} + \cdots + sx^2 + tx + u = 0$$

Substitute $-y$ for x. Since y has the same value as x with the sign changed, the equation becomes

$$(-y)^n + p(-y)^{n-1} + q(-y)^{n-2} + \cdots + s(-y)^2 + t(-y) + u = 0 \quad (1)$$

When n is *even*, the first term is positive, the second negative, the third positive, etc., and equation (1) becomes

$$y^n - py^{n-1} + qy^{n-2} - \cdots + sy^2 - ty + u = 0 \quad (2)$$

When n is *odd*, the first term is negative, the second positive, the third negative, etc., and equation (1) becomes

$$-y^n + py^{n-1} - qy^{n-2} + \cdots + sy^2 - ty + u = 0 \quad (3)$$

Changing all the signs, equation (3) becomes

$$y^n - py^{n-1} + qy^{n-2} - \cdots - sy^2 + ty - u = 0 \quad (4)$$

By comparing equations (2) and (4), the truth of the following principle is apparent.

465. PRINCIPLE. — *An equation may be transformed into another having the same roots, but with opposite signs, by changing the signs of the alternate terms, beginning with the second.*

If the equation is incomplete, the missing terms must be supplied by writing the missing quantity with 0 for its coefficient.

EXAMPLES.

1. Transform the equation $x^5 + x^3 - 4x^2 + 3x - 6 = 0$ into another having the same roots with contrary signs.

SOLUTION.

Supplying the missing term, $\quad x^5 \pm 0\,x^4 + x^3 - 4x^2 + 3x - 6 = 0$

Changing the signs, $\quad\quad\quad x^5 \mp 0\,x^4 + x^3 + 4x^2 + 3x + 6 = 0$

$\quad\quad\quad\quad\quad\quad\quad = x^5 + x^3 + 4x^2 + 3x + 6 = 0$

In like manner transform the following:

    2. $x^5 + 3x^4 - 2x^2 + 3x - 6 = 0.$

    3. $x^6 - 3x^4 - 4x^2 + 7x + 13 = 0.$

    4. $x^5 + 2x^4 - 5x^3 - 3x + 4 = 0.$

466. To transform an equation into another whose roots are some multiple of the roots in the given equation.

Let the given equation be

$$x^n + px^{n-1} + qx^{n-2} + \cdots + sx^2 + tx + u = 0 \qquad (1)$$

Let $y = mx$, a multiple of x; then $x = \dfrac{y}{m}$.

For x substitute $\dfrac{y}{m}$. Equation (1) becomes

$$\frac{y^n}{m^n} + p\frac{y^{n-1}}{m^{n-1}} + q\frac{y^{n-2}}{m^{n-2}} + \cdots + s\frac{y^2}{m^2} + t\frac{y}{m} + u = 0$$

Since m^n is the L.C.M. of the denominators, we multiply the equation by it, obtaining

$$y^n + pmy^{n-1} + qm^2y^{n-2} + \cdots + sm^{n-2}y^2 + tm^{n-1}y + um^n = 0 \quad (2)$$

Hence,

467. Principle. — *An equation may be transformed into another whose roots are a multiple of those in the given equation by multiplying the successive terms by* 1, m, m^2, m^3, *etc.*, m *being the given factor of the multiple.*

EXAMPLES.

Transform the following equations into others whose roots are *three times* the roots of the given equation:

    1. $x^3 + 4x - 16 = 0.$ **3.** $x^3 - 2x^2 - 9 = 0.$

    2. $x^3 + 2x^2 - 3 = 0.$ **4.** $x^3 - 3x - 2 = 0.$

360   HIGH SCHOOL ALGEBRA.

468. To transform a given equation into another whose coëfficients are all integers, 1 being the coëfficient of the first term.

By the principle (Art. 467), when m is the L.C.M. of the denominators of the coëfficients, the transformed equation will have integral coëfficients, but often a smaller number will remove the denominators.

Thus, transforming $x^3 - \frac{2}{3}x^2 - \frac{1}{4}x - 5\frac{5}{6} = 0$ by putting $\dfrac{y}{12}$ for x, we have

$$\frac{y^3}{1728} - \frac{2}{3}\cdot\frac{y^2}{144} + \frac{y}{48} - \frac{35}{6} = 0$$

and $$y^3 - 8y^2 + 36y - 10080 = 0$$

But $\dfrac{y}{6} = x$ will transform the equation, and it becomes

$$y^3 - 4y^2 + 9y - 1260 = 0$$

In general, take for the denominator of y the smallest number whose power corresponding to the highest exponent of x is a multiple of the denominators.

Thus, 216 is the smallest multiple of 12 which is a cube, and its root is 6, the denominator of y.

Transform the following equations into others whose coëfficients are all integral, 1 being the coëfficient of the first term.

1. $\dfrac{x^4}{2} - \dfrac{x^3}{4} + \dfrac{x}{8} - 9 = 0.$

2. $x^3 - \frac{7}{3}x^2 + \frac{11}{36}x - \frac{25}{72} = 0.$

3. $x^3 - 3x^2 + \dfrac{3x}{4} - \dfrac{5}{9} = 0.$

4. $x^3 - \frac{11}{2}x^2 - 3x + 3\frac{7}{8} = 0.$

5. $x^4 - \frac{4}{5}x^3 + \frac{2}{3}x^2 - 4x + \frac{7}{15} = 0.$

6. $x^4 - \frac{13}{12}x^3 + \frac{13}{40}x^2 - \frac{17}{24}x + \frac{1}{80} = 0.$

7. $x^4 + \frac{2}{5}x^3 - \frac{3}{4}x^2 + \frac{1}{2}x = 0.$

8. $x^5 - \frac{1}{4}x^3 + \frac{2}{3}x - 24 = 0.$

TRANSFORMATION OF EQUATIONS.

469. To transform an equation into another whose roots are greater or less than the roots of the given equation by a given quantity.

It is obvious that the equation $x^4 - 3x^2 + 5x - 6 = 0$ can be transformed into another whose roots are less by 2, by substituting $y + 2$ for x. The equation then becomes

$$y^4 - 8y^3 + 21y^2 - 15y + 4 = 0$$

The process of expanding the binomials and reducing the equation is so tedious that a shorter and simpler method has been devised. It is determined as follows:

Assuming the general equation with one unknown quantity

$$x^n + px^{n-1} + qx^{n-2} + \cdots + sx^2 + tx + u = 0 \qquad (1)$$

For x substitute $y + r$. It will then be transformed into another equation whose roots are r less than those of the given equation. It will become

$$(y+r)^n + p(y+r)^{n-1} + \cdots + s(y+r)^2 + t(y+r) + u = 0 \quad (2)$$

Expanding and reducing, (2) becomes

$$\left. \begin{array}{l} y^n + nr \left| \begin{array}{l} y^{n-1} + \dfrac{n(n-1)r^2}{2} \\ \phantom{y^{n-1}} + (n-1)pr \\ \phantom{y^{n-1}} + q \end{array} \right| \begin{array}{l} y^{n-2} + \cdots + r^n \\ \phantom{y^{n-2}} + pr^{n-1} \\ \phantom{y^{n-2}} + \cdots \\ \phantom{y^{n-2}} + tr \\ \phantom{y^{n-2}} + u \end{array} \\ +p \end{array} \right\} = 0 \quad (3)$$

Indicating the coëfficient of y^{n-1} by p', the coëfficient of y^{n-2} by q' and the absolute term by u', (3) becomes

$$y^n + p'y^{n-1} + q'y^{n-2} + \cdots s'y^2 + t'y + u' = 0 \qquad (4)$$

Restoring the value of y, which is $x - r$, (4) becomes

$$(x-r)^n + p'(x-r)^{n-1} + q'(x-r)^{n-2} + \cdots s'(x-r)^2 \\ + t'(x-r) + u' = 0 \qquad (5)$$

The first member of (5) is *identical* with the first member of (1), for the process by which (4) was deduced from (1) is simply reversed. Hence,

$$x^n + px^{n-1} + qx^{n-2} + \cdots + sx^2 + tx + u$$
$$= (x-r)^n + p'(x-r)^{n-1} + q'(x-r)^{n-2} + \cdots s'(x-r)^2$$
$$+ t'(x-r) + u' \qquad (6)$$

is an *identical equation*.

Dividing the second member by $x - r$, the remainder u' is obtained, which is the absolute term in (4).

Dividing the quotient by $x - r$, the remainder t' is obtained, which is the coëfficient of x in (4).

Hence, if we continue to divide by $x - r$, the successive coëfficients of x in (4) will be determined from the remainders.

Since the first member of (6) is identical with the second member, the coëfficients of (4), the transformed equation, may be found in the same way. Hence,

470. Principle. — *To transform an equation into another whose roots are greater or less than the roots of the given equation by a given quantity,*

Divide the first member of the equation by $x \pm r$, and continue the process until a remainder is found which is independent of x; then divide this quotient by the same divisor until n divisions have been performed. The successive remainders will be the coëfficients of the transformed equation.

The result may be obtained more readily and more simply by the use of *Detached Coëfficients* and *Synthetic Division*.

471. Division of polynomials may be readily performed by the use of the coëfficients alone, provided care is taken to restore the proper literal quantities in the terms of the quotient.

Let it be required to divide $x^3 - 5x^2 + 13$ by $x - 2$.

Since the coëfficients of the quotient depend upon the coëfficients of the dividend and divisor, and not upon the literal factors of the terms, the process by *detached coëfficients* may be employed.

TRANSFORMATION OF EQUATIONS. 363

$$\begin{array}{r|l} 1-5\pm 0+13 & \underline{1-2} \\ \underline{1-2} & 1-3-6 \\ -3\pm 0 & \\ \underline{-3+6} & \\ -6+13 & \\ \underline{-6+12} & \\ 1 & \end{array}$$

∴ The quotient is $x^2 - 3x - 6 +$ Rem. 1.

1. Since the first term of the divisor is 1, the first term of the quotient is the same as the first term of the dividend.

2. The other terms of the quotient are the same as the first terms of the successive partial dividends.

3. Each remainder is found by subtracting the product of the first term of the partial dividend by the second term of the divisor from the successive terms of the dividend.

4. Since the first term of the divisor is 1, it may be omitted, and if the sign of the second term of the divisor be changed, the products may be added.

The process given above may, therefore, be abridged as follows:

Dividend, $\quad\quad\quad 1-5\pm 0+13\ \underline{|+2}$
Partial products, $\quad\quad 2-6-12$
Quotients, $\quad\quad\quad \overline{1-3-6+\ \ 1}$, Remainder.

∴ The quotient is $x^2 - 3x - 6 +$ Rem. 1.

472. The method of division illustrated above is called *Synthetic Division*.

Divide by synthetic division:

1. $x^4 - 14x^3 + 71x^2 - 154x + 120$ by $x - 5$.
2. $x^5 + 6x^4 - 10x^3 - 112x^2 - 207x - 110$ by $x + 2$.
3. $x^4 - 12x^3 + 47x^2 - 72x + 36$ by x.

EXAMPLES.

1. Transform the equation $x^3 - 7x + 7 = 0$ into another whose roots are 1 less than those of the given equation.

Solution.

Using synthetic division, and applying the principle (Art. 470), the dividend and successive quotients are divided by $x - 1$, or changing the sign of the second term, by $+1$.

The coëfficients of terms of the polynomial that are wanting must be supplied by ± 0.

364　HIGH SCHOOL ALGEBRA.

$$1 \pm 0 - 7 + 7 \;\underline{|\,+1\,}$$
$$+1 + 1 - 6$$

Dividing by $+1$, $\;+1-6+1\;\;\therefore +1 =$ 1st Rem.
$$+1+2$$

Dividing by $+1$, $\;+2-4\;\;\;\;\;\;\therefore -4 =$ 2d Rem.
$$+1$$
$$+3\;\;\;\;\;\;\;\;\;\;\therefore +3 =$$ 3d Rem.

Hence, the required coëfficients are $1, +3, -4,$ and $+1$.
$\therefore y^3 + 3y^2 - 4y + 1 = 0$ is the transformed equation.

2. Transform the equation $x^3 - 27x - 36 = 0$ into another whose roots are less by 3.

SOLUTION.

$$1 \pm 0 - 27 - 36\;\underline{|\,+3\,}$$
$$+3+\;\;9-54$$
$$+3-18-90\;\;\therefore 90 =$$ 1st Rem.
$$+3+18$$
$$+6 \pm \;0\;\;\;\;\therefore \;\;0 =$$ 2d Rem.
$$+3$$
$$+9\;\;\;\;\;\;\;\;\therefore \;\;9 =$$ 3d Rem.

Hence, the required coëfficients are $1, +9, \pm 0,$ and -90.
$\therefore y^3 + 9y^2 - 90 = 0$ is the transformed equation.

3. Transform $y^4 + 8y^3 + 21y^2 + 25y + 8 = 0$ into an equation whose roots are 2 greater than the roots of this equation.

4. Transform the equation $y^3 - 15y^2 + 9 = 0$ into an equation whose roots are greater than the roots of the given equation by 5.

5. Find the equation whose roots are less by 10 than those of the equation $x^5 + 5x^3 - 3x - 5 = 0$.

6. Find the equation whose roots are less by 3 than the roots of the equation $x^4 - 3x^3 - 15x^2 + 49x - 12 = 0$.

7. Find the equation whose roots are less by 10 than the roots of the equation $x^4 + 3x^3 - 3x^2 + 7x - 317 = 0$.

COMMENSURABLE ROOTS.

8. Find the equation whose roots are less by 2 than the roots of the equation $y^5 + 37y^3 - 94y + 13 = 0$.

9. Transform the following equation into another equation, whose roots are less by 3 than the roots of the given equation $x^4 - 22x^3 + 179x^2 - 638x + 840 = 0$.

10. Transform the following equations into others, whose roots are less by 2 than those of the given equations:

1. $x^5 - 4x^4 + 3x^3 - 2x^2 + x - 1 = 0$.
2. $x^4 + 40x^3 - 20x^2 + 35 = 0$.
3. $y^5 - 1 = 0$.
4. $x^3 - 80x + 90 = 0$.

COMMENSURABLE ROOTS.

473. Commensurable Roots are the integers or rational fractions which are roots of an equation.

They may be either positive or negative.

EXAMPLES.

What are the commensurable roots of:

1. $x^3 - 9x^2 + 26x - 24 = 0$.

SOLUTION.

First. There are three roots (Art. 453).
Second. All roots are positive (Art. 460).
Third. Their product is 24 (Art. 455).

The moderate increase in the coëfficients as we pass from left to right indicates that the roots are not the largest possible factors of 24.

Try 2 as one of the roots. Then by synthetic division

$$\begin{array}{r} 1 - 9 + 26 - 24 (+2 \\ 2 - 14 - 24 \\ \hline -7 + 12 \end{array}$$

∴ $x^2 - 7x + 12 = 0$ is the equation which will give the other roots.

Solving, $\qquad x = 4 \text{ or } 3$

∴ The roots are 2, 4, and 3.

366 HIGH SCHOOL ALGEBRA.

2. $6x^3 - 11x^2 + 6x - 1 = 0.$

SOLUTION.

Transforming so that the first term shall have 1 for a coëfficient by substituting $\frac{y}{6}$ for x,

$$y^3 - 11y^2 + 36y - 36 = 0$$

First. There are three roots (Art. 453).
Second. All roots are positive (Art. 460).
Third. Their product is 36 (Art. 455).

Solving, as in Example 1, the roots of the transformed equation are 2, 3, and 6. Consequently the roots of the original equation are $\frac{1}{3}, \frac{1}{2}$, and 1

3. $x^4 - 14x^3 + 73x^2 - 170x + 150 = 0.$

SOLUTION.

Try 3 as one of the roots. Then, by synthetic division, the quotient is $x^3 - 11x^2 + 40x - 50 = 0.$
Try 5 as another root. Then the quotient is $x^2 - 6x + 10 = 0.$
Solving, the other roots are found.

$$\begin{array}{r} 1 - 14 + 73 - 170 + 150 \; \underline{|+3} \\ 3 - 33 + 120 - 150 \\ \hline -11 + 40 - 50 \underline{|+5} \\ 5 - 30 - 50 \\ \hline -6 + 10 \end{array}$$

$$\therefore x^2 - 6x + 10 = 0$$

Solving, $\qquad x = 3 \pm \sqrt{-1}$

$\therefore$ The roots of the equation are 3, 5, $3 + \sqrt{-1}$ and $3 - \sqrt{-1}$.

Find the commensurable roots of the following equations:

4. $x^3 - 10x^2 + 29x = 20.$

5. $6x^3 - 11x^2 - 45x + 72 = 0.$

6. $x^4 + 5x^3 - 28x^2 - 122x + 60 = 0.$

7. $x^4 - 5x^3 - 46x^2 + 152x + 192 = 0.$

8. $6x^4 - 25x^3 + 31x^2 - 25x + 25 = 0.$

9. $x^4 - 16x^3 + 89x^2 - 206x + 168 = 0.$

10. $x^4 - 48x^3 + 827x^2 - 6048x + 15876 = 0.$

REVIEW EXERCISES.

474. Find the value of x in the following equations:

1. $\dfrac{9x-7}{2} + \dfrac{11}{x} + \dfrac{4x+12}{2} = \dfrac{14}{x} + \dfrac{75x}{6} - \dfrac{7x+6}{2x}.$

2. $\dfrac{x-bx}{x-1} + \dfrac{x^3}{b(x-1)} = \dfrac{x^2-2b}{b} + 2.$

3. $ax + \dfrac{bx+4a}{4} - ab^2 = bx + \dfrac{a^2-3bx}{a} + \dfrac{6bx-5a^2}{2a}.$

4. $\tfrac{1}{2}(x+1) + \tfrac{1}{3}(x+2) = 16 - \tfrac{1}{4}(x+3).$

5. $\dfrac{x-7}{x+7} + \dfrac{1}{2(x+7)} = \dfrac{2x-15}{2x-6}.$

6. $\dfrac{5x-4}{9} + \dfrac{12x+2}{11x-8} = \dfrac{10x+17}{18}.$

7. $\sqrt{x+3} + \sqrt{x} = \dfrac{6}{\sqrt{x+3}}.$

8. $\sqrt{x+5} + \sqrt{x+12} = 7.$

9. $\sqrt{x} + \sqrt{a+x} = \dfrac{2a}{\sqrt{a+x}}.$

10. $\dfrac{2x-1}{x+1} + \dfrac{3x-1}{x+2} = \dfrac{5x-11}{x-1}.$

11. $\dfrac{\sqrt{a+x}+\sqrt{a-x}}{\sqrt{a+x}-\sqrt{a-x}} = c.$

12. $\dfrac{\sqrt{ax}+\sqrt{b}}{\sqrt{ax}-\sqrt{b}} = \dfrac{\sqrt{a}-\sqrt{b}}{\sqrt{b}}.$

13. $\dfrac{\sqrt{b}-\sqrt{b-\sqrt{b^2-bx}}}{\sqrt{b}+\sqrt{b-\sqrt{b^2-bx}}} = c.$

14. $a + x = \sqrt{a^2 + x\sqrt{b^2+x^2}}.$

15. $\dfrac{3x-1}{\sqrt{3x}+1} = 1 + \dfrac{\sqrt{3x}-1}{2}$.

16. $x^2 - 2x + 6\sqrt{x^2 - 2x + 5} = 11$.

17. $\sqrt{x^5} - \dfrac{40}{\sqrt{x}} = 3x$.

18. $(3x + .5)^2 + (5x - .5)^2 = 6(3x - .5)^2 + 4x$.

19. Reduce to its simplest form the expression

$\sqrt{-25} + \sqrt{-49} + \sqrt{-121} - \sqrt{-64} + \sqrt{-1} - \sqrt{-36} \pm 0\sqrt{-2}$.

Solve the following equations:

20. $\begin{cases} 3x - 2y + 3z = 26. \\ 5y + 2z = 22. \\ 6u - 4x = 14. \\ 8y + 3u = 31. \end{cases}$

21. $\begin{cases} 4x - 5y + 3z + 4v - 5u = 12. \\ 7y - 2z + u - 2v = 10. \\ 9z + 3u + 4v - 4x = 6. \\ 4x - 8y + 2z - 3u = 10. \\ 5u - 4v + 3x - 2y = 7. \end{cases}$

22. $\dfrac{\sqrt{a^2 + x^2} + x}{\sqrt{a^2 + x^2} - x} = \dfrac{b}{c}$.

23. $\dfrac{\sqrt{9x} - 4}{\sqrt{x} + 2} = \dfrac{15 + \sqrt{9x}}{\sqrt{x} + 40}$.

24. $x + \sqrt{a^2 + x^2} = \dfrac{na^2}{\sqrt{a^2 + x^2}}$.

25. $\dfrac{x-4}{12} - \dfrac{(x-5)(x+5)}{x+4} = x - 4$.

26. $\dfrac{\sqrt{a^2 - x^2} - \sqrt{b^2 + x^2}}{\sqrt{a^2 - x^2} + \sqrt{b^2 + x^2}} = \dfrac{c}{d}$.

REVIEW EXERCISES.

27. $x + \sqrt{a^2 + x^2} = \dfrac{a^2 + 2}{\sqrt{a^2 + x^2}}.$

28. $\dfrac{1}{1 - \sqrt{1 - x^2}} - \dfrac{1}{1 + \sqrt{1 - x^2}} = \dfrac{\sqrt{3}}{x^2}.$

29. $\begin{cases} 4x^2 + xy + 4y^2 = 58. \\ 5x^2 + 5y^2 = 65. \end{cases}$

30. $\begin{cases} \dfrac{a}{a+x} + \dfrac{b}{b+y} = 1. \\ x + y = a + b. \end{cases}$

31. $\begin{cases} x^3 + y^3 = (x+y)xy. \\ x + y = 4. \end{cases}$

32. $\begin{cases} x^2 - y^2 - (x+y) = 8. \\ (x+y)(x-y)^2 = 32. \end{cases}$

33. $\begin{cases} \dfrac{1}{x} + \dfrac{1}{y} = \dfrac{1}{2}. \\ \dfrac{2}{xy} = \dfrac{1}{9}. \end{cases}$

34. $\begin{cases} y\sqrt{x} + \sqrt{y} = 21. \\ xy^2 + y = 333. \end{cases}$

35. $\begin{cases} x^{\frac{3}{2}} + y^{\frac{2}{3}} = 3x. \\ x^{\frac{1}{2}} + y^{\frac{1}{3}} = x. \end{cases}$

36. $\begin{cases} x^3 - y^3 = 56. \\ x - y = \dfrac{16}{xy}. \end{cases}$

37. $\begin{cases} y^2 - 2xy = 12. \\ y^4 - 12xy^2 = 432. \end{cases}$

38. $\begin{cases} x^2 + 3x + y = 73 - 2xy. \\ y^2 + 3y + x = 44. \end{cases}$

39. $\begin{cases} x^2 + 2x^3y = 441 - x^4y^2. \\ xy = 3 + x. \end{cases}$

40. $\begin{cases} (2x - 4y)^2 + x - 2y = 5. \\ x^2 - y^2 = 8. \end{cases}$

41. $\begin{cases} 2x + 3y = 19. \\ 5x^2 - 7y^2 = 62. \end{cases}$

42. $\begin{cases} 4xy + x^2y^2 = 96. \\ x + y = 6. \end{cases}$

43. $\begin{cases} x^3 + y^3 = 35. \\ x^2y + xy^2 = 30. \end{cases}$

44. $\begin{cases} xy + xy^2 = 12. \\ y + xy^2 = 10. \end{cases}$

45. $\begin{cases} x^2 + y^2 = 20. \\ x^3y + y^3x = 160. \end{cases}$

46. $\begin{cases} x^4 - y^4 = 65. \\ x^3y + xy^3 = 78. \end{cases}$

47. $\begin{cases} x^2 + y^2 - x - y = 98. \\ xy + x + y = 71. \end{cases}$

48. $\begin{cases} x^2 + 2xy + y^2 + 2x + 2y = 120. \\ xy - y^2 = 8. \end{cases}$

ALGEBRA.—24.

49. $\begin{cases} x^2 + y^2 : x^2 - y^2 :: 25 : 7. \\ xy = 48. \end{cases}$
50. $\begin{cases} x + y + \sqrt{x+y} = 12. \\ x^3 + y^3 = 189. \end{cases}$

51. $\begin{cases} (3x+4y)(7x-2y) + 3x + 4y = 44. \\ (3x+4y)(7x-2y) - 7x + 2y = 30. \end{cases}$

52. $\begin{cases} 12 : x :: y : 3. \\ \sqrt{x} + \sqrt{y} = 5. \end{cases}$

53. $\begin{cases} \dfrac{x^2}{y} + \dfrac{y^2}{x} = 9. \\ x + y = 6. \end{cases}$

54. $\begin{cases} (x^2 + y^2)xy = 300. \\ x^4 + y^4 = 337. \end{cases}$

55. $\begin{cases} \dfrac{x^2 + xy + y^2}{x + y} = 14. \\ \dfrac{x^2 - xy + y^2}{x - y} = 18. \end{cases}$

56. $\begin{cases} x^2 + \dfrac{x^4}{y^2} + y^2 = 84. \\ x + \dfrac{x^2}{y} + y = 14. \end{cases}$

57. $\begin{cases} x^2 + y^2 - x - y = 8. \\ 2x + xy + 2y = 16. \end{cases}$

58. $\begin{cases} 2(x^2 + y^2)(x + y) = 15xy. \\ 4(x^4 - y^4)(x^2 - y^2) = 45x^2y^2. \end{cases}$

59. $\begin{cases} (x+y)^2 - 2x^2 = 49. \\ 3x^2 + 4(x+y)^2 = 372. \end{cases}$

60. $\begin{cases} (x+y)xy = 30. \\ (x^2 + y^2)x^2y^2 = 468. \end{cases}$

61. $\begin{cases} 5(x^2 + y^2) + 4xy = 356. \\ x^2 + y^2 + x + y = 62. \end{cases}$

62. $\begin{cases} x + y = 5. \\ (x^2 + y^2)(x^3 + y^3) = 455. \end{cases}$

63. $\begin{cases} 2xy - 24(x+y) = -240. \\ x^2 + y^2 = 100. \end{cases}$

64. $\begin{cases} x(x+y+z) = 60. \\ y(x+y+z) = 75. \\ z(x+y+z) = 90. \end{cases}$

65. $\begin{cases} x^2 + y^2 + z^2 = 125. \\ x^2 + y + z^2 = 69. \\ (x+y+z)y = 152. \end{cases}$

REVIEW EXERCISES.

66. $\begin{cases} x+y=a. \\ x^3+y^3=b. \end{cases}$
67. $\begin{cases} x+\sqrt{xy}=a. \\ y+\sqrt{xy}=b. \end{cases}$

68. $\begin{cases} (x-y)(x^2-y^2)=a. \\ (x+y)(x^2+y^2)=b. \end{cases}$

69. Insert 6 arithmetical means between $\frac{1}{2}$ and $\frac{2}{3}$.

70. Insert 3 geometrical means between 9 and $\frac{1}{9}$.

71. Find the arithmetical mean between $\frac{8}{5}$ and $-\frac{10}{7}$.

72. Find the geometrical mean between $23\frac{1}{3}$ and $4\frac{2}{7}$.

73. Expand by the binomial theorem $(a-x)^{\frac{1}{2}}$ to five terms.

74. Expand by the binomial theorem $(a^3+b^2)^{n+1}$ to five terms.

75. Expand $(1+2x-x^2)^3$ by the binomial theorem.

76. Find the fifth term of $(a^{-1}+2x^{\frac{1}{2}})^{10}$ by means of the binomial theorem.

77. Find the approximate value of $\sqrt[4]{125}$ to five decimal places, by the binomial theorem.

78. Expand $\dfrac{1-x}{1-3x-2x^2}$ to five terms in ascending powers of x.

79. Separate $\dfrac{7x^4+1}{x^8-1}$ into partial fractions.

80. Separate $\dfrac{5x^2+10x+4}{x^3+3x^2+2x}$ into partial fractions.

81. Show that 2 is a root of the equation $x^3-7x^2+14x-8=0$.

82. Show that 3 is not a root of the equation $x^3-3x^2+6x+8=0$.

83. One root of the equation $x^3-11x^2+37x-35=0$ is 5. Find the other roots.

84. Form the equation whose roots are 1, -2, and -4.

85. Given $3x + 5y + 7z = 100$ and $8x + 4y + 5z = 100$. Find all the different possible values of x, y, and z in whole numbers.

86. Separate 590 into two parts, so that their product shall be 80464.

87. Divide the number 327 into three parts proportional to the numbers $\frac{2}{3}$, 6, and $\frac{7}{8}$.

88. A gentleman bought a horse for a certain sum. He afterward sold him for $144, and gained as much per cent. as the horse cost him. How much did he pay for the horse?

89. An engraving, whose length was twice its breadth, was mounted on Bristol board, so as to have a margin 3 inches wide, and equal in area to the engraving, lacking 36 inches. What was the width of the engraving?

90. There is a stack of hay whose length is to its breadth as 5 to 4, and whose height is to its breadth as 7 to 8. It is worth as many cents per cubic foot as it is feet in breadth; and the whole is worth at that rate 224 times as many cents as there are square feet on the bottom. What are the dimensions of the stack?

91. I let fall a stone into a deep pit, and I hear it strike the bottom in seven seconds. Granting that a falling body falls 16 feet the first second, four times as far in two seconds, nine times as far in three seconds, and so on, also, that sound travels 1100 feet per second, how deep is the pit?

92. The hypotenuse of a right triangle is 24 feet more than the base, and 3 feet more than the perpendicular. Determine the sides of the triangle.

93. A farmer had two flocks of sheep, each containing 147. From the first flock he sold a certain number, and from the second a number expressed by the same digits inverted. One flock then contained twice as many sheep as the other. How many did he sell from each flock?

REVIEW EXERCISES.

94. There are two vessels, one holding a gallons of wine, the other holding b gallons of water. Determine the capacity of a third vessel, such that if it be separately filled from each of the first two, and the liquor from each be poured into the other, the resulting mixtures shall contain like proportions of wine and water.

95. The base of a right-angled triangle is 20 inches, and is a geometrical mean between the hypotenuse and the perpendicular. What is the hypotenuse?

96. It is required to find three numbers, such that the difference of the first and second shall exceed the difference of the second and third by 6, the sum of the numbers shall be 33, and the sum of the squares 441.

97. A piece of cloth, by being wet in water, shrinks one eighth in its length, and one sixteenth in its breadth. If the perimeter of the piece is diminished $4\frac{1}{4}$ feet, and the surface $5\frac{3}{4}$ square feet by wetting, what were the length and breadth of the piece?

98. A rectangular lot is 119 feet long and 19 feet broad. How much must be added to the breadth, and how much taken from the length, in order that the perimeter may be increased by 24 feet, and the contents of the lot remain the same?

99. From two towns, 396 miles apart, two persons, A and B, set out at the same time, and traveled toward each other. After as many days as are equal to the difference of miles they traveled per day, they met, when it appeared that A had traveled 216 miles. How many miles did each travel per day?

100. Find two numbers such that their sum, their product, and the difference of their squares may be all equal to one another.

101. Find two numbers, such that their product shall be equal to the difference of their squares, and the sum of their squares shall be equal to the difference of their cubes.

102. A miner bought two cubical masses of ore for $820. Each of them cost as many dollars per cubic foot as there were feet in a side of the other; and the base of the greater contained a square yard more than the base of the less. What was the price of each?

103. A May-pole was broken by the wind in such a manner that 4 times the upper or broken part, added to 6 times the remaining part, was equal to 5 times the whole, and 28 over; and the proportion of the former to the latter part was 9 to 16. Required the height at first.

104. Two messengers start from the two towns, A and B, to travel toward each other, but one started 2 hours earlier than the other. They meet each other $2\frac{1}{12}$ hours after the starting of the second messenger, and they reach the towns, A and B, at the same instant. In how many hours did each messenger perform the journey?

105. A ship, with a crew of 175 men, set sail with a supply of water sufficient to last to the end of the voyage; but in 30 days the scurvy made its appearance, and carried off 3 men every day; and at the same time a storm arose, which protracted the voyage 3 weeks. They were, however, just enabled to arrive in port without any diminution in each man's daily allowance of water. Required the time of the passage, and the number of men alive when the vessel reached the harbor.

106. What are the dimensions of a rectangular court, which, if lengthened 7 feet and made 4 feet broader, would contain 363 feet more; but if made 4 feet shorter and 3 feet narrower, would be diminished 208 feet?

107. A vintner has two casks of wine, from each of which he draws 6 gallons, and finds the remainders in the proportion of 4 to 7. He then puts into the less 3 gallons, and into the greater 4 gallons, and the proportion is that of 7 to 12. How many gallons were there in each at first?

REVIEW EXERCISES. 375

108. A and B traveled on the same road, and at the same rate, from Cumberland to Baltimore. At the 50th milestone from Baltimore, A overtook a drove of geese, which was proceeding at the rate of 3 miles in 2 hours, and 2 hours afterward met a wagon which was moving at the rate of 9 miles in 4 hours. B overtook the same drove of geese at the 45th milestone, and met the same wagon 40 minutes before he came to the 31st milestone. Where was B when A reached Baltimore?

109. In one of the corners of a rectangular garden there is a fish-pond of similar shape, whose area is $\frac{1}{9}$ of the whole garden, the periphery of the garden exceeding that of the pond by 200 yards. If the greater side be increased by 3 yards, and the other by 5 yards, the garden will be enlarged by 645 square yards. What is the length of each side of the garden?

110. The number of deaths in a besieged garrison amounted to 6 daily; and, allowing for this diminution, their stock of provisions was sufficient to last 8 days. But on the evening of the sixth day 100 men were killed in a sally, and afterward the mortality increased to 10 daily. Supposing the stock of provisions unconsumed at the end of the sixth day to support 6 men for 61 days, it is required to find how long it would support the garrison, and the number of men alive when the provisions were exhausted.

111. A wall was built round a rectangular court to a certain height. The length of one side of the court was 2 yards less than 8 times the height of the wall, and the length of the adjacent side was 5 yards less than 6 times the height of the wall. The number of square yards in the court was greater than the number in the inside face of the wall by 178. Required the dimensions of the court, and the height of the wall.

112. Find the value of $\sqrt{\frac{3}{7}}$ to within .001.

113. Find the value of $\sqrt{\dfrac{\sqrt{355}+\sqrt{113}}{\sqrt{355}-\sqrt{113}}}$ to within .001.

ANSWERS.

Page 10.—2. Vest, $6; coat, $24. **3.** Henry, $9; James, $27. **4.** 8 bu.; 16 bu. **5.** B, $200; A, $600. **6.** 1st, 50; 2d, 100; 3d, 300. **7.** Charles, 70; William, 280. **8.** 130. **9.** Cow, $50; horse, $200. **10.** B, 105; A, 315.

Page 11.—11. Sister, 120; brother, 360. **12.** Less, 90; greater, 450. **13.** B, $350; A, $1400. **14.** Wheat, 220 bu.; corn, 1100 bu. **15.** Rye, 150 bu.; corn, 300 bu.; wheat, 900 bu. **16.** A, $80; B, $160; C, $320. **17.** 1st, 13; 2d, 39; 3d, 117. **18.** $7260. **19.** 1st yr., $3450; 2d yr., $6900. **20.** 1st, $75; 2d, $225. **21.** A owns 2000; B, 6000; C, 2000. **22.** 3.

Page 12.—23. 2 ducks. **24.** Younger daughter, $1000; elder, $2000; son, $3000. **25.** 15 slate pencils. **26.** 1st part, 2; 2d, 16; 3d, 6; 4th, 12. **27.** 15. **28.** 4. **29.** 13. **30.** B, $3100; A, $12,400. **31.** Daughter, $1200; son, $3600; widow, $9600. **32.** Barley, 4; oats, 12; wheat, 16. **33.** 1st, 12; 2d, 36; 3d, 96.

Page 13.—34. Cherry, 20; peach, 60; apple, 480. **35.** John, 6¢; James, 36¢. **36.** Fiction, 4500. **37.** Sarah, 10¢; Mary, 50¢. **38.** 1st, 62; 2d, 124; 3d, 31. **39.** 1st yr., $1000; 4th yr., $8000. **40.** A, $3000; B, $2000; C, $1000. **41.** 1st class, 50; 2d class, 100; 3d class, 150; 4th class, 300.

Page 18.—1. 5. **2.** 4. **3.** 11. **4.** 15. **5.** 5. **6.** 1. **7.** 1. **8.** 9. **9.** 3. **10.** $1\frac{3}{5}$. **11.** 4. **12.** 70. **13.** 240. **14.** 36. **15.** 6. **16.** 18. **17.** 2. **18.** 8. **19.** 9. **20.** 32. **21.** 47. **22.** 34. **23.** 9. **24.** 8. **25.** $2\frac{8}{11}$. **26.** 5. **27.** 6. **28.** 12. **29.** 13. **30.** $27\frac{9}{15}$. **31.** $33\frac{1}{3}$. **32.** 16. **33.** 37. **34.** 31. **35.** 2. **36.** 0. **37.** 9. **38.** 4.

Page 21.—3. $26b$. **4.** $19ax$. **5.** $25x^2y$. **6.** $-23z^2y^2$. **7.** $-13cx^3$. **8.** $30ax$. **9.** $26mn$.

Page 22.—10. $-26x^2y^2$. **11.** $19x^3y^3$. **12.** $4a$. **13.** a^3x. **14.** $6\sqrt{xy}$. **15.** $-4(xy)^3$. **16.** $(x+y)^4$. **17.** $11(x-y)$. **18.** $3(a+b)$. **19.** $9(a-x)$. **20.** $5(a-b)^2$. **21.** $16(x+y)^3$. **22.** $10\sqrt{a^2-x^2}$.

Page 23.—3. $6a+4b-6c$. **4.** $7x-6xy+z$. **5.** $8x+9z-2xz$. **6.** $-2y+6z$. **7.** $2xy+6z-6y+4x$. **8.** $4ac+4ay$. **9.** $16b+5cd-9e-4c$. **10.** $-3x^2y+6xy+z$. **11.** $-3a+8c+8d$. **12.** $-5y+5w+z$. **13.** $10a^2b^2-7c^3y^3+d^2$.

Page 24.—14. $7ab+9\sqrt{xy}+25$. **15.** $12x^3+3x+6$. **16.** $5\frac{13}{16}ax^2+5\frac{2}{3}a^2+5\frac{3}{8}x^3y-2\frac{1}{15}b^3$. **17.** $2ab^2+\frac{1}{4}a^3+3\frac{3}{4}abc+1\frac{1}{3}a^2c+b^3+1\frac{1}{2}b^2c+c^3$. **18.** $6(a-x)+3x^2$. **19.** $20(a+b)^2+b^2c^2$. **20.** $4(a-x)+4ax-(ab+c)$. **21.** $-6(a+c)^2-4a(x+y)$. **22.** $6a(x+1)-2(y-2)+6a^2$. **23.** $\sqrt{a-x}+9x^2$. **24.** $10a-15(b+c)-4$. **25.** $5(x+3)+y^4+2ab^2$. **26.** $2ax(a-1)-2(b^2-2)-4y^2$.

Page 25.—27. $12\sqrt{a+b}-2x\sqrt[3]{a-3b}-7abc$. **28.** $3a\sqrt[5]{b-c}+5\sqrt{x}+4y$. **30.** $(2a-3b+4c+3d)x$. **31.** $(2a+4b+3c+4)x^2$.

(376)

ANSWERS.

32. $(6+5a)(a+b)$. **33.** $(3a+2b+7)(a+3)$. **34.** $(5a+5)\sqrt{x+y}$.
35. $(a+b+1)(x+y)$. **36.** $(b+b^2+b^3)(x-y)$. **37.** $(4a+6c)\sqrt{a-b}$.
38. $(a+b+c)(x^2+y^2)$. **39.** $(11+5x)\sqrt{a^2-c^2}$. **40.** $(2+5b)(x+y+1)$. **41.** $(2a+2b+2c)\sqrt[3]{x-y}$.

Page 26.—**2.** 2. **3.** 4. **4.** 2. **5.** 4. **6.** 4. **7.** 4.
8. 9. **9.** 2. **10.** 4. **11.** Harvey, 7; Henry, 21; James, 42.
12. C, $20; B, $40; A, $80. **13.** Samuel, 5; Henry, 15; William, 30. **14.** B, $200; C, $400.

Page 27.—**15.** 10. **16.** Fiction, 2000; reference, 20000; historical, 6000. **17.** A, $20000; B, $2000; C, $6000; D, $8000. **18.** A, 612; B, 306; C, 204. **19.** Board, $36; wages, $60. **20.** 1st man, $60; 2d, $90; 3d, $120; 4th, $150. **21.** Men, 11; women, 22; children, 44. **22.** Library, $10000; hospital, $20000; school, $40000.

Page 29.—**3.** $9a$. **4.** $5xy$. **5.** $-2x^3y^2$. **6.** $-3xyz$.
7. $-12x^2y^3z$. **8.** $-3a^2b^3c$. **9.** $4x+4y$. **10.** $a-b$. **11.** $-2xy+2z$.

Page 30.—**12.** $2x^2y^2-3z$. **13.** $-2xy^3z-xy$. **14.** $p^2qs+3pq^2s$.
15. $2m^2nx-mnx$. **16.** $4x^2y+5y^2$. **17.** $-2xy^2-3z$. **18.** $-4p^2q^2-pq$.
19. $14x^2y^2z^2-2y^2$. **20.** $-5yz^4+2y^4z$. **21.** $6p^2q^2-2qs$.
22. $-9xyz^3-3xyz$. **24.** $6a^3x$. **25.** $8x^3y^3$. **26.** $4x+3y$.
27. $3y-6z$. **28.** $4ax+5by$. **29.** $a+6b-8c$. **30.** $2x+7y+2z$.
31. $2xy+6z-3x^2+y$.

Page 31.—**32.** $-b+2c$. **33.** $x+5y-7z$. **34.** $3a^2+5b^2+5c^2$.
35. $6a^3-6c^3-6d^3$. **36.** $11x^4-7y^2$. **37.** $13p^2+6q^2-2r^3$.
38. $-ax+4ay$. **39.** $8yz+7xz$. **40.** $-6x^3y^2+24xy^3+14xy$.
41. $9x^3y^3-11yz^3$. **42.** $x^2+5xy+5z^2+w$. **43.** $15x^3+5y^3+4z^3-7r^3$. **44.** $3x^5y+7x+4$. **45.** $5bx^3+3ay^2-2cy+9$. **46.** $x^2y^4+5xy-9x+5$. **47.** $2x^5y^2-9xy^5-9z^4-9$. **48.** $4ar^2-6bs^3+3rs-p-7$. **49.** $14x^5-39x^3y^3-11y^4-4z$. **50.** $-x^m-6x^ny^m+4y^m+4x^{2m}$. **51.** $7x^{2n}-4x^{3n}y^m-4y^{m-1}$. **52.** $\sqrt{xy}+5z+\sqrt[3]{y^2}$. **53.** $6(a+b)^2-4a+6c$. **54.** $5\sqrt{a+b^2}-9\sqrt[3]{x+y}+7\sqrt{x+y}$. **55.** $\sqrt{a+b^2}-5\sqrt[3]{c+d}$.

Page 32.—**57.** $(a-c)y+(d+2)x$. **58.** $(c-3a)x+(2a-3d)z$.
59. $(2c-2a)d-6ab+(c+2d)e^2$. **60.** $(a-b)x+(a+b)y+(c-1)z$.
61. $(5a-c)y+(a+2c)z+(d-6)x$. **62.** $(a-2b)x^2+(3a+2c)y+(3+c)x^3y$. **63.** $(2p-r)x^2+(r+p)y^2-(3q+s)xy$. **64.** $(c-9)x+(7a^2-15)b^2$. **65.** $(3-bc)x+(7+a)y-(8+q)z$. **66.** $(c-b)x+(a+c)y$. **67.** $(d-c)x+2ay$. **68.** $(4b-4c)(x-y)+(4cd-a)x$.
69. $2\sqrt{x}+(3-2a)\sqrt{y}$. **70.** $(a^2+c^2)x+(b^2+c^2)y$. **71.** $(a-b)\sqrt{x+y}+(a+b)\sqrt{x-y}+(c+d)x$. **72.** $(2n+3)\sqrt[m]{x^2y^2}-2m\sqrt[n]{xy}-y^2$.

Page 33.—**1.** $-b$. **2.** y. **3.** $2a+b$. **4.** $2a+b$. **5.** b.
5. $x+2y$. **7.** $2a-y$. **8.** $x+4y$.

Page 34.—**9.** $7x-7y$. **10.** $6x-y+2z$. **11.** $x-z^2$. **12.** $-xy+3x^3y-x^2$. **13.** $7x^2+4y^2+z^2$. **14.** $6ab^2+4ac^2$. **15.** $2b$.
16. $x-a$. **17.** 4. **18.** $2a^2x-19$. **19.** $8ab$. **20.** $3x+1-x^2$.
21. $2bx+11$. **22.** $a-b+2c-d$. **23.** $4c$. **24.** a. **25.** $x+c$.
26. $3a-2b+c-2d$. **27.** $6y-5x^2+7x^3$. **28.** $-5x^2y+4y+1$.
29. $ab-2bc-4bd-6c$.

Page 35.—**30.** $6xy+11z$. **31.** $5x$. **32.** y^2-c^2. **33.** b.
34. a. **35.** $a-10$. **36.** 1.

Page 38.—**2.** 4. **3.** 8. **4.** 12. **5.** 7. **6.** 10. **7.** 4.
8. 2. **9.** 5. **10.** 3. **11.** 4. **12.** 2. **13.** 3. **14.** 4. **15.** 5.
16. $1\frac{1}{2}$. **17.** 6. **18.** 4. **19.** 7. **20.** 8. **21.** 3. **22.** 11. **23.** 2.
24. 17. **25.** 8. **26.** 25. **27.** 46. **28.** 18. **29.** 43. **30.** 15.
31. 18. **32.** 10. **33.** 12. **34.** 11.

378   HIGH SCHOOL ALGEBRA.

Page 39.—35. 13.   **37.** John, 20; James, 30; Henry, 35.   **38.** 1st, 110; 2d, 130; 3d, 155.   **39.** 48.   **40.** $1500; $1650; $1800; $1950.   **41.** $2000; $2250; $2500; $2750.

Page 43.—3. -24.   **4.** -12.   **5.** $21\,a$.   **6.** $-12\,x$.   **7.** $-20\,x$.   **8.** $6\,x^5$.   **9.** $6\,x^7$.   **10.** $8\,x^5y^2$.   **11.** $6\,x^5y^6$.   **12.** $-12\,x^3m^2y^3$.   **13.** $-40\,x^3y^5z^3$.   **14.** $16\,x^4y^3z^3$.   **15.** $-20\,a^3b^3$.   **16.** $-24\,a^3b^3x^4$.   **17.** $6\,x^5y^3$.   **18.** $15\,a^6b^4x^2y^2$.   **19.** $-12\,c^4d^2y$.   **20.** $-20\,ax^5y^4$.   **21.** $-15\,a^4x^6y^3z$.   **22.** $-20\,x^3y^4z$.   **23.** $24\,x^4y^4z^2$.   **24.** $-12\,a^2c^2y^4z^2$.   **25.** $-15\,abx^4yz$.   **26.** $2(x+y)$.   **27.** $-12(a+b)$.

Page 44.—28. $15(y+z)^5$.   **29.** $4(a-b)^6$.   **30.** $6(c+d)^5$.   **31.** $-10(x+y+z)^7$.   **32.** $12\,x^{2n}$.   **33.** $-20\,a^{3n}$.   **34.** $-15\,a^{3}x^{3n+1}$.   **35.** $8\,a^{2}x^{m+n}$.   **36.** $-15\,a^{3n}x^{4n}$.   **37.** $20\,x^{m+n}y^{m+n}$.   **38.** $-15\,x^{2n+2}$.   **39.** $3\,x^2y - 6\,y^2$.   **40.** $2\,x^2yz - 4\,z^2$.   **41.** $12\,x^3y - 6\,x^2y^2$.   **42.** $-6\,x^4y - 4\,x^2y^3$.   **43.** $-16\,x^4y^2z^2 - 8\,x^2z^4$.   **44.** $9\,x^3y^3z - 6\,xy^2z^2$.   **45.** $4\,x^4y + 2\,xy^2 + 3\,xyz$.   **46.** $6\,x^3yz + 2\,xyz - 6\,x^2z^2$.   **47.** $18\,x^3y^4 + 12\,xy^4$.   **48.** $12\,a^2bcd - 9\,a^2c^2d$.   **49.** $-25\,a^2c^2x + 30\,a^2cx^2$.   **50.** $-20\,a^2b^2c^2d + 12\,a^2bc^2d^2$.   **51.** $-6\,a^3x^3y + 4\,a^3bcx^2$.   **52.** $6\,x^3y^2z + 14\,x^2yz^2$.   **53.** $12\,ab^3c^2 - 9\,bc^4$.   **54.** $15\,a^4x^5 - 12\,a^3x^4$.   **55.** $20\,x^{n+4}y^3 + 15\,x^ny^4$.   **56.** $18\,a^4x^2 - 27\,a^3x^3$.   **57.** $6\,a^{2n+5}x^3 - 9\,a^{2n+1}x^{n+1}$.   **58.** $35\,a^2bc^2dy^2 + 20\,d^2xy^2$.   **59.** $\tfrac{1}{4}a^5x^7 - \tfrac{9}{20}a^4x^4$.   **60.** $14\,a^{2n}b^{m+1} + 21\,a^{n+1}b^{m+2}$.

Page 45.—4. $x^2 - y^2$.   **5.** $3\,a^2 + 10\,ac + 3\,c^2$.   **6.** $12\,a^2 - 18\,ab + 6\,b^2$.   **7.** $6\,y^2 + yz - 12\,z^2$.   **8.** $4\,x^2 + 6\,xy + 2\,y^2$.   **9.** $9\,x^2 - 24\,xy + 16\,y^2$.   **10.** $15\,a^2 - 29\,ac - 14\,c^2$.   **11.** $a^2c^2 + 2\,abxy + b^2y^2$.   **12.** $4\,a^2c^2 - 9\,b^2c^2$.   **13.** $6\,b^2d^2 + b^2cd - 12\,b^2c^2$.   **14.** $6\,x^4y^4 + x^2y^2z^2 - 12\,z^4$.   **15.** $6\,x^3y^2z^2 + 4\,xy^3 + 3\,x^2z^3 + 2\,yz$.   **16.** $8\,a^2b^3 + 6\,ab^2c^2 + 8\,ab^3c^2 + 6\,b^2c^4$.   **17.** $25\,x^4y^4 - 15\,ax^2y^3 - 10\,ax^4y + 6\,a^2x^2$.   **18.** $a^3 + 3\,a^2b + 3\,ab^2 + b^3$.   **19.** $x^3 + 6\,x^2 + 12\,x + 8$.   **20.** $a^3 - 2\,ay^2 + y^3$.   **21.** $6\,a^3 - 3\,a^2b - 9\,ab^2 + 6\,b^3$.   **22.** $a^8 - a^2$.   **23.** $x^6 - y^6$.   **24.** $6\,x^2 - 5\,xy + 2\,xz - 6\,y^2 + 23\,yz - 20\,z^2$.   **25.** $6\,a^4 + 11\,a^3b + a^2c^2 - 10\,a^2b^2 + 31\,abc^2 - 15\,c^4$.   **26.** $21\,x^4 - 34\,x^3y + 34\,x^2y^2 + 2\,xy^3 - 15\,y^4$.   **27.** $1 - 5\,x + 11\,x^2 - 12\,x^3 + 6\,x^4$.   **28.** $a^4 + a^2c^2 + x^4$.   **29.** $x^2 + 3\,xy - 3\,xz + 2\,y^2 - 5\,yz + 2\,z^2$.   **30.** $a^{2m} - b^{2m}$.   **31.** $x^{2n} + 2\,x^ny^n + y^{2n}$.   **32.** $x^{m+n} + x^ny^m + x^my^n + y^{m+n}$.   **33.** $x^{2m+2n} + 2\,x^{m+n}y^{m+n} + y^{2m+2n}$.   **34.** $a^{2m-2n} - b^{2m-2n}$.

Page 46.—35. $x^2 + 2\,xy + y^2$.   **36.** $4\,x^2 - 4\,xy + y^2$.   **37.** $9\,x^2 - 16\,y^2$.   **38.** $16\,x^2 - 36\,y^2$.   **39.** $9\,a^2x^2 + 6\,axy + 6\,axz + 4\,yz$.   **40.** $4\,x^2 - 8\,x^2y - 4\,xz + 8\,xyz$.   **41.** $9\,a^4 - 4\,b^2c^2$.   **42.** $a^3 + ab + a^2b^2 + b^3$.   **43.** $a^2 - b^2 - 2\,bc - c^2$.   **44.** $a^3 + 3\,a^2b + 3\,ab^2 + b^3$.   **45.** $a^4 - 2\,a^2b^2 + b^4$.   **46.** $x^4 - 2\,x^2 + 1$.   **47.** $a^4 - 2\,a^2b^2 + b^4$.   **48.** $1 + 2\,a - 2\,a^3 - a^4$.   **49.** $x^8 - 4\,x^6y^2 + 6\,x^4y^4 - 4\,x^2y^6 + y^8$.   **50.** $a^{16} - 2\,a^{12}b^4 + 2\,a^4b^{12} - b^{16}$.   **51.** $a^{16}b^{16} - 2\,a^{12}b^{16} + 2\,a^4b^{16} - b^{16}$.   **52.** $-x^4 + 2\,x^2y^2 + 2\,x^2z^2 - y^4 + 2\,y^2z^2 - z^4$.   **53.** $1 - x^{16}$.   **54.** $16\,x^4 - 81$.   **55.** $64\,y^{2p} - 9\,y^{4p}$.   **56.** $m^7 - m^6n + 2\,m^5n^2 - m^4n^3 + m^3n^2 - m^2n^3 + m^3n^4 + mn^4 + n^7$.   **57.** $x^5 - x^4 - 27\,x^3 + 10\,x^2 - 30\,x - 200$.   **58.** $-x^2 - 3\,x^4 + 8 - x^3 + 6\,x + 4\,x^5 - 4\,x^6$.   **59.** $a^3 + 3\,ab + b^3 - 1$.   **60.** $x^6 - 3\,x^5 + 2\,x^4 + x^3 - 2\,x^2 + x + 1$.   **61.** $3\,m^5 - 5\,m^4p + 8\,m^3p^2 - 6\,m^2p^3 + 4\,mp^4 - p^5$.   **62.** $a^2 - 4\,b^2 + 12\,bc - 9\,c^2$.   **63.** $a^4 - 3\,a^2c - b^2 + bc + 2\,c^2$.   **64.** $m^4 - 2\,m^2n^2 - 4\,mnp^2 + n^4 - p^4$.

Page 47.—65. $r^4 - 4\,r^2t^2 - 4\,rt^3 - t^4$.   **66.** $m^p - mn^{p-1} - m^{p-1}n + n^p$.   **67.** $x^{2p+2} - x^{2p}y^{p-3c+2} + x^{p+2c}y^{p+4-3c} - x^{p+2-2c}y^{p+3c-2} + x^{p-2c}y^{2p} - y^{3p+2}$.   **68.** $x^y - xy^b + x^{1-a}y^{b-1} - x^{a-1}y^{1-b} + x^ay - y^0$.   **69.** $c^{3p} - d^{3q}$.   **70.** $a^4 + 2\,a^2c^2 - b^4 - 2\,b^2d^2 + c^4 - d^4$.   **71.** $a^{m+n} + a^nb^m + a^mb^n + b^{m+n}$.   **72.** $x^4 - 2\,x^3y + 2\,xy^3 - y^4$.   **73.** $a^5 + y^5$.

2. 6.   **3.** 4.   **4.** 2.   **5.** 17.   **6.** $1\tfrac{7}{9}$.   **7.** 2.   **8.** 12.   **9** -1.   **10.** 8.   **11.** $2\tfrac{1}{2}$.

Page 48.—12. 38.   **13.** 4.   **14.** 10.   **15.** 14.   **16.** 25.   **17.** $1\tfrac{1}{2}$.   **18.** 6.   **19.** 3.   **20.** 1.   **21.** 17.   **23.** 4.   **24.** A, 10; B, 10.   **25.** Henry, 9; John, 12.   **26.** 1st, 20; 2d, 35.

ANSWERS.

Page 49.—27. C, $5; B, $10; A, $20. **28.** Smaller, 10; larger, 40. **29.** B, $1400; A, $2800. **30.** Amount wanted, 46 lb.; in 1st firkin, 40 lb.; in 2d firkin, 60 lb. **31.** $120. **32.** 1st, 24; 2d, 25. **33.** 8 doz. at 25¢; 5 doz. at 30¢. **34.** B, 12; A, 36.

Page 50.—1. $c^2+2cd+d^2$. **2.** $m^2+2mn+n^2$. **3.** $r^2+2rs+s^2$. **4.** x^2+4x+4. **5.** a^2+6a+9. **6.** $9a^2+6ax+x^2$. **7.** $4b^2+4bc+c^2$. **8.** $4y^2+4y+1$. **9.** $m^2+4mn+4n^2$. **10.** $4c^2+8cd+4d^2$. **11.** $4x^2+12x+9$. **12.** $4x^2+16xy+16y^2$. **13.** $9a^2+12ab+4b^2$. **14.** $x^4+2x^2y^2+y^4$. **15.** $16x^2+24xy+9y^2$. **16.** $9p^2+12pq+4q^2$. **17.** $4x^4+20x^2y^2+25y^4$. **18.** $4y^6+12x^3y^3+9x^6$. **19.** $x^{2n}+2x^ny^n+y^{2n}$. **20.** $x^{6n}+2x^{3n}y^{2n}+y^{4n}$. **21.** $x^{2p-2q}+2x^{p-q}y^p+y^{2p}$. **22.** $x^{2m+2n}+2x^{m+n}y^{2n}+y^{4n}$.

Page 51.—24. 484. **25.** 529. **26.** 1681. **27.** 2704. **28.** 5041. **29.** 6724. **30.** 8281. **31.** 10201. **32.** 10609. **33.** 40804. **34.** 42849. **35.** 91809. **37.** $30\frac{1}{4}$. **38.** $72\frac{1}{4}$. **39.** $56\frac{1}{4}$. **40.** $156\frac{1}{4}$. **41.** $110\frac{1}{4}$. **42.** $90\frac{1}{4}$. **43.** 6.25. **44.** 12.25. **45.** 20.25. **46.** 2025. **47.** 56.25. **48.** 5625. **49.** 72.25. **50.** 7225. **51.** 1225.

Page 52.—1. $a^2-2ac+c^2$. **2.** $y^2-2yz+z^2$. **3.** $r^2-2rs+s^2$. **4.** $b^2-2bc+c^2$. **5.** x^2-2x+1. **6.** $x^2-4xy+4y^2$. **7.** $x^2-4xyz+4y^2z^2$. **8.** $4x^2-12xz+9z^2$. **9.** $4a^2-4ac+c^2$. **10.** $9y^2-12yz+4z^2$. **11.** $9x^2-24xy+16y^2$. **12.** $4a^2+8ad+4d^2$. **13.** $4r^2-12rs+9s^2$. **14.** $4s^2-4sq+q^2$. **15.** $9m^2-24mn+16n^2$. **16.** $4v^2-4vw+w^2$. **17.** $4x^4-8x^2y^2+4y^4$. **18.** $4x^2-12x+9$. **19.** $9a^2x^2-12ax^3+4x^4$. **20.** $x^{2m}-2x^my^n+y^{2n}$. **21.** $x^{2m-2n}-2x^{m-n}y^{m+n}+y^{2m+2n}$. **22.** $4x^4-12x^2+ny^n+9x^{2n}y^{2n}$. **24.** 324. **25.** 841. **26.** 1521. **27.** 1444. **28.** 2401. **29.** 2304. **30.** 3481. **31.** 3364. **32.** 6241. **33.** 6084. **34.** 9801. **35.** 9604. **36.** 994009. **37.** 996004. **38.** 998001.

Page 53.—1. c^2-d^2. **2.** r^2-s^2. **3.** m^2-n^2. **4.** c^2-a^2. **5.** x^2-1. **6.** $4-x^2$. **7.** c^2-4d^2. **8.** $4x^2-9$. **9.** $9m^2-16n^2$. **10.** $4x^2-25y^2$. **11.** $a^2b^2-c^2d^2$. **12.** $4x^2-16$. **13.** $4x^4-y^2$. **14.** x^4-y^4. **15.** x^8-y^8. **16.** $9v^2-4w^2$. **17.** $25x^2y^2-9$. **18.** $4a^4-9b^4$. **19.** $9a^2b^4-25b^2c^4$. **20.** $16x^6y^6-25$. **21.** $25x^{2m}-16y^{2n}$. **22.** $49a^4y^2x^6-36z^{2n}$. **24.** $x^2-y^2+2yz-z^2$. **25.** $a^2-b^2-2bc-c^2$. **26.** $a^2-2ab+b^2-c^2$. **27.** $m^2-n^2-2nr-r^2$. **28.** x^4-2x^2+1. **29.** $16-x^2+2xy-y^2$. **30.** $x^4+x^2y^2+y^4$. **31.** $16a^2-9b^2+6bc-c^2$. **32.** $4x^2+12xz+9z^2-16$. **33.** $4m^4+12m^2n-m^2n^2+9n^2$.

Page 54.—35. 399. **36.** 899. **37.** 891. **38.** 3584. **39.** 4884. **40.** 8099. **41.** 6396. **42.** 9991. **43.** 884. **44.** 1596. **45.** 3591. **46.** 2475. **47.** 9999. **48.** 9996. **49.** 9984. **50.** 9964. **53.** 361. **54.** 841. **55.** 961. **56.** 1521. **57.** 3481. **58.** 2401. **59.** 9604. **60.** 9216. **61.** 8836. **62.** 6084. **63.** 6241. **64.** 4624. **65.** 10609. **66.** 11449. **67.** 12544. **68.** 994009.

Page 55.—1. $x^2+7x+12$. **2.** $x^2-2x-15$. **3.** x^2-x-12. **4.** $x^2-10x+24$. **5.** $a^2+(b+c)a+bc$. **6.** $a^2+(m+n)a+mn$. **7.** $4x^2-2x-20$. **8.** $9y^2-9x-10$. **9.** x^4+4x^2-21. **10.** $x^6+ax^3-2a^2$. **11.** $9x^2-33x+30$. **12.** $4a^2+(x+y)2a+xy$. **13.** $25b^2+10bc-3c^2$. **14.** $9a^4-3a^2-6$. **15.** $16d^2+28d+10$. **16.** $49y^2-49y+12$. **17.** $9x^2-6x-35$. **18.** $4y^2-14y+12$. **19.** $16a^2+(b+c)4a+bc$. **20.** $25a^2+(2b-2c)5a-4bc$. **21.** $9a^2x^2-9ax-28$. **22.** $4a^4x^2-8a^2x-12$. **23.** $3x^4y^6+22x^2y^3+28$. **24.** $9a^2c^4-6ac^2-15$. **25.** $25c^4d^6+(x-y)5c^2d^3-xy$. **26.** $9a^2x^6+33ax^3+28$. **27.** $25c^2d^2x^2-20cdx-5$. **28.** $16c^6+(ab-d)4c^3-abd$. **29.** $a^2x^2-4ax-45$. **30.** $4a^2x^2+(4-b)2ax-4b$. **31.** $9x^{2n}+(m^n+n^m)3x^n+m^nn^m$. **32.** $a^2d^4x^6-13ad^2x^3+30$.

Page 56.—**1.** $x^2+y^2+z^2+2xy-2xz-2yz$. **2.** $x^2+y^2+z^2-2xy+2xz-2yz$. **3.** $a^2+b^2+c^2-2ab-2ac+2bc$. **4.** $a^2+b^2+c^2+2ab-2ac-2bc$. **5.** $x^2+y^2+9+2xy+6x+6y$. **6.** $4x^2+y^2+49+4xy-28x-14y$. **7.** $4x^2+y^2+z^2-4xy-4xz+2yz$. **8.** $9x^2+y^2+16+6xy-24x-8y$. **9.** $4x^2+9y^2+36+12xy-24x-36y$. **10.** $x^2+36y^2+25-12xy-10x+60y$. **11.** $9x^2+4y^2+9z^2-12xy+18xz-12yz$. **12.** $a^2+b^2+c^2+d^2+2ab+2ac+2ad+2bc+2bd+2cd$. **13.** $a^2+b^2+c^2+d^2-2ab-2ac-2ad+2bc+2bd+2cd$. **14.** $x^2+y^2+z^2+d^2+2xy-2xz-2dx-2yz-2dy+2dz$. **15.** $x^2+y^2+z^2+16+2xy+2xz+8x+2yz+8y+8z$. **16.** $9x^2+4y^2+9z^2+9+12xy+18xz+18x+12yz+12y+18z$. **17.** $4x^2+9y^2+4z^2+25-12xy-8xz+20x+12yz-30y-20z$. **18.** $4x^2+25y^2+z^2+w^2+20xy+4xz+4wx+10yz+10wy+2wz$. **19.** $9x^2+y^2+4z^2+25+6xy+12xz+30x+4yz+10y+20z$. **20.** $4x^2+9y^2+25+4z^2+12xy-20x+8xz-30y+12yz-20z$. **21.** $9x^2+49+4y^2-42x+12xy-30xz-28y+70z-20yz$. **22.** $16x^2+4y^2+4z^2+36-16xy-16xz+48x+8yz-24y-24z$. **23.** $4a^2+9b^2+4c^2-12ab+8ac+12bc$. **24.** $16a^2n^2+9a^2b^2+36+24a^2bn+48an+36ab$. **25.** $9a^2x^4+4b^2y^4+49-12abx^2y^2+42ax^2-28by^2$. **26.** $4x^2+9y^2+4z^2+16+12xy-8xz+16x-12yz+24y-16z$. **27.** $x^{2n}+y^{2n}+z^{2n}+w^{2n}+2x^ny^n+2x^nz^n+2x^nw^n+2y^nz^n+2y^nw^n+2z^nw^n$. **28.** $x^{4n}+y^{2n}+z^{6n}+64+2x^2ny^n-2x^{2n}z^{3n}-16x^{2n}-2y^nz^{3n}-16y^n+16z^{3n}$.

Page 59.—**4.** 2. **5.** -4. **6.** $-3ay$. **7.** $4xy$. **8.** $-3z^2$.
Page 60.—**9.** $-5xyz$. **10.** $2a^4b^4$. **11.** $2f$. **12.** $2x^2$. **13.** $-2xz$. **14.** $-3w$. **15.** $-3rz^2$. **16.** $7n$. **17.** $2y^2$. **18.** $-2ax^2y^2$. **19.** $4s$. **20.** $-18x$. **21.** -24. **22.** $-9n^ny^n$. **23.** $-5x^{-1}y^{-1}z$. **24.** $-\dfrac{7z^{2m}}{z^3x}$. **25.** $-\dfrac{5n^2}{m^2}$. **26.** $4xyz$. **27.** $ax-2y$. **28.** $3y-3x$. **29.** $2x+y$. **30.** $ab-2b^2$. **31.** $ab-c$. **32.** $3xy+z$. **33.** $a-3b+c^2$. **34.** $x-y+xy^2$. **35.** $x-2y+\dfrac{y^2}{x}$. **36.** $z-3x+3z$. **37.** $m+2-\dfrac{3m}{n}$. **38.** $c-3d+\dfrac{4d^2}{c}$. **39.** $-a(b+c)-b(b+c)^2$. **40.** $3-2(a-c)^2$. **41.** $-2+5a-10x$. **42.** $10y^2-7x^{-1}y+4$. **43.** $7a^{-1}x^5+9bx^4-8ab^2x^6$. **44.** $3d^{-1}x^2+4cx-5c^2dx^3$. **45.** $y^{-1}+cx+dx^2y$. **46.** $a(b-c)^2+b(b-c)^3-c(b-c)$.

Page 61.—**47.** $-3(x+y)^{-1}+9-6(x+y)^3$. **48.** $\tfrac{1}{2}x^3-\tfrac{1}{4}x^2-\tfrac{1}{6}x-1-\dfrac{1}{2x}$. **49.** $x^{m-3}+x^{m-2}+x^{m-1}+x^m$. **50.** $1-y-y^2-y^3$.

Page 63.—**7.** $a-b$. **8.** $x+2$. **9.** $3+x$. **10.** x^2+y^2. **11.** a^3+y^3. **12.** $x^2+2xy+y^2$. **13.** $r+s$. **14.** $x^3+3x^2y+3xy^2+y^3$. **15.** $c^2+2cd+d^2$. **16.** x^2+5x+7. **17.** $a+x$. **18.** $a+b-c$. **19.** $x-2ay+y^2$. **20.** x^2-ax-b. **21.** $5a^2+2ab-3b^2$. **22.** $3x^2-5y^2+3z^2$. **23.** $2a^2-3a+1$. **24.** $2a+3b$. **25.** $27a^2+12ab+6b^2-1$. **26.** $5a^3+4a^2+3a+2$. **27.** $x^2-xy+y^2-xz-yz+z^2$. **28.** $6x^2-7x+8$.

Page 64.—**29.** $8x^3+12ax^2-18a^2x-27a^3$. **30.** $a^2-2ax+4x^2$. **31.** x^2+xz+z^2. **32.** $x^3+x^2y+xy^2+y^3$. **33.** $x^4-x^3y+x^2y^2-xy^3+y^4$. **34.** $x^5-x^5+x^4-x^3+x^2-x+1$. **35.** $x^3+3x^2y+9xy^2+27y^3$. **36.** $27a^3-18a^2b+12ab^2-8b^3$. **37.** $x^{n-1}-x^{n-2}y+x^{n-3}y^2$, etc. **38.** $a^2+2ab+3b^2$. **39.** $4x^2-6xy+9y^2$. **40.** x^2+ax-a^2. **41.** $a^2-2ax+x^2$. **42.** x^2-2x+2. **43.** $4m^6+6m^3n^2+9n^4$. **44.** $x^3-3x^2y+3xy^2-y^3$. **45.** $x^4-x^3+x^2-x+1$. **46.** $a-b$. **47.** $a^3+2a^2b+2ab^2+b^3$.

Page 65.—**3.** $c+3$. **4.** $a-4$. **5.** $2a+3$. **6.** $d-3a$.
Page 66.—**7.** $a+b$. **8.** $2a+3b$. **9.** a^2+c. **10.** $3a+2b$.

ANSWERS. 381

11. $7+5b$. **12.** $5a+2b$. **13.** $2c^2-d$. **14.** $b-3c$. **15.** $2m^2-3m+1$. **16.** a^2-6a+9. **17.** $2m^2+3mn+n^2$. **18.** $3a^2-b+b^2$. **19.** $110. **20.** 1st, $1000; 2d, $2000; 3d, $4000; 4th, $3500. **21.** 5 beggars; 19 cents. **22.** 6th, 10 years; 5th, 14 years; 4th, 18 years; 3d, 22 years; 2d, 26 years; 1st, 30 years. **23.** 7 gallons.

Page 67.—24. $\frac{4}{7}$. **25.** B, $1000; A, $1100; C, $1300. **26.** In 1st, 200; 2d, 160. **27.** $600. **28.** 1st, $28; 2d, $48; 3d, $76. **29.** 1st, 12; 2d, 15; 3d, 10. **30.** To Denver, 100; Omaha, 110; St. Paul, 335; Chicago, 670.

1. $20ax+20+8\sqrt{x}+10x^2$. **2.** $2am+5x+3\sqrt{y}+z-x^2$. **3.** $4\sqrt{a^2-b^2}-\sigma(x+y)-6$.

Page 68.—4. $3\sqrt{x}-\sqrt{y}-3z+22+y$. **5.** $(a^2-b^2)x^2-ay+(c-3)y^5-3z^6$. **6.** $-4x^{2n}-6x^2y^2+6x^2y^n-6yz+4az+4x^m+6z$. **7.** $x^6+2x^4y+x^3y^3+2x^5y+4x^3y^2+2x^2y^4-x^4y^2-2x^2y^3-xy^5$. **8.** $x^{2n}+4x^{2m}y^n+2xn^ny^n+4x^{2n}y^{2n}+4x^ny^{2n}+y^{2n}$. **9.** $3x^{2n}+2x^{3n}y^{2n}-xn^ny^n-4xn^ny^{2n}-2x^{2n}y^{1n}+y^{3n}+3x^{2m}y^n+2x^{3n}y^{3n}$. **10.** $9x^{n+4}+6x^2y^{n+m}+3x^2z^m-6x^{n+2}y^{n-m}-4y^{2n}-2y^{n-m}z^m+3x^{n+2}z^{2m}+2y^{n+m}z^{2m}+z^{3m}$. **11.** x^2-3x+2. **12.** x^2+x-6. **13.** $x^2-3x-18$. **14.** x^2-x-90. **15.** $a^2-3a-28$. **16.** x^2-5x+6. **17.** x^2+x-30. **18.** a^2+a-2. **19.** y^4-3y^2-28. **20.** $c^2+7c-44$. **21.** m^2+5m-4. **22.** $p^2+23p+132$. **23.** $y^2+7y-18$. **24.** $d^2+35d+300$. **25.** $324-16x^2$. **26.** 99. **27.** $x^2+4xy+4y^2$. **28.** $a^{2n}-b^{2n}$. **29.** $9m^4-4m^2$. **30.** $x^{2n}-x^2$. **31.** $a^2+2ab+b^2-c^2$. **32.** $x^4-2x^2y^2+y^4$. **33.** $1-x^{16}$. **34.** $x^5+5x^4y+10x^3y^2+10x^2y^3+5xy^4+y^5$. **35.** a^4-8a^2+16. **36.** $81a^4-648a^2+1296$.

Page 69.—37. $p^4-2p^2q+q^2$. **38.** $x^6+2x^3y^2+y^4$. **39.** 625. **40.** 4096. **41.** 324. **42.** $4x^2+20xy+25y^2$. **43.** $9x^4-12x^2y^2+4y^4$. **44.** $9a^2-12ab+4b^2$. **45.** $4a^2x^2+12a^2xy+9a^2y^2$. **46.** $16a^4-24a^3y+9a^2y^2$. **47.** $4a^{2m}+4a^mb^{2n}+b^{4n}$. **48.** $x^{4n}+4x^{2n}y^{2n}+4y^{4n}$. **49.** $x^{4+2n}-2x^2+ny^2+y^4$. **50.** $a^2+b^2+c^2+2ab+2ac+2bc$. **51.** $a^2+b^2+c^2-2ab+2ac-2bc$. **52.** $a^2+b^2+c^2+2ab-2ac-2bc$. **53.** $4x^2-y^2$. **54.** $9x^2-49y^2$. **55.** $16x^4-4y^4$. **56.** $9a^4b^2-4b^4$. **57.** $49a^4-4b^2$. **58.** $25b^4-36c^4$. **59.** $64x^2y^2-9x^4$. **60.** $81a^2m^2-64x^2y^2$. **61.** $a^2x^{2n}-y^{2n}$. **62.** $a^2x^{4n}-a^2y^{4n}$. **63.** a^8-x^8. **64.** $x^{16}-y^{16}$. **65.** $256x^8-2592x^4+6561$. **66.** $2a^2+3ab+b^2$. **67.** $x+y$. **68.** m^2-4m-4. **69.** $x^3-2x^2+10x-24$. **70.** x^3+2x^2-x. **71.** $a^4-a^3+a^2-a+1$. **72.** $x^3-3x^2y+3xy^2-y^3$. **73.** $2x^2-4xy+2y^2$. **74.** $x^5-4x^4y+11x^3y^2-x^2y^3-xy^4+3y^5$. **75.** $a-b-c$.

Page 70.—76. -7. **77.** $2a^3+a^2b-2ab^2+2b^3$. **78.** a^2+3b^2. **79.** $1+a$. **80.** $18+3x$. **81.** $2a+3c$. **82.** $a+3b-4c$. **83.** $5x$. **84.** $-5a$. **85.** $-4x$. **86.** $11x-36y$. **87.** $a-2b+3c$. **88.** x. **89.** $3c-4a-3b$. **90.** $4z$. **91.** $(a-c)x-(a+b)y+(b-c)z$. **92.** $(2a-5)x+(4c-1)y+(3a-3b)z$. **93.** $(4a-5b)x+(b+c)y+(3a-3d)z$. **94.** $(3a-6b)x+(2a-4)y+(5a-3b)z$. **95.** $(4b-4c-a)y-(3b+2c)x+(a+2c)z$. **96.** $(a-4b-2c)x-(a+2b+3c)y+(a+3c)z$. **97.** $(6a-5c)x+(5a+2b+6c)y+(3c-6b)z$. **98.** $(3a-4b-2c)x-(2b+2c)y+4cz$.

Page 71.—99. $x^6+2x^3y^3+y^6$. **100.** $x^{10}-2x^5y^5+y^{10}$. **101.** $4x^2+4x+1$. **102.** $9x^4+24x^2y^3+16y^6$. **103.** $25a^{2n}-20a^nc^2+4c^4$. **104.** $4a^2d^2+12acd^2+9c^2d^2$. **105.** $36x^4-36x^2y^3+9y^6$. **106.** $9a^4y^{2n}+12a^2c^3xny^n+4c^6x^{2n}$. **107.** $4x^4-9y^4$. **108.** $4a^5-b^4$. **109.** $9x^2-4a^2$. **110.** $25x^4-9$. **111.** $49b^2c^2-d^4$. **112.** $4x^2+14x+6$. **113.** $4x^2-16x+15$. **114.** $a^2x^2+abx-6b^2$. **115.** $x^2+y^2+z^2+2xy+2xz+2yz$. **116.** $x^2+4y^2+9z^2+4xy-6xz-12yz$. **117.** $x^2+9y^2+4z^2-6xy+4xz-12yz$. **118.** $m^2+n^2+p^2+q^2-2mn+2mp-2mq-2np+2nq-2pq$.

119. 5. **120.** 1. **121.** 5. **122.** 6. **123.** 2. **124.** 10.
125. 2. **126.** $\frac{2}{5}$. **127.** 2. **128.** $a+b$. **129.** $a^2-2ab+b^2$.
130. $a+b$. **131.** $\dfrac{7bc-3b^2}{b-3c}$. **132.** $b-a$. **133.** $2a-3b$. **134.** $3+2a+c$. **135.** $2a+3c+d$.

Page 73.—**2.** $a^2(5b+6c)$. **3.** $4x^2(2y^2+3z^2)$. **4.** $6xyz(1+2xy)$.
5. $9xy^2z(x^2+2z^2)$. **6.** $a^2xyz(xy+z)$. **7.** $c(a^2+b^2+cd^2)$. **8.** $xy(4x+cy+3y^2)$. **9.** $2ax(2b+3ax+4)$. **10.** $3ay(a^2-2ay+3y^2)$.
11. $ac(2a-2ac+3)$. **12.** $cd(5a-2cd+b)$. **13.** $c(4b^2c-12ab-9c)$.
14. $ab(3a+c-d)$. **15.** $5a^2x^2(a-x+2z)$. **16.** $xy(6xy^2z-3yz^2+1)$.
17. $3acx(4a+5c-4)$. **18.** $3a^2b^3(2a^2b^2+7a-8b)$. **19.** $5c^2x(4cx-3c^2+x)$. **20.** $8x^3y^6(7+14xy^2-27x^3y^3)$. **21.** $5x^4y^6(13xy^3-17x^3+51y^2)$. **22.** $75ay^4(a^2+2ay^2-3y^3)$. **23.** $48b^2c^6(1-3b^4c^2-4b^6c^3)$.

Page 74.—**1.** $(a+b)(a+b)$. **2.** $(x+y)(x+y)$. **3.** $(b-c)(b-c)$.
4. $(r+s)(r+s)$. **5.** $(x+1)(x+1)$. **6.** $(x+2)(x+2)$. **7.** $(y-1)(y-1)$. **8.** $(2y-1)(2y-1)$. **9.** $(3x+1)(3x+1)$. **10.** $(4x+2)(4x+2)$. **11.** $(3y-3)(3y-3)$. **12.** $(z^3+8)(z^3+8)$. **13.** $(4ab^2-bc^2)(4ab^2-bc^2)$. **14.** $(3m+3n)(3m+3n)$. **15.** $(3+x)(3+x)$.
16. $(1-x^2)(1-x^2)$. **17.** $(4n-1)(4n-1)$. **18.** $(4+2a)(4+2a)$.
19. $(6+a^2)(6+a^2)$. **20.** $(7-x^3)(7-x^3)$. **21.** $(9x-a)(9x-a)$.
22. $(2a^n+3b^n)(2a^n+3b^n)$. **23.** $(x-15)(x-15)$. **24.** $(2x^2-16)(2x^2-16)$. **25.** $(7x-8y)(7x-8y)$. **26.** $(14abc+4bcd)(14abc+4bcd)$.

Page 75.—**29.** $(3x-2y)(3x-2y)$. **30.** $(2+x-y)(2+x-y)$.
31. $(a+b+3)(a+b+3)$. **32.** $(4+a-b)(4+a-b)$. **33.** $(3x-y)(3x-y)$. **34.** $(2x-y)(2x-y)$. **35.** $(5x-3y)(5x-3y)$.
36. $(2x+y)(2x+y)$. **37.** $(a+2c)(a+2c)$. **38.** $(2a+c)(2a+c)$.
39. $(x+z)(x+z)$. **40.** $(3x+z)(3x+z)$. **41.** $(a+5c)(a+5c)$.
42. $(2a+6c)(2a+6c)$. **43.** $(4a+4y)(4a+4y)$. **44.** $(5x+5z)(5x+5z)$.

Page 76.—**1.** $(a+b)(a-b)$. **2.** $(c+d)(c-d)$. **3.** $(m+n)(m-n)$.
4. $(2x+2y)(2x-2y)$. **5.** $(3x+y)(3x-y)$. **6.** $(x+3y)(x-3y)$.
7. $(4x+4y)(4x-4y)$. **8.** $(3c+4d)(3c-4d)$. **9.** $(5a+3b)(5a-3b)$.
10. $(3y+1)(3y-1)$. **11.** $(5x^2+4y^3)(5x^2-4y^3)$. **12.** $(6y+7z^3)(6y-7z^3)$. **13.** $(2x+3y)(2x-3y)$. **14.** $(xy+2yz)(xy-2yz)$.
15. $(m^2+n^2)(m+n)(m-n)$. **16.** $(a^4+b^4)(a^2+b^2)(a+b)(a-b)$.
17. $(m^n+n^m)(m^n-n^m)$. **18.** $(3a^n+2b^{2n})(3a^n-2b^{2n})$. **19.** $(a^8+b^4)(a^4+b^2)(a^2+b)(a^2-b)$. **20.** $(3ab+2c^2)(3ab-2c^2)$. **21.** $(a^2x^2+3cd^2)(a^2x^2-3cd^2)$. **22.** $(x^6+y^5)(x^6-y^5)$. **23.** $(2bc^2+5d^3)(2bc^2-5d^3)$.
24. $(9b^2+4)(3b+2)(3b-2)$. **25.** $(11a+6b^2)(11a-6b^2)$.
26. $(13c^2+7d^3)(13c^2-7d^3)$. **27.** $(10ad^2+8)(10ad^2-8)$. **28.** $(8+5x^2y^2)(8-5x^2y^2)$. **29.** $(7+6xy^2)(7-6xy^2)$. **30.** $(4m^n+n^{2m})(4m^n-n^{2m})$.
31. $(17x^3+y^{3n})(17x^3-y^{3n})$. **32.** $(16a^2+6z^4)(16a^2-6z^4)$.
33. $(14x^{2n}+3)(14x^{2n}-3)$. **34.** $(15xy^n+11z^2)(15xy^n-11z^2)$.
35. $(18xy^2z^3+9)(18xy^2z^3-9)$. **36.** $(15xy^4+12)(15xy^4-12)$.

Page 77.—**39.** $(c+a+b)(c-a-b)$. **40.** $(b+x-y)(b-x+y)$.
41. $(2c+x+y)(2c-x-y)$. **42.** $(3c+x-y)(3c-x+y)$.
43. $(x+a+2b)(x-a-2b)$. **44.** $(c+3a+b)(c-3a-b)$.
45. $(b+2a+3c)(b-2a-3c)$. **46.** $(2c+2a+3b)(2c-2a-3b)$.
47. $(3c+2a-3b)(3c-2a+3b)$. **48.** $(2a)(2b)$. **49.** $(3x+y+3)(y-x-3)$. **50.** $(3x+y-4)(y-x+4)$. **51.** $(5x-1)(7-x)$.
52. $(5x-y)(7y-x)$. **53.** $(9x-y)(x+5y)$. **54.** $(9x-2y)(5x-4y)$. **55.** $(11x+3y)(5x+7y)$. **56.** $(11x-2y)(7x+8y)$.

Page 78.—**59.** $(b+c+d)(b+c-d)$. **60.** $(a+x+2y)(a+x-2y)$.

ANSWERS. 383

61. $(2a+x-y)(2a-x+y)$. **62.** $(a-3x+2c)(a-3x-2c)$.
63. $(2x+y+2z)(2x+y-2z)$. **64.** $(x-y+b)(x-y-b)$.
65. $(x+2y+2z)(x+2y-2z)$. **66.** $(x+3+3y)(x+3-3y)$.
67. $(2x+3y+3c)(2x+3y-3c)$. **68.** $(2a-3b+4c)(2a-3b-4c)$.
69. $(3x+4y+5z)(3x+4y-5z)$. **70.** $(a+c+b-d)(a+c-b+d)$.
71. $(a+2b+x-2y)(a+2b-x+2y)$. **72.** $(d-2c+a-2x)$.
$(d-2c-a+2x)$.

Page 79.—**2.** $(x+2)(x+1)$. **3.** $(x+4)(x+3)$. **4.** $(x-7)(x+3)$.
5. $(x-9)(x+2)$. **6.** $(x+4)(x+2)$. **7.** $(x+8)(x+4)$. **8.** $(b-5)$
$(b-3)$. **9.** $(b+4)(b-3)$. **10.** $(b-4)(b+3)$. **11.** $(b+7)(b-5)$.
12. $(y+8)(y-7)$. **13.** $(y+8)(y-5)$. **14.** $(b+16c)(b+3c)$.
15. $(c^4d^2+4)(c^4d^2+3)$. **16.** $(xy+5z)(xy+2z)$. **17.** $(x+12)(x-3^\circ)$.
18. $(x-56)(x-1)$. **19.** $(x-13)(x+3)$. **20.** $(x-16)(x+4)$.
21. $2(2x-3)(x-1)$. **22.** $9(x-2)(x-1)$. **23.** $2(x+3a)(x+a)$.
24. $3(a+2b)(3a+4b)$. **25.** $(2a+3)(2a-1)$. **26.** $2(2b+7)(b-1)$.
27. $(3a-7)(3a+2)$. **28.** $(4c-7)(4c-5)$. **29.** $3(3x+5y)(x-y)$.
Page 80.—**30.** $(4y+7z)(4y-z)$. **31.** $2(2x-5y)(x-y)$.
32. $(5y-3z)(5y-2z)$. **33.** $(7x^3+5y)(7x^3-3y)$.
Page 81.—**2.** $(x+2)(2x-1)$. **3.** $3(x+1)(x+1)$. **4.** $(x-5)$
$(3x-2)$. **5.** $(5x-3)(x-7)$. **6.** $(3x+2)(x-1)$. **7.** $(3x-2)(x+3)$.
8. $(3x-4)(x+5)$. **9.** $(2x+3)(x+4)$. **10.** $(4x-3y)(x+y)$.
11. $(4x-y)(x-y)$. **12.** $(5x-9y)(x-4y)$. **13.** $(7x-3)(x+18)$.
14. $(5x-y)(2x+y)$. **15.** $(3x-2y)(x+3y)$. **16.** $(4x-5)(x+3)$.
17. $(3x-2y)(2x+y)$.
3. $(b+x)(c+x)$. **4.** $(a+b)(x-y)$. **5.** $(a+b)(b-c)$.
6. $(a-y)(b+y)$. **7.** $(3a-b)(x-y)$. **8.** $(ac+b)(ac+d)$.
9. $(a+b)(x^2+2)$. **10.** $(x-3)(x-y)$. **11.** $(b-x)(c+x)$. **12.** $(cx+y)$
$(dx-y)$. **13.** $(ax+y)(bx-y)$. **14.** $(bd-ac)(dx-by)$. **15.** $(2a+b)$
$(3c-2d)$. **16.** $(5a+3b)(3x-4y)$. **17.** $9(2a-b)(2b-a)$.
18. $(3a-2b)(4x-3y)$. **19.** $(a+b)(c+d)(x+y)$. **20.** $(2a-3b)$
$(x+2y)(z-v)$.

Page 82.—**3.** $(b^2+bc+c^2)(b^2-bc+c^2)$. **4.** $(x^2+x+1)(x^2-x+1)$.
5. $(3b^2+3bc+5c^2)(3b^2-3bc+5c^2)$. **6.** $(3b^2+3bx+2x^2)$
$(3b^2-3bx+2x^2)$. **7.** $(5x^2+7xy+4y^2)(5x^2-7xy+4y^2)$.
8. $(4a^2+5ab+b^2)(4a^2-5ab+b^2)$; or, $(4a^2-3ab-b^2)(4a^2+3ab-b^2)$.
9. $(6x^2+5xy+4y^2)(6x^2-5xy+4y^2)$. **10.** $(7b^2+9bc+5c^2)$
$(7b^2-9bc+5c^2)$. **11.** $(8a^2+4ab+9b^2)(8a^2-4ab+9b^2)$.
12. $7x^2+6xy+5y^2)(7x^2-6xy+5y^2)$. **13.** $(3c^2+2cd+7d^2)$
$(3c^2-2cd+7d^2)$. **14.** $(6a^2+10ab+7b^2)(6a^2-10ab+7b^2)$.
15. $(5b^2+7bc+8c^2)(5b^2-7bc+8c^2)$. **16.** $(5b^2+9bc+8c^2)$
$(5b^2-9bc+8c^2)$. **17.** $(9x^2+4xy+2y^2)(9x^2-4xy+2y^2)$.
18. $(9x^2+10xy+2y^2)(9x^2-10xy+2y^2)$.

Page 83.—**1.** $x^7+x^6y+x^5y^2+x^4y^3+x^3y^4+x^2y^5+xy^6+y^7$. **2.** x^8+
$x^7y+x^6y^2+x^5y^3+x^4y^4+x^3y^5+x^2y^6+xy^7+y^8$. **3.** x^3+x^2+x+1. **4.** x^3+
$2x^2+4x+8$. **5.** $x^6+x^4y^2+x^2y^4+y^6$. **6.** $(a-3)(a^2+3a+9)$.
7. $(x-2)(x^2+2x+4)(x^3+8)$. **8.** $(2a-b)(4a^2+2ab+b^2)$.
9. $(3x-y)(9x^2+3xy+y^2)$. **10.** $(7-x)(49+7x+x^2)$.
Page 84.—**1.** $x^7-x^6y+x^5y^2-x^4y^3+x^3y^4-x^2y^5+xy^6-y^7$. **2.** x^9-
$x^8y+x^7y^2-x^6y^3+x^5y^4-x^4y^5+x^3y^6-x^2y^7+xy^8-y^9$. **3.** x^3-x^2+x-1.
4. x^3-2x^2+4x-8. **5.** $x^6-x^4y^2+x^2y^4-y^6$. **6.** $(x+3)$
$(x-3)(x^2+9)$. **7.** $(x+2)(x-2)(x^2+2x+4)(x^2-2x+4)$.
8. $(2a+2b)(8a^3-8a^2b+8ab^2-8b^3)$. **9.** $(ax+y)(ax-y)$
$(a^2x^2+axy+y^2)(a^2x^2-axy+y^2)$. **10.** $(ab+cd)(ab-cd)$
$(a^2b^2+c^2d^2)(a^4b^4+c^4d^4)$.

384   HIGH SCHOOL ALGEBRA.

Page 85.—**1.** $x^6-x^5y+x^4y^2-x^3y^3+x^2y^4-xy^5+y^6$. **2.** $x^8-x^7y+x^6y^2-x^5y^3+x^4y^4-x^3y^5+x^2y^6-xy^7+y^8$. **3.** $x^4-x^3+x^2-x+1$. **4.** $x^2y^2-xyab+a^2b^2$. **5.** $x^8-x^3yz+x^4y^2z^2-x^2y^3z^3+y^4z^4$. **6.** $x^8-x^6+x^4-x^2+1$.
Page 86.—**7.** $(a+b^2c^2)(a^4-a^3b^2c^2+a^2b^4c^4-ab^6c^6+b^8c^8)$. **8.** $(x+yz)(x^5-x^5yz+x^4y^2z^2-x^3y^3z^3+x^2y^4z^4-xy^5z^5+y^6z^6)$. **9.** $(mn+rs)(m^4n^4-m^3n^3rs+m^2n^2r^2s^2-mnr^3s^3+r^4s^4)$. **10.** $(ab^2+cd^2)(a^4b^8-a^3b^6cd^2+a^2b^4c^2d^4-ab^2c^3d^5+c^4d^8)$. **11.** $(ax^2+b^2y^2)(a^4x^8-a^3x^6b^2y^2+a^2x^4b^4y^4-ax^2b^6y^6+b^8y^8)$. **12.** $(x^2y+zy^2)(x^8y^4-x^6y^5z+x^4y^6z^2-x^2y^7z^3+z^4y^8)$.

1. $(a-4)(a+2)$. **2.** $(3a+b)(3a-b)$. **3.** $(2a)(2b)$. **4.** $(y+m)(y+m)$. **5.** $(b-15)(b+3)$. **6.** $(x^2+y^2)(x+y)(x-y)$. **7.** $(n+10)(n-8)$. **8.** $(m^2+n^2)(m+n)(m-n)$. **9.** $(a+b)(c+d)$. **10.** $(q-2r)(q-2r)$. **11.** $(3x-4c)(3x-4c)$. **12.** $(x+y)(x^4-x^3y+x^2y^2-xy^3+y^4)$. **13.** $(a-10)(a+3)$. **14.** $(a-7)(a-2)$. **15.** $(x^2-y)(x^4+x^2y+y^2)(x^{12}+x^6y^3+y^6)$. **16.** $(a-b)(x-y)$. **17.** $(a-1)(z+2)$. **18.** $(p-5)(p+1)$. **19.** $(a-b)(a^2+ab+b^2)(a+b)(a^2-ab+b^2)(a^2+b^2)(a^4-a^2b^2+b^4)$. **20.** $(3a-5)(2a-7)$. **21.** $(x-y+z)(x-y-z)$. **22.** $(x+9y)(x+4y)$. **23.** $(a^m+b^n)(a^m-b^n)$. **24.** $c^2(2a-3c)(2a-3c)$. **25.** $(c+7)(c-2)$. **26.** $(y+4)(y+2)$. **27.** $(x-1)(x^2+x+1)(x+1)(x^2-x+1)$. **28.** $(x-y)(y+2)$. **29.** $(3a+2z^2)(3a+2z^2)$. **30.** $4(a-b)(a^2+ab+b^2)(a+b)(a^2-ab+b^2)$. **31.** $(x^m+1)(x^m+1)$. **32.** $(2x^{3m}+y^m)(2x^{3m}+y^m)$. **33.** $(2p-2r)(2p-2r)$. **34.** $(4x^{2n}+9y^{4n})(4x^{2n}-9y^{4n})$. **35.** $(m-10)(m+9)$. **36.** $(x+7)(x-5)$. **37.** $(a+b)(c-d^2)$. **38.** $(m-7)(m+4)$. **39.** $(x+10)(x+4)$. **40.** $(17-y)(5+y)$.
Page 87.—**41.** $(2x+3)(2x+1)$. **42.** $(x^2+x+1)(x^2-x+1)$. **43.** $(a-b+c)(a-b-c)$. **44.** $(2c^2+5c)(2c^2-5c)$. **45.** $(c-2)(c+2)(c-2)(c+2)$. **46.** $(c-2)(c+2)(c^2+3)$. **47.** $(y-2)(y-1)$. **48.** $(z^2-12)(z^2+2)$. **49.** $(p+5)(p+2)$. **50.** $(x^4+y^4)(x^2+y^2)(x+y)(x-y)$. **51.** $(3b+1)(2b-3)$. **52.** $(20x+y)(-y)$. **53.** $(3a-n)(2a-m)$. **54.** $4(c+5d)(c+5d)$. **55.** $(2x+3y^2)(2x-3y^2)$. **56.** $(2a+b)(a+2)$.

57. $(x-20)(x+18)$. **58.** $(5p^{\frac{m}{2}}+6x^p)(5p^{\frac{m}{2}}-6x^p)$. **59.** $(3c+4)(4c-3)$. **60.** $(z-x+y)(z-x+y)$. **61.** $(x^2-2xy+1)(x^2-2xy+1)$. **62.** $(x-y)(x-y)$. **63.** $4(a+c)(b+d)$. **64.** $(x-2y)(x+2z)$. **65.** $x(x-y)(y-z)$. **66.** $(x+a)(x-a)(z^2+a)$. **67.** $(ax+by-c)(by+c)$. **68.** $(a-y)(a-y)(y-b)$. **69.** $(a-b+m-n)(a-b-m+n)$. **70.** $(a-b+c+d)(a-b-c-d)$. **71.** $(a^2-b^2+m+n)(a^2-b^2-m-n)$. **72.** $3(a-b+2c)(a-b)$. **73.** $(x+y)(x+y)(x-y)(x-y)$. **74.** $(8ab)(a^2+b^2)$. **75.** $2(x+y-z)(a+b)$.
Page 89.—**3.** $6m^2nx^2$. **4.** $4r^2s^2x^2$. **5.** $7x^2y^2z^3$. **6.** $5x^5yz^2$. **7.** axy. **8.** a^2xy^2.
Page 90.—**9.** $2b^2c$. **10.** $2xy^2$. **11.** $2r^2s^2t$. **12.** $5a^2xy^2$. **13.** $3x^2y^2z^2$. **14.** $a-b$. **15.** $x-2$. **16.** $4x-y$. **17.** $x+3$. **18.** $x+5$. **19.** $x+6$. **20.** $x+3$. **21.** $x+7$. **22.** $x+6$. **23.** $x+y$. **24.** $2a-3x$. **25.** $b-3a$. **26.** $4a(b^2-c^2)$. **27.** $z(y-1)$. **28.** $a-1$. **29.** $3y+2$. **30.** $a-b$. **31.** $2(x-y)$. **32.** $a-b$. **33.** $z+3$. **34.** $x+2$. **35.** $1+a$.
Page 94.—**4.** $x-7$. **5.** $x+4$. **6.** $x-3$. **7.** $x+6$. **8.** $3x-2$. **9.** $x-4y$. **10.** $x-y$. **11.** $(x-1)^2$. **12.** $2(x+1)^2$. **13.** $3x+9$. **14.** $5x^2-1$. **15.** $x+3$. **16.** $x-7$. **17.** $x+3$. **18.** a^2-b^2. **19.** $x+2$. **20.** $a+b$. **21.** $3x^2+2x+1$.
Page 95.—**22.** x^2+4. **23.** a^2+4. **24.** $5x^2-2x+1$. **25.** $ax-a$. **26.** $2a^2-2b^2$. **27.** $x-5$. **28.** $3x+33$. **29.** x^2-x-1. **30.** $a^2-13a+5$. **31.** $2x^2+x+1$. **32.** $x-2$.

ANSWERS.

Page 97.—3. $40\ a^2b^2c^3$. 4. $100\ x^3y^3z^3$. 5. $70\ a^2b^2c^2x^2y$. 6. $72\ m^2n^3y^3$. 7. $36\ r^3s^3z^4$. 8. $x^3-x^2y-xy^2+y^3$. 9. $x^3+x^2y-xy^2-y^3$. 10. $x^4-2x^2y^2+y^4$. 11. $x^4+x^3y-xy^3-y^4$. 12. $a^2y^2(x^2-z^2)$. 13. x^4-1. 14. $12\,xy^2(x^2-y^2)$. 15. $x(x^6-1)$. 16. $x(x^4+x^3-x-1)$. 17. $8(1-x^2)$. 18. $x^3+9\,x^2+26\,x+24$. 19. $a^3-4\,a^2-17\,a+60$. 20. $x^3-11\,x^2-4\,x+44$. 21. $x^4-6\,x^3-6\,x^2+70\,x-75$. 22. x^4-y^4.

Page 98.—23. $x^5-xy^4+x^4y-y^5$. 24. $a^5+2\,a^4-16\,a-32$. 25. x^5-xy^4. 26. $x^3-3\,x^2-4\,x+12$. 27. $x^4-2\,ax^3+a^2x^2-10\,x^3+20\,ax^2-10\,a^2x+25\,x^2-50\,ax+25\,a^2$. 28. $x^5+2\,x^4-16\,x-32$. 29. $m^3-5\,m^2+8\,m-4$. 30. $p^3-3\,p^2-18\,p+40$. 31. $8\,c^2-c^3-19\,c+12$. 32. $1-2\,p^2+2\,p^3-2\,p^5+p^6$. 33. $4\,bc(a-c-1)$. 34. $3\,x^3-31\,x^2+82\,x-24$. 35. $m^4+6\,m^3-127\,m^2-600\,m+2700$. 36. $y^3-3\,y^2-33\,y+35$. 37. $2\,x^4-14\,x^3+31\,x^2-31\,x+12$. 38. $a^6+2\,a^5-a^4-a^2-2\,a+1$. 39. $(c-1)(c-2)(c-4)(c-4)$. 40. $(a+4)(a-1)(a-5)$. 41. $(x+y+z)(x-y-z)(y-x-z)(z-x-y)$. 42. $(a-1)(a-2)(a-3)(a-4)$. 43. $(2\,c+3)(2\,c-3)(3\,c+2)$. 44. $1+x^2+x^4$. 45. $x^4+2\,x^3y-2\,xy^3-y^4$. 46. $(a+b)(a+c)(b+c)$. 47. a^3-b^3. 48. $4\,a^3(a^2-x^2)(a^2+x^2)$.

Page 102.—3. $\dfrac{12\,a}{28}$. 4. $\dfrac{30\,x^2}{36}$. 5. $\dfrac{10\,a+20\,b}{15}$. 6. $\dfrac{15\,x+35}{30}$. 7. $\dfrac{6\,x}{18\,x+9}$. 8. $\dfrac{9\,x}{18\,x-24}$. 9. $\dfrac{4\,ax^2}{6\,x+4\,xy}$. 10. $\dfrac{(a-b)(2\,a+x)}{a^2-b^2}$. 11. $\dfrac{(a+b)(3\,x-y)}{a^2+2\,ab+b^2}$.

Page 103.—12. $\dfrac{x}{5}$. 13. $\dfrac{3}{4\,y}$. 14. $\dfrac{2}{5\,xy}$. 15. $\dfrac{2\,z^2}{3\,xym}$. 16. $\dfrac{7\,n^2}{4\,z}$. 17. $\dfrac{2\,z}{y}$. 18. $\dfrac{5\,x}{7\,yz}$. 19. $\dfrac{2\,az^2}{3\,y}$. 20. $\dfrac{a+b}{a-b}$. 21. $\dfrac{a-b}{a+b}$. 22. $\dfrac{x}{3-2\,x}$. 23. $\dfrac{2(a+1)}{a-1}$. 24. $\dfrac{3\,a-b}{a}$. 25. $\dfrac{x+z}{x^2+xz+z^2}$. 26. $\dfrac{a}{a-2\,b}$. 27. $\dfrac{2}{3(x^4y^4-1)}$. 28. $\dfrac{3\,c}{7\,d}$. 29. $\dfrac{x+y}{x-y}$. 30. $\dfrac{m^2-mn+n^2}{m^2+2\,mn+n^2}$. 31. $\dfrac{(1-x)^2}{1+x}$. 32. $\dfrac{a-6}{a+3}$. 33. $\dfrac{x-2}{x+7}$. 34. $\dfrac{a-9}{a-2}$. 35. $\dfrac{2\,x-3}{x+2}$. 36. $\dfrac{m-8}{m+6}$. 37. $\dfrac{3\,a}{2\,x}$. 38. $\dfrac{7\,ab}{a-x}$. 39. $\dfrac{x-1}{x+1}$. 40. $\dfrac{x-1}{2\,y}$. 41. $\dfrac{x(x+a)}{x-a}$.

Page 104.—42. $\dfrac{x^3}{x^2+y^2}$. 43. $\dfrac{x+3}{x(x-4)}$. 44. $\dfrac{x+4}{x-4}$. 45. $\dfrac{m+n}{m^2+mn+n^2}$. 46. $\dfrac{a^4-a^2b^2+b^4}{a^2-b^2}$. 47. $\dfrac{m-n}{m^2-mn+n^2}$. 48. $\dfrac{(m^4+n^4)(m^2+n^2)}{m^4+m^2n^2+n^4}$. 49. $\dfrac{3\,ab}{a+b}$. 50. $\dfrac{x^2+xy+y^2}{x-y}$. 51. $\dfrac{y(y-b)}{y+b}$. 52. $\dfrac{a+2\,b}{a-2\,b}$. 53. $\dfrac{a-1}{a+2}$. 54. $\dfrac{a+2}{a+4}$. 55. $\dfrac{3\,c+2}{4\,c+5}$. 56. $\dfrac{a-b-c}{a}$. 57. $\dfrac{5\,x+2\,y}{3\,x+2\,y}$. 58. $\dfrac{a-5}{a-8}$. 59. $\dfrac{x^2+4\,x+1}{x^2-x+1}$.

Page 105.—2. $\dfrac{10\,x+4\,y}{5}$. 3. $\dfrac{20\,x-3\,y}{4}$. 4. $\dfrac{8\,x-6\,z}{2}$. 5. $\dfrac{4\,x+4\,y+3}{4}$. 6. $\dfrac{8\,a+3\,x+4}{4}$. 7. $\dfrac{16\,x+3\,y-4}{8}$. 8. $\dfrac{18\,x-2\,y-3}{6}$. 9. $\dfrac{10\,a-3\,x-4}{2}$.

ALGEBRA—25.

HIGH SCHOOL ALGEBRA.

10. $\dfrac{24a-3y-7}{4}.$ **11.** $\dfrac{3cd+4a+b}{d}.$ **12.** $\dfrac{4acd+3c-d}{cd}.$ **13.** $\dfrac{3ax^2+6a-x}{ax}.$

14. $\dfrac{5x+20+2c-d}{5}.$ **15.** $\dfrac{a^2-2ac+c^2}{a}.$ **16.** $\dfrac{x^2-9x+6}{x-2}.$ **17.** $\dfrac{2a^2}{a-x}.$

18. $\dfrac{a^2+2ac-2c^2}{a-c}.$ **19.** $\dfrac{2x^2-2y^2}{x+y}.$ **20.** $\dfrac{-14}{x-4},$ or $\dfrac{14}{4-x}.$ **21.** $\dfrac{a^2-4ax+4x^2}{a-x}.$

22. $\dfrac{-2ab}{a-b},$ or $\dfrac{2ab}{b-a}.$ **23.** $\dfrac{m^2-2mn-2n^2}{m-n}.$

Page 106. —**2.** $a+\dfrac{c^2}{a}.$ **3.** $x+\dfrac{cd}{b}.$ **4.** $2b-\dfrac{b^2}{a+b}.$ **5.** $a+x.$ **6.** $a-x.$

7. $x^2-x+1.$ **8.** $x^2+x+1+\dfrac{2}{x-1}.$ **9.** $1+\dfrac{x}{1-x}.$ **10.** $x^2-xy+y^2-\dfrac{y^3}{x+y}.$

11. $7a+7b+\dfrac{7b^2}{a-b}.$ **12.** $1-\dfrac{2y}{x+y}.$ **13.** $2x+8+\dfrac{39}{x-4}.$ **14.** $b^2-a.$

15. $x^5+x^2y^3+\dfrac{x^2y^6-y^8}{x^3-y^3}.$ **16.** $2x.$ **17.** $a^2+ab+b^2+\dfrac{2b^3}{a-b}.$ **18.** $1+\dfrac{5ay+x}{ax}.$ **19.** $2a-2b.$ **20.** $x+y+1+\dfrac{y^2-y}{x+y}.$ **21.** $a^2+ab+b^2.$

22. $x+\dfrac{y^2}{x+y}.$ **23.** $m+3n+\dfrac{4n^2}{m-n}.$ **24.** $x^2+xz+z^2-\dfrac{y^3}{x-z}.$ **25.** $2x^2+3xy^2+\dfrac{3xy^4-6y^4}{2x-y^2}.$ **26.** $x^2+2x+1-\dfrac{4}{x^2-2}.$ **27.** $a-1+\dfrac{5a+4}{5a^2+4a-1}.$

28. $a+\dfrac{b^2}{a+b+1}.$ **29.** $a-b+\dfrac{2b^3}{a^2-2ab+b^2}.$

Page 108. —**2.** $\dfrac{9x}{12},\ \dfrac{10x}{12}.$ **3.** $\dfrac{21a}{24},\ \dfrac{20a}{24}.$ **4.** $\dfrac{6x^2y}{8},\ \dfrac{xy^2}{8}.$

5. $\dfrac{4x}{6a},\ \dfrac{4y}{6a}.$ **6.** $\dfrac{2by}{3y^2},\ \dfrac{c}{3y^2}.$ **7.** $\dfrac{9acz}{6x^2yz},\ \dfrac{4bdy}{6x^2yz}.$ **8.** $\dfrac{4x-8y}{10x^2},\ \dfrac{3x^2-8xy}{10x^2}.$

9. $\dfrac{16a+20b}{12a^2},\ \dfrac{9a^2+12ab}{12a^2}.$ **10.** $\dfrac{6ax-4ay}{10a^2c},\ \dfrac{4x-3y}{10a^2c}.$ **11.** $\dfrac{18xz}{12x^2yz},\ \dfrac{12az}{12x^2yz},\ \dfrac{20c}{12x^2yz}.$ **12.** $\dfrac{2cxy^2}{8x^2y^2},\ \dfrac{2dxy}{8x^2y^2},\ \dfrac{d}{8x^2y^2}.$ **13.** $\dfrac{3cd}{3a^2c^2},\ \dfrac{ac}{3a^2c^2},\ \dfrac{12a^2c^2}{3a^2c^2}.$ **14.** $\dfrac{acx+acy}{4ac},\ \dfrac{2ax-2ay}{4ac},\ \dfrac{2cx^2+2cy^2}{4ac}.$ **15.** $\dfrac{x^2+3x+2}{x^2-1},\ \dfrac{x^2-3x+2}{x^2-1},\ \dfrac{x+3}{x^2-1}.$ **16.** $\dfrac{ax^2y-bx^2y}{a^2-b^2},\ \dfrac{axy+bxy}{a^2-b^2},\ \dfrac{xy^2}{a^2-b^2}.$

17. $\dfrac{x^2+2xy+y^2}{x^2-y^2},\ \dfrac{x^2-2xy+y^2}{x^2-y^2},\ \dfrac{x^2+y^2}{x^2-y^2}.$ **18.** $\dfrac{x^4-2x^2+1}{x^4-1},\ \dfrac{x^4+2x^2+1}{x^4-1},\ \dfrac{x^4+1}{x^4-1}.$ **19.** $\dfrac{a-c}{(a-b)(b-c)(a-c)},\ \dfrac{b-c}{(a-b)(b-c)(a-c)}.$

Page 109. —**20.** $\dfrac{a^3-9a}{a^3+a^2-9a-9},\ \dfrac{a^2-2a-3}{a^3+a^2-9a-9},\ \dfrac{2a+2}{a^3+a^2-9a-9},\ \dfrac{a^3+3a^2-a-3}{a^3+a^2-9a-9}.$ **21.** $\dfrac{18x-6}{6x+12},\ \dfrac{6x+15}{6x+12},\ \dfrac{4x-1}{6x+12}.$ **22.** $\dfrac{32x^2-200}{4x^3-20x^2-25x+125}.$

ANSWERS.

$$\frac{25x-125}{4x^3-20x^2-25x+125}, \quad \frac{6x^3-45x^2+75x}{4x^3-20x^2-25x+125}.$$ **23.** $\frac{m^3+2m^2n+2mn^2+n^3}{m^4-mn^3+m^3n-n^4}$,

$\frac{2m^3+2m^2n+2mn^2}{m^4-mn^3+m^3n-n^4}, \quad \frac{m^2+2mn+n^2}{m^4-mn^3+m^3n-n^4}$. **24.** $\frac{x^4+2x^3-5x^2-18x-36}{x^5-5x^3-12x^2-8x+24}$,

$\frac{x^4+2x^3+3x^2-2x-4}{x^5-5x^3-12x^2-8x+24}, \quad \frac{x^4+2x^3-8x-16}{x^5-5x^3-12x^2-8x+24}, \quad \frac{x^3-2x^2-5x+6}{x^5-5x^3-12x^2-8x+24}$

Page 110.—**3.** 20. **4.** 18. **5.** 12. **6.** 10.
Page 111.—**7.** 14. **8.** 30. **9.** 72. **10.** 5. **11.** 24.
12. 24. **13.** 12. **14.** 24. **15.** 12. **16.** $17\frac{10}{19}$. **17.** 24.
18. 48. **19.** 7. **20.** 1. **21.** 12. **22.** 40. **23.** 5. **24.** 8.
25. $20\frac{2}{3}$. **26.** $15\frac{5}{7}$. **27.** 23. **28.** $2\frac{1}{3}$.
Page 112.—**29.** $5\frac{1}{2}$. **30.** 1. **31.** 11. **32.** 3. **33.** 2.
34. 7. **35.** $-5\frac{7}{5}$. **36.** 3. **37.** $\frac{30}{49}$. **38.** 28. **39.** 6. **40.** 9.
Page 113.—**42.** 12. **43.** 18. **44.** 30 yr. **45.** 1st, $1000; 2d, $300; 3d, $200. **46.** A's, $1500; B's, $2000. **47.** Horse, $180; carriage, $240. **48.** A, 16\frac{8}{37}$; B, 8\frac{4}{37}$; C, 21\frac{23}{37}$; D, 4\frac{2}{37}$. **49.** 90. **50.** $630.
Page 114.—**51.** 150. **52.** 24, 76. **53.** 5, 6. **54.** A, 12 yr.; B, 8 yr. **55.** 15, 35. **56.** 60. **57.** 4, 12.

Page 116.—**4.** $\frac{a^2+xz}{a^2-az+ax-xz}$. **5.** $\frac{2+2x^2}{1-x^2}$. **6.** $\frac{2+2x^4}{1-x^4}$. **7.** $\frac{1+x^2}{1-x^2}$.

8. $\frac{x+xy-y^2}{x^2-y^2}$. **9.** $\frac{a+1}{a}$. **10.** $\frac{a(15d-4b)}{6xy}$. **11.** $\frac{mn(2y-3)}{4y^2}$.

12. $\frac{a-5b}{6}$. **13.** $\frac{5b-a}{a^2-b^2}$. **14.** $\frac{x-9}{x^2-1}$. **15.** $\frac{-4x}{x^2-1}$ or $\frac{4x}{1-x^2}$.

16. $\frac{x^2+2x}{x^2-1}$. **17.** $3a-b+\frac{1}{a^2-b^2}$. **18.** $\frac{2b(2a-b)}{a^2-b^2}$. **19.** $\frac{y(y-2x)}{x^2-xy}$.

20. $\frac{2x^2-1}{x^2(x^2-1)}$. **21.** $\frac{2}{1+x^2+x^4}$.

Page 117.—**22.** $\frac{2b(3a^2+b^2)}{a^2-b^2}$. **23.** $3x+\frac{27a-8b}{30a}$. **24.** $4x+$

$\frac{3x-3z}{y}$. **25.** $b+\frac{a+b}{ax}$. **26.** $3x+\frac{10x-1}{12}$. **27.** $a-b+\frac{4ax}{a^2-x^2}$.

28. $\frac{7y-3x+4x^2-3xy}{x^2y}$. **29.** $\frac{b^2-8a^2b^2-4ab-7a^2}{a^3b^3}$. **30.** $\frac{2x}{1-x^4}$.

31. $\frac{x^2+2ax-8a^2}{x^3-3ax^2+3a^2x-a^3}$. **32.** $\frac{2x-3}{4x^3-x}$. **33.** 1. **34.** $\frac{2(3x^2+2x-3)}{x^3-7x-6}$.

35. 0. **36.** $\frac{7x+2}{2x(x^2+x-2)}$.

Page 118.—**37.** $\frac{2x(5x+7)}{x^3+2x^2-x-2}$. **38.** $\frac{2}{a^4+a^2+1}$. **39.** $\frac{4y^2}{x^4-y^4}$.

40. $\frac{4a^5+2a^2b^3+6b^5}{a^6-b^6}$. **41.** $\frac{x^3+2x^2y-2xy^2+2y^3}{x^4-y^4}$. **42.** $\frac{3x^2+2x+4}{x^2+x-2}-x-2$.

43. $\frac{13x+2}{x^3-5x^2-18x+72}$. **44.** $\frac{11p-2n}{6}-3m$. **45.** $\frac{2}{x^4+x^2+1}$.

46. $\frac{3ax+3x^2-a^2}{2a+x}$. **47.** $\frac{2a+3}{a^3-7a-6}$. **48.** $\frac{m(m-1)}{m^3-10m^2+11m+70}$.

49. $\dfrac{x^2-8x+16}{x^2+5x+6}$. **50.** $\dfrac{2x^2y(y-1)}{x^4-y^4}$. **51.** $\dfrac{4xy}{x^2-2xy+y^2}$.

52. $\dfrac{x^3+4x^2-8x+1}{6x^3-11x^2-5x+12}$.

Page 119. — **53.** $\dfrac{-2}{x^3+9x^2-x-105}$. **54.** $\dfrac{3x^2-20}{4x^4-10x^3-14x^2+20x+12}$

55. $\dfrac{x(5x+y)}{x^3+x^2y-xy^2-y^3}$.

Page 120. — **3.** $\dfrac{3cx}{2by}$. **4.** $\dfrac{15x^2}{2a^2}$. **5.** $\dfrac{a^4b^4}{2y^{2n}}$. **6.** $\dfrac{axy-ay^2}{2x}$.

7. $\dfrac{ax}{30}$. **8.** $\dfrac{2a+3b}{4b}$. **9.** $x-a$. **10.** $\dfrac{ab}{x^2-y^2}$.

Page 121. — **11.** x. **12.** 1. **13.** $\dfrac{x^3-xy^2}{a}$. **14.** $\dfrac{cd}{x^4-y^4}$. **15.** $\dfrac{9a}{2}$.

16. $\dfrac{3a}{x}$. **17.** $\dfrac{3x}{c}$. **18.** $\dfrac{1}{(x+y)^2}$. **19.** $1\tfrac{1}{2}$. **20.** $\dfrac{x-6}{x-3}$.

21. $\dfrac{x^2-7x+6}{x^2}$. **22.** $\dfrac{x^2-2x-8}{x^2-2x}$. **23.** $\dfrac{x^2y^2}{x^2-y^2}$. **24.** $\dfrac{72c^2-2x^2}{9c^2}$.

25. $\dfrac{y^2-2y+1}{y^3+11y^2+28y+20}$. **26.** $\dfrac{a^4+a^2+1}{a^2}$. **27.** $\dfrac{x^2-13x+40}{x^2+14x+49}$.

28. $\dfrac{x^3+y^3}{xy^3}$. **29.** $\dfrac{x^2+11x-12}{x^3}$.

Page 123. — **3.** $\dfrac{8a^3}{3dyx}$. **4.** $\dfrac{7ax}{2y}$. **5.** $\dfrac{2cx}{3az}$. **6.** $\dfrac{4}{3ayz}$. **7.** $\dfrac{dz}{8cx}$.

8. $\dfrac{abc}{m^2n}$. **9.** $\dfrac{a+x}{2}$. **10.** $\dfrac{2a+x}{a^2+ax+x^2}$. **11.** $\dfrac{2(m^2-n^2)}{3m+n}$. **12.** $\dfrac{a-c}{1+x}$.

13. $\dfrac{a^2+4x+4}{ad+ac+dz+cz}$. **14.** $\dfrac{5(x+y)}{x-y}$. **15.** $\dfrac{ab+cd}{a^2-c^2}$. **16.** $\dfrac{3an+cm}{x^4-y^4}$.

17. $\dfrac{4(ax+by)}{ac^2+ad^2+bc^2+bd^2}$. **18.** $\dfrac{x(y+z)}{ax+nx+az+nz}$. **19.** $\dfrac{1}{x-2}$.

20. $\dfrac{ax}{b(x-6)}$. **21.** $\dfrac{mp}{ac}$. **22.** $\dfrac{16a}{c^2}$. **23.** $a+b$. **24.** $\dfrac{a^2+b^2}{a^2-b^2}$.

25. $\dfrac{a^3-b^3}{a^3+b^3}$. **26.** $\dfrac{x+4}{x^2-x-12}$. **27.** $\dfrac{a^2-4a+4}{a^2-4a-5}$. **28.** $\dfrac{x^2+12x+20}{x^2+10x+21}$.

Page 124. — **29.** $\dfrac{x+y-z}{z+y-x}$. **30.** $\dfrac{x^2-xy+y^2}{x^2+y^2}$. **31.** $a+1$. **32.** $\dfrac{6(m-9)}{a(m-7)}$.

2. $\dfrac{d(a+cx)}{c(b+dx)}$. **3.** $\dfrac{5(3a^2+x)}{3(20+x)}$. **4.** $9(a-y)$. **5.** $\dfrac{4xy(x-y)}{ab(5x-3y)}$.

6. $\dfrac{2x}{x-y}$. **7.** $4(a-x)^2$. **8.** $\dfrac{2(3acx+2d)}{3(2acx+3d)}$. **9.** $\dfrac{2x^2-y^2}{x-3y}$.

10. $\dfrac{x^2(acy-3)}{ac^2(a+2x)}$.

Page 125. — **11.** $\dfrac{cz(ay+bx)}{by(ac-xz)}$. **12.** $\dfrac{y}{y-x}$. **13.** $\dfrac{x-1}{x-5}$. **14.** y.

ANSWERS.

15. $\dfrac{a+5}{a-5}$. **16.** $\dfrac{2+8x-6x^2}{3x(x+1)}$. **17.** $\dfrac{1}{a+1}$. **18.** $\dfrac{a^2+1}{a(a^2+2)}$.

19. $\dfrac{x(x^2+2xy+2y^2)}{x^2+2xy+y^2}$. **20.** $\dfrac{a^5-ab^4+a^4b-b^5}{2a^3b^2}$. **21.** $\dfrac{3x^2-2x+1}{3x-1}$. **22.** $\dfrac{1+c}{1+c^2}$.

1. $\dfrac{x-3}{x+1}$. **2.** $\dfrac{m-1}{m+1}$. **3.** $\dfrac{x^2-x+3}{x^2-4x+9}$. **4.** $\dfrac{a-2}{a+2}$. **5.** $\dfrac{c+5}{c^2-6c+8}$.

6. $\dfrac{x^3-2x^2-2x+1}{4x^2-7x-1}$. **7.** $\dfrac{a^2-5a+6}{3a^2-8a}$. **8.** $\dfrac{x+1}{3x-7}$. **9.** $\dfrac{x^2-x+2}{x^3+x+1}$.

10. $\dfrac{2a^2-6a-7}{6a^2-17a-20}$. **11.** $\dfrac{3x^2}{x^2-1}$.

Page 126.—12. $\dfrac{1}{x+2}$. **13.** $\dfrac{1}{(x-y)(y-z)}$. **14.** $\dfrac{1}{abc}$. **15.** $\dfrac{4xy}{x^2-y^2}$.

16. $\dfrac{4x^4}{(x+y)^2}$. **17.** $a^2+1+\dfrac{1}{a^2}$. **18.** $\dfrac{4a}{x(a+1)}$. **19.** x. **20.** $\dfrac{x}{a}$.

21. $-1\tfrac{1}{5}$. **22.** x. **23.** $\dfrac{4(5x+4)}{9x+20}$. **24.** $\dfrac{xy}{x^2+y^2}$.

Page 127.—25. $\dfrac{2x^3(x+y)}{x^3-xy^2-x^4-4x^2y-4x^2}$. **26.** $\dfrac{10x^3+47x^2-17x-10}{17(x-6)}$.

27. $\dfrac{c}{a}$. **28.** $\dfrac{15x^2}{(x-1)^2}$. **29.** $\dfrac{2y^3}{x}$. **30.** $\dfrac{x^4-y^4+2xy^5-2y^6}{x^2y^2+y^4}$.

31. $\dfrac{a^2b}{2}$. **32.** $\dfrac{ab-bc^2-ac^2-abc}{c(a^2-b^2)}$. **33.** $\dfrac{y(x-12)}{x(y+12)}$.

Page 128.—34. $\dfrac{a^2+b^2}{(a-b)^2}$. **35.** $\dfrac{a+a^3+a^5}{4}$. **36.** $\dfrac{32-14x}{x^2-5x+4}$.

37. $\dfrac{6x(x-4)}{x^2-36}$. **38.** $\dfrac{x}{a+x}$. **39.** $\dfrac{1}{x-y}$. **40.** $\dfrac{x^6+2x^5y-2xy^5+y^6}{x^6-y^6}$.

41. $\dfrac{c}{ax+c}$. **42.** $\dfrac{a(a+3x)}{x(a-3x)}$. **43.** a. **44.** $\dfrac{1}{c^3}$. **45.** $\dfrac{a-x^2-y^2}{x(a-x+y)(a-x-y)}$.

46. $\dfrac{1}{ax}$. **47.** $\dfrac{ax(2ax-a^2-x^2)}{a^4+a^2x^2+x^4}$.

Page 130.—1. 8. **2.** $2\tfrac{2}{5}$. **3.** $\tfrac{1}{2}$. **4.** 6. **5.** $\tfrac{3}{13}$. **6.** 12.

7. $22\tfrac{5}{16}$. **8.** $\dfrac{b}{a-1}$. **9.** $\dfrac{1+6a}{2(2a+b)}$. **10.** $\dfrac{a(c+bd)}{d(1+a)}$. **11.** 5.

12. $\dfrac{b(a^2-1)}{2}$. **13.** $\dfrac{-3a}{4}$. **14.** $1\tfrac{1}{23}$. **15.** $\dfrac{5(a+27b)}{24}$.

16. $\dfrac{a^2(c-a)(c+1)}{c^2}$. **17.** 12. **18.** 4. **19.** 2. **20.** 9. **21.** −7.

22. 4. **23.** $\tfrac{1}{2}$.

Page 131.—24. 9. **25.** 8. **26.** $-2\tfrac{1}{43}$. **27.** $\dfrac{ab(2c-a-b)}{c(a+b)-(a^2+b^2)}$.

28. $-7\tfrac{6}{7}$. **29.** $\dfrac{a^2}{b-a}$. **30.** 4. **31.** 66. **32.** 8. **33.** 6. **34.** $4\tfrac{5}{7}$.

Page 132.—35. 3. **36.** 65. **37.** 6. **38.** 2. **39.** 3. **40.** −2.
41. 20. **42.** 22. **43.** 13. **44.** 7. **45.** −10. **46.** 5. **47.** 1. **48.** 9.

HIGH SCHOOL ALGEBRA.

Page 133.—**49.** 10.   **50.** 6.   **51.** $\dfrac{2a^3}{b-1}$.   **52.** $\dfrac{ac}{b}$.   **53.** 2.
54. 12.   **55.** 6.   **56.** 3.   **57.** 9.   **58.** 20.   **59.** 29.   **60.** 1.   **61.** $\dfrac{a+b}{a-b}$.
62. b.   **63.** $\dfrac{8a}{a+c^2-2}$.   **64.** $\dfrac{m^2+n^2}{m+n}$.

Page 134.—**65.** 1.   **66.** $\dfrac{2(a-b)+d(a+b)+c}{2b}$.   **67.** $(a+b)^2$.
68. 50.   **69.** 12, greater; 4, less.   **70.** $2400.   **71.** 47, greater; 23, less.   **72.** 75.

Page 135.—**74.** $4\tfrac{14}{19}$.   **75.** $2\tfrac{29}{143}$.   **76.** $4\tfrac{4}{17}$.   **77.** 12.   **78.** 72.
79. 2280.

Page 136.—**81.** 64, A's; 80, B's; 144, C's; 104, D's.   **82.** 150, corn; 100, oats; 50, rye.   **83.** $103\tfrac{1}{33}$, greater; $100\tfrac{32}{33}$, less.   **84.** $84.
85. $5000.

Page 137.—**87.** 1024.   **88.** 760.   **89.** 90.   **91.** 42, 30.   **92.** 12, A's age; 32, B's age.

Page 138.—**93.** $8000, E's; $7000, Y's.   **94.** $450, A's; $270, B's.
95. $60.   **96.** 51.   **97.** 60.   **98.** $1400.   **99.** $360.

Page 139.—**100.** 21, 15.   **101.** 18, 1st; 22, 2d; 10, 3d; 40, 4th.
102. $4000.   **103.** 10 yd.   **104.** 6.   **105.** 720.   **106.** $24, A; $36, B; $80, C; $175, D.   **107.** 100, 150.

Page 140.—**108.** 7¢; 12¢.   **109.** 12, A; 24, B; 18, C.
110. $90, poorer horse; $120, better horse.   **111.** 13, 40.
113. $27\tfrac{3}{11}$ min.

Page 141.—**114.** $43\tfrac{7}{11}$ min.   **115.** $54\tfrac{6}{11}$ min.   **116.** 6 o'clock.
117. $16\tfrac{4}{11}$ min.   **118.** 30 min., or 9 o'clock.   **119.** $300, better horse; $160, poorer horse.   **120.** $2, laborer; $4, 1st mechanic; $6, 2d mechanic.   **121.** 28 miles.   **122.** 43, larger; 37, smaller.
123. $100.   **124.** 80, 1st; 40, 2d; 110, 3d.

Page 142.—**125.** 18 days.   **126.** 10 rd. long; 5 rd. wide.
127. 80 miles.   **128.** 7 dimes; 28 dollar pieces.   **129.** $7500.   **130.** 7 hours.
131. 23¢.

Page 143.—**133.** $20, saddle; $180, horse.   **134.** $\dfrac{b}{1+a}=\$15$, B;
$\dfrac{ab}{1+a}=\$60$, A.   **135.** $\dfrac{b}{1+a}=3$, 1st; $\dfrac{ab}{1+a}=21$, 2d.   **136.** $\dfrac{mn}{m+n}$;
$2\tfrac{11}{12}$; $4\tfrac{4}{9}$.   **137.** $\dfrac{d(a+b)}{b}=3$.   **138.** $\dfrac{b(c-b)}{1+b}=64$.

Page 146.—**2.** $x=3$; $y=2$.   **3.** $x=1$; $y=1$.   **4.** $x=2$; $y=1$.
5. $x=2$; $y=1$.   **6.** $x=2$; $y=2$.   **7.** $x=1\tfrac{2}{7}$; $y=5\tfrac{2}{7}$.
Page 147.—**8.** $x=5$; $y=6$.   **9.** $x=5$; $y=4$.   **10.** $x=2\tfrac{12}{17}$; $y=4\tfrac{7}{17}$.   **11.** $x=5$; $y=4$.   **12.** $x=4$; $y=3$.   **13.** $x=6$; $y=4$.
14. $x=12$; $y=10$.   **15.** $x=8$; $y=9$.   **16.** $x=5$; $y=3$.   **17.** $x=5$; $y=4$.   **18.** $x=5\tfrac{20}{77}$; $y=15\tfrac{55}{77}$.   **19.** $x=16$; $y=18$.
Page 148.—**2.** $x=2$; $y=3$.   **3.** $x=4$; $y=5$.   **4.** $x=2\tfrac{3}{5}$; $y=7\tfrac{4}{5}$.
5. $x=8$; $y=10$.   **6.** $x=2$; $y=3$.   **7.** $x=1$; $y=2$.   **8.** $x=5$; $y=3$.   **9.** $x=4$; $y=5$.   **10.** $x=5$; $y=9$.   **11.** $x=7$; $y=3$.
Page 149.—**12.** $x=4$; $y=10$.   **13.** $x=3$; $y=2$.   **14.** $x=6$; $y=3$.   **15.** $x=15$; $y=6$.   **16.** $x=6$; $y=6$.   **17.** $x=15$; $y=12$.
18. $x=7$; $y=5$.   **19.** $x=6\tfrac{69}{236}$; $y=-3\tfrac{53}{59}$.   **20.** $x=15\tfrac{2}{5}$; $y=18\tfrac{18}{25}$.
21. $x=15$; $y=20$.   **22.** $x=12$; $y=14$.

ANSWERS.

Page 150. — **2.** $x=4$; $y=3$. **3.** $x=3$; $y=4$. **4.** $x=16$; $y=5$. **5.** $x=1$; $y=3$. **6.** $x=\frac{1}{3}$; $y=\frac{1}{5}$. **7.** $x=4$; $y=3$. **8.** $x=9$; $y=6$. **9.** $x=2$; $y=5$. **10.** $x=60$; $y=36$. **11.** $x=11$; $y=6$.
Page 151. — **12.** $x=6$; $y=12$. **13.** $x=12\frac{3}{7}$; $z=15\frac{3}{7}$. **14.** $x=10$; $y=5$. **15.** $x=42$; $y=35$. **16.** $x=\frac{1}{2}$; $y=\frac{1}{4}$. **17.** $x=\frac{61}{92}$; $y=\frac{61}{103}$. **18.** $x=\frac{1}{3}$; $y=\frac{1}{5}$. **19.** $x=\frac{2a}{m+n}$; $y=\frac{2b}{m-n}$. **20.** $x=40$; $y=60$. **21.** $x=6$; $y=12$. **22.** $x=4\frac{123}{236}$; $y=4\frac{42}{59}$. **23.** $x=5\frac{1}{2}$; $y=4\frac{1}{2}$. **24.** $x=5$; $y=7$. **25.** $x=2$; $y=3$.

Page 152. — **26.** $x=a$; $y=b$. **27.** $x=\frac{ac}{a+b}$; $y=\frac{bc}{a+b}$. **28.** $x=\frac{2b^2-6a^2+d}{3a}$; $y=\frac{3a^2-b^2+d}{3b}$. **29.** $x=\frac{an-bm}{cn-be}$; $y=\frac{an-bm}{ae-cm}$. **30.** $x=\frac{bs-cr}{a(b-c)}$; $y=\frac{r-s}{b-c}$. **31.** $x=\frac{3m}{2}$; $y=\frac{n}{2}$. **32.** $x=\frac{1}{ab}$; $y=\frac{1}{cd}$. **33.** $x=\frac{c(n-d)}{an-dm}$; $y=\frac{c(m-a)}{an-dm}$. **34** $x=a^2b$; $y=ab^2$.

Page 153. — **2.** $x=\frac{1}{3}$; $y=\frac{1}{2}$. **3.** $x=\frac{1}{2}$; $y=\frac{1}{3}$. **4.** $x=2$; $y=3$. **5.** $x=\frac{13}{3b-2a}$; $y=\frac{13}{5a-b}$. **6.** $x=4$; $y=6$. **7.** $x=1$; $y=3$. **8.** $x=2$; $y=3$. **9.** $x=\frac{a^2-b^2}{ac-bd}$; $y=\frac{a^2-b^2}{ad-bc}$. **10.** $x=-2$; $y=2$. **11.** $x=\frac{1}{3}$; $y=\frac{1}{4}$. **12.** $x=0$; $y=3$. **13.** $x=\frac{ab(aq-bp)}{a^2-b^2}$; $y=\frac{ab(ap-bq)}{a^2-b^2}$.

Page 155. — **4.** $x=16$; $y=8$. **5.** $x=17$; $y=12$. **6.** $x=41$; $y=7$. **7.** $4, men; $2, boys. **8.** $\frac{4}{21}$. **9.** $56, $33. **10.** $180, $120. **11.** $250, A; $320, B. **12.** $\frac{5}{18}$. **13.** 24. **14.** 1 and 5.
Page 156. — **15.** 12 persons; $5, each paid. **16.** $x = £3$; $y = £2$. **17.** 53. **18.** $4800, A's; $5000, B's. **19.** 55 at $20, 45 at $30. **20.** 50, father's; 30, son's. **21.** 60 ct., A.; 40 ct., B. **22.** 65 at $45, 35 at $37. **23.** 72 apples, 60 pears.

Page 157. — **24.** 58¢, corn; $1, wheat. **25.** 12 and 48. **26.** 50¢, 1st boy; $2, 2d boy. **27.** 20, in 1st field; 30, in 2d field. **28.** 12, Paul's; 6, Mabel's. **29.** $\frac{3}{4}$ and $\frac{3}{8}$. **30.** $6000; 2 years. **31.** 9 ft., length; 4 ft., breadth. **32.** 21 and 18.

Page 158. — **33.** 82. **34.** 4 days. **35.** $2000 in 3% bonds; $2400 in 2½% bonds. **36.** 300 yd., A's rate per min.; 275 yd., B's rate. **37.** 30 miles per hour; 90 miles.

Page 160. — **2.** $x=3$; $y=4$; $z=5$. **3.** $x=5$; $y=6$; $z=8$. **4.** $x=1$; $y=2$; $z=3$. **5.** $x=4\frac{1}{6}$; $y=4\frac{1}{6}$; $z=4\frac{5}{6}$. **6.** $x=20$; $y=10$; $z=5$. **7.** $x=2$; $y=10$; $z=14$. **8.** $x=6$; $y=4$; $z=2$. **9.** $x=40$; $y=30$; $z=24$; $u=26$. **10.** $x=3$; $y=5$; $z=7$. **11.** $x=35$; $y=30$; $z=25$. **12.** $x=15$; $y=20$; $z=25$. **13.** $x=3$; $y=7$; $z=4$. **14.** $x=12$; $y=20$; $z=8$. **15.** $x=2$; $y=3$; $z=1$. **16.** $x=5$; $y=6$; $z=7$. **17.** $x=6$; $y=8$; $z=10$.

Page 161. — **18.** $x=2$; $y=-2$; $z=0$. **19.** $x=\frac{a}{11}$; $y=\frac{5a}{11}$; $z=\frac{7a}{11}$. **20.** $x=a+b-c$; $y=a-b+c$; $z=b-a+c$. **21.** $x=\frac{r-s+v}{2a}$; $y=\frac{r+s-v}{2b}$; $z=\frac{s-r+v}{2c}$. **22.** $x=2$; $y=3$; $z=4$;

$w = 5$.   **23.** $x = 4$; $y = 5$; $z = 6$; $u = 2$; $v = 3$.   **24.** $x = \frac{1}{2}$; $y = \frac{1}{3}$; $z = \frac{1}{4}$.   **25.** $x = 3$; $y = 6$; $z = 9$.   **26.** $x = a$; $y = b$; $z = c$.   **27.** $x = \dfrac{a+b-c}{2}$; $y = \dfrac{a-b+c}{2}$; $z = \dfrac{b+c-a}{2}$.   **28.** $x = 4$; $y = 5$; $z = 6$; $u = 8$; $w = 7$.   **29.** $x = 1\frac{3}{35}$; $y = 1\frac{3}{30}$; $z = 1\frac{3}{56}$.

Page 162.—30. $x = \dfrac{2}{a+b-c}$; $y = \dfrac{2}{a-b+c}$; $z = \dfrac{2}{b-a+c}$.
31. $x = 3$; $y = 5$; $z = 6$; $u = 7$; $t = 8$.   **32.** $x = 3a$; $y = 2a$; $z = a$.
33. $x = \dfrac{b+c}{2}$; $y = \dfrac{a+b}{2}$; $z = \dfrac{a+c}{2}$.   **34.** $x = \dfrac{a+2b-c+3d}{5}$;
$y = \dfrac{3a+b+2c-d}{5}$; $z = \dfrac{3b-a+c+2d}{5}$; $v = \dfrac{2a-b+3c+d}{5}$.

35. $x = \dfrac{2a}{a+c}$; $y = \dfrac{2b}{b-c}$; $z = \dfrac{2c}{a-b}$.   **36.** $x = \dfrac{a-2b+2c}{3}$; $y = \dfrac{a+b-c}{3}$; $z = \dfrac{2b-a+c}{6}$.   **37.** $x = \dfrac{5a-11b+13c}{56}$; $y = \dfrac{5b+13a-11c}{56}$; $z = \dfrac{13b-11a+5c}{56}$.   **38.** $x = a - c$; $y = b + c$; $z = 0$; $v = a - b$.   **39.** $x = \dfrac{2}{b+c}$; $y = \dfrac{2}{a+c}$; $z = \dfrac{2}{a+b}$.

1. 10, 30, 20.   **2.** 50, 58, 80.

Page 163.—3. A, $300; B, $420; C, $780.   **4.** 10¢, sugar; 25¢, coffee; 75¢, tea.   **5.** 16, 1st; 24, 2d; 5, 3d; 80, 4th.   **6.** $14\frac{34}{43}$, A; $17\frac{23}{41}$, B; $23\frac{7}{31}$, C.   **7.** 361.   **8.** 4 horses; 16 cows; 60 sheep.
9. 20 dollars; 24 half-dollars; 40 quarters.

Page 164.—10. A, 40 yr.; B, 20 yr.; C, 80 yr.   **11.** $200, fine gold watch; $120, plain gold watch; $60, open face watch; $20, chain.   **12.** $150, horse; $5, sheep; 20, cows.   **13.** 1000, A; 600, B; 100, C.   **14.** $\frac{9}{15}$, larger; $\frac{4}{15}$, smaller.   **15.** $40, eldest; $30, second; $24, third; $26, fourth.

Page 165.—16. 40 sheep; 8 cows; 6 horses.   **17.** $981\frac{13}{19}$ in 1st; $295\frac{15}{19}$ in 2d; $352\frac{12}{19}$ in 3d.   **18.** A, $280; B, $300; C, $320.
19. 60, 120, 180.   **20.** A, $13; B, $7; C, $4.   **21.** A, $490; B, $770; C, $1190.

Page 167.—3. $36 x^4y^2$.   **4.** $-64 a^6b^6$.   **5.** $-27 a^3b^9$.   **6.** $9 c^4d^4$.
7. $32 a^5x^{15}y^{10}$.   **8.** $64 x^{12}y^6z^{18}$.   **9.** $256 a^4b^{16}d^{12}$.   **10.** $-a^{35}b^{14}c^{42}$.
11. $256 a^8b^8c^{16}$.   **12.** $32 x^{10}y^5z^{20}$.   **13.** $-125 a^9b^3c^6$.   **14.** $a^{32}b^{16}c^{24}$.
15. $256 a^8b^{12}c^{16}$.   **16.** $512 a^{18}b^{18}c^{27}$.   **17.** $16 x^8y^{12}z^4$.   **18.** $27 x^6y^{12}$.
19. $-64 a^9z^6y^3$.   **20.** $32 x^{15}y^{20}z^{15}$.   **21.** $-32 a^{10}y^{15}$.   **22.** $16 a^4x^8y^{12}$.
23. $x^{12}y^6z^{3n}$.   **24.** $x^{5m}y^{5n}z^{5m}$.   **25.** $a^{8n}z^{4n}v^{4m}$.   **26.** $a^{8y^{10n}}z^{9n}$.
27. $\pm x^{in}y^{3n}z^{n^2}$.   **28.** $\pm a^{20n}b^{10n}c^{5n^2}d^{10n^2}$.   **29.** $a^{3n-6}b^{4n-8}c^{3n-6}d^{(n-2)^2}$.

Page 168.—31. $\dfrac{4 a^2}{9 b^2}$.   **32.** $\dfrac{4 x^4}{9 y^2}$.   **33.** $-\dfrac{216 x^3y^3}{125 a^3b^3}$.   **34.** $\dfrac{4096 a^8b^4}{2401 x^8y^4}$.

35. $\dfrac{a^{12}b^{6n}}{c^{24}d^{18}}$.   **36.** $\dfrac{x^{7n}y^{14n}}{a^{7n}z^{28}}$.   **37.** $\dfrac{a^{2n}b^{2n}c^{n^2-n}}{x^{n^2-2n}y^{n^2-3n}}$.   **38.** $\dfrac{a^{4n}b^{6n}c^{8n}}{x^{8n}y^{8n}z^{8n}}$.   **39.** $\dfrac{a^{9n}c^{12n}}{b^{12n}y^{9n}}$.

40. $\dfrac{a^nb^{2n}c^{n^2+n}}{x^{mn}y^{n^2-n}}$.   **41.** $\dfrac{a^{16}b^{12}c^8}{x^8y^{16}}$.   **42.** $\dfrac{a^{2n}b^{6n}c^{2n^2}}{x^{10n}y^{4n}z^{6mn}}$.   **1.** $4 a^2 + 4 ab + b^2$.

ANSWERS.

2. $9a^2 - 12ac + 4c^2$. **3.** $4a^2 + 12ab + 9b^2$. **4.** $16a^2 - 16ab + 4b^2$. **5.** $9a^2 + 24ac + 16c^2$. **6.** $25a^2 - 40ac + 16c^2$. **7.** $4x^4 - 4x^2y + y^2$. **8.** $9x^4 + 12x^2y + 4y^2$. **9.** $9x^4 - 18x^2y^2 + 9y^4$. **10.** $16x^4 + 40x^2y + 25y^2$. **11.** $25x^6 - 30x^3y^2 + 9y^4$. **12.** $9x^4 + 36x^2y^2 + 36y^4$. **13.** $a^2 + b^2 + c^2 + d^2 + 2ab + 2ac - 2ad + 2bc - 2bd - 2cd$. **14.** $4x^2 + 9y^2 + 4z^2 - 12xy + 8xz - 12yz$. **15.** $x^4 + y^4 + 4z^4 - 2x^2y^2 + 4x^2z^2 - 4y^2z^2$. **16.** $x^4 + 4y^2 + 9z^4 + 4x^2y + 6x^2z^2 + 12yz^2$. **17.** $4x^2 + 9y^4 + z^4 + 12xy^2 - 4xz^2 - 6y^2z^2$. **18.** $4x^4 + 9y^4 + 4z^4 - 12x^2y^2 - 8x^2z^2 + 12y^2z^2$.

Page 169.— 19. $x^3 + 3x^2y + 3xy^2 + y^3$. **20.** $a^5 - 5a^4b + 10a^3b^2 - 10a^2b^3 + 5ab^4 - b^5$. **21.** $a^4 + 4a^3b + 6a^2b^2 + 4ab^3 + b^4$. **22.** $x^6 + 6x^5y + 15x^4y^2 + 20x^3y^3 + 15x^2y^4 + 6xy^5 + y^6$. **23.** $a^7 - 7a^5c + 21a^3c^2 - 35a^4c^3 + 35a^3c^4 - 21a^2c^5 + 7ac^6 - c^7$. **24.** $16b^4 + 32b^3c + 24b^2c^2 + 8bc^3 + c^4$. **25.** $8a^3 - 12a^2b + 6ab^2 - b^3$. **26.** $81a^4 + 108a^3b + 54a^2b^2 + 12ab^3 + b^4$. **27.** $256x^4 - 256x^3y + 96x^2y^2 - 16xy^3 + y^4$. **28.** $3125x^5 + 3125x^4z + 1250x^3z^2 + 250x^2z^3 + 25xz^4 + z^5$. **29.** $8x^6 - 36x^4y^2 + 54x^2y^4 - 27y^6$. **30.** $27x^6 + 54x^4y^2 + 36x^2y^4 + 8y^6$. **31.** $256x^8 + 768x^5y + 864x^4y^2 + 432x^2y^3 + 81y^4$. **32.** $125x^5 - 150x^4y^2 + 60x^2y^4 - 8y^6$. **33.** $256x^8 - 512x^6y^2 + 384x^4y^4 - 128x^2y^6 + 16y^8$.

Page 171.— 2. $x^3 + 3x^2y + 3xy^2 + y^3$. **3.** $a^3 - 3a^2b + 3ab^2 - b^3$. **4.** $a^3 + 3a^2c + 3ac^2 + c^3$. **5.** $a^4 + 4a^3x + 6a^2x^2 + 4ax^3 + x^4$. **6.** $a^3 - 3a^2x + 3ax^2 - x^3$. **7.** $a^3 + 3a^2b + 3ab^2 + b^3$. **8.** $x^3 - 3x^2y + 3xy^2 - y^3$. **9.** $b^4 + 4b^3c + 6b^2c^2 + 4bc^3 + c^4$. **10.** $x^3 + 3x^2 + 3x + 1$. **11.** $x^3 - 3x^2 + 3x - 1$. **12.** $1 + 3a + 3a^2 + a^3$. **13.** $1 - 4a + 6a^2 - 4a^3 + a^4$. **14.** $x^4 + 4x^3a + 6x^2a^2 + 4xa^3 + a^4$. **15.** $x^3 + 3x^2b + 3xb^2 + b^3$. **16.** $x^4 - 4x^3c + 6x^2c^2 - 4xc^3 + c^4$. **17.** $x^4 + 4x^3y + 6x^2y^2 - 4xy^3 + y^4$. **18.** $a^4 + 4a^3b + 6a^2b^2 + 4ab^3 + b^4$. **19.** $a^5 - 5a^4c + 10a^3c^2 - 10a^2c^3 + 5ac^4 - c^5$. **20.** $a^4 - 4a^3x + 6a^2x^2 - 4ax^3 + x^4$. **21.** $a^6 + 6a^5x + 15a^4x^2 + 20a^3x^3 + 15a^2x^4 + 6ax^5 + x^6$. **22.** $a^9 - 9a^8c + 36a^7c^2 - 84a^6c^3 + 126a^5c^4 - 126a^4c^5 + 84a^3c^6 - 36a^2c^7 + 9ac^8 - c^9$. **23.** $x^7 - 7x^6y + 21x^5y^2 - 35x^4y^3 + 35x^3y^4 - 21x^2y^5 + 7xy^6 - y^7$. **24.** $x^{10} + 10x^9y + 45x^8y^2 + 120x^7y^3 + 210x^6y^4 + 252x^5y^5 + 210x^4y^6 + 120x^3y^7 + 45x^2y^8 + 10xy^9 + y^{10}$. **25.** $x^5 + 5x^4 + 10x^3 + 10x^2 + 5x + 1$. **26.** $x^6 - 6x^5 + 15x^4 - 20x^3 + 15x^2 - 6x + 1$. **27.** $1 + 5a + 10a^2 + 10a^3 + 5a^4 + a^5$. **28.** $1 - 7a + 21a^2 - 35a^3 + 35a^4 - 21a^5 + 7a^6 - a^7$. **29.** $x^4 + 4x^3ac + 6x^2a^2c^2 + 4xa^3c^3 + a^4c^4$. **30.** $x^5 + 5x^4bc + 10x^3b^2c^2 + 10x^2b^3c^3 + 5xb^4c^4 + b^5c^5$. **31.** $x^6 - 6x^5ac + 15x^4a^3c^2 - 20x^3a^3c^3 + 15x^2a^4c^4 - 6xa^5c^5 + a^6c^6$.

Page 172.— 33. $a^3 + 6a^2b + 12ab^2 + 8b^3$. **34.** $27a^3 - 27a^2b + 9ab^2 - b^3$. **35.** $8a^3 + 36a^2b + 54ab^2 + 27b^3$. **36.** $27a^3 - 81a^2c + 81ac^2 - 27c^3$. **37.** $81a^4 + 216a^3c + 216a^2c^2 + 96ac^3 + 16c^4$. **38.** $8a^6 - 12a^4c + 6a^2c^2 - c^3$. **39.** $8a^3 + 12a^2c^2 + 6ac^4 + c^6$. **40.** $27x^6 + 54x^4y^2 + 36x^2y^4 + 8y^6$. **41.** $256a^8 - 256a^6c + 96a^4c^2 - 16a^2c^3 + c^4$. **42.** $27a^6 + 108a^4c^2 + 144a^2c^4 + 64c^6$. **43.** $81x^5 - 540x^6c^2 + 1350x^4c^4 - 1500x^2c^6 + 625c^8$. **44.** $625a^8 + 2000a^6c^2 + 2400a^4c^4 + 1280a^2c^6 + 256c^8$. **46.** $a^3 + 3a^2b + 3ab^2 + b^3 + 3a^2c + 6abc + 3b^2c - 3a^2d - 6abd - 3b^2d + 3ac^2 - 6acd + 3ad^2 + 3bc^2 - 6bcd + 3bd^2 + c^3 - 3c^2d + 3cd^2 - d^3$. **47.** $x^3 - 3x^2y + 3xy^2 - y^3 + 3x^2z - 6xyz + 3y^2z + 3xz^2 - 3yz^2 + z^3$. **48.** $x^3 - 3x^2y + 3xy^2 - y^3 - 3x^2z + 6xyz - 3y^2z + 3xz^2 - 3yz^2 - z^3$. **49.** $a^3 + 3a^2b + 3ab^2 + b^3 - 3a^2c - 6abc - 3b^2c + 3ac^2 + 3bc^2 - c^3$. **50.** $a^3 - 3a^2b + 3ab^2 - b^3 - 3a^2c + 6abc - 3b^2c - 3a^2d + 6abd - 3b^2d + 3ac^2 + 6acd + 3ad^2 - 3bc^2 - 6bcd - 3bd^2 - c^3 - 3c^2d - 3cd^2 - d^3$. **51.** $a^3 + 3a^2b + 3ab^2 + b^3 - 3a^2c - 6abc - 3b^2c + 3a^2d + 6abd + 3b^2d + 3ac^2 - 6acd + 3ad^2 + 3bc^2 - 6bcd + 3bd^2 - c^3 + 3c^2d - 3cd^2 + d^3$. **52.** $1 + 3x + 3x^2 + x^3 - 3y - 6xy - 3x^2y - 3z - 6xz - 3x^2z + 3y^2 + 6yz + 3z^2 + 3xy^2 + 6xyz + 3xz^2 - y^3 - 3y^2z - 3yz^2 - z^3$.

394 HIGH SCHOOL ALGEBRA.

Page 174. — **3.** $8a^3+36a^2c+54ac^2+27c^3$. **4.** $8a^3-48a^2b+96ab^2-64b^3$. **5.** $81a^4+216a^3x+216a^2x^2+96ax^3+16x^4$. **6.** $32a^5-400a^4x+2000a^3x^2-5000a^2x^3+6250ax^4-3125x^5$. **7.** $625a^4+1500a^3c+1350a^2c^2+540ac^3+81c^4$. **8.** $1024a^5+3840a^4x+5760a^3x^2+4320a^2x^3+1620ax^4+243x^5$. **9.** $243a^5-2025a^4x+6750a^3x^2-11250a^2x^3+9375ax^4-3125x^5$. **10.** $2401a^4-5488a^3c+4704a^2c^2-1792ac^3+256c^4$. **11.** $46656a^6+233280a^5x+486000a^4x^2+540000a^3x^3+337500a^2x^4+112500ax^5+15625x^6$. **12.** $243x^5-405x^4+270x^3-90x^2+15x-1$. **13.** $16x^4+160x^3+600x^2+1000x+625$. **14.** $a^5+\tfrac{5}{2}a^3+\tfrac{5}{2}a+\dfrac{5}{4a}+\dfrac{5}{16a^3}+\dfrac{1}{32a^5}$.
15. $1-\tfrac{15}{2}x+\tfrac{45}{2}x^2-\tfrac{135}{4}x^3+\tfrac{405}{16}x^4-\tfrac{243}{32}x^5$. **16.** $\tfrac{81}{16}+\tfrac{45}{4}c+\tfrac{75}{2}c^2+\tfrac{250}{9}c^3+\tfrac{625}{81}c^4$. **17.** $a^{10}+10a^8c^2+40a^6c^4+80a^4c^6+80a^2c^8+32c^{10}$.

Page 176. — **2.** $\pm 4ab^2c^2$. **3.** $-2ab^2c$. **4.** $\pm 2a^2c^2x$. **5.** $3xy^2z$.

Page 177. — **6.** $\pm 2ab^2c^2$. **7.** $-2a^{\tfrac{1}{3}}bc^{\tfrac{2}{3}}$. **8.** $-ac^2x^{\tfrac{2}{5}}y^{\tfrac{1}{5}}$. **9.** ax^3y.
10. $axy^{\tfrac{2}{3}}z^{\tfrac{5}{3}}$. **11.** $\pm xy^2z^{\tfrac{3}{4}}w^{\tfrac{1}{2}}$. **12.** $\pm 2x^ny^{\tfrac{n}{2}}z^{\tfrac{n}{4}}$. **13.** $\pm a^2xy^2z^4$.
14. $a^2x^4y^3$. **15.** $\pm\dfrac{4a}{5y^2}$. **16.** $\dfrac{2x}{3y^2}$. **17.** $\dfrac{5xy}{6a^2y}$. **18.** $\pm\dfrac{9x^5}{17y^2z^3}$.
19. $-\dfrac{3a^2b^3}{7x^4y^6}$. **20.** $x^{\tfrac{8}{5}}yz^{\tfrac{n}{5}}$. **21.** xy^2z^3. **22.** $ax^2y^3z^4$. **23.** $-2a^2x^3y^4$.

Page 179. — **3.** $x+2$. **4.** $b+x$. **5.** $2x+1$. **6.** $a+\tfrac{1}{2}b$. **7.** $3a-2b$. **8.** $a+b-c$. **9.** $2x^2-3x+1$. **10.** $2a^2+a-2$. **11.** x^3-2x^2+3x. **12.** $4a^2-3ax+5x^2$. **13.** $a-b-c$. **14.** $3x^2-2x+6$. **15.** $2x^3+3x^2-x-1$. **16.** $7x^2-2x-\tfrac{3}{4}$. **17.** $3a+c-1$.

Page 180. — **18.** $2m-3x+2$. **19.** $a-b-6c$. **20.** $a^m-b^2+c^{3n}$.
21. $1-\dfrac{a}{2}+\dfrac{a}{3}$. **22.** m^3-m^2-m-1. **23.** $a^2+\dfrac{a}{2}+\dfrac{1}{4}$. **24.** $c^2-c+\tfrac{1}{4}$.
25. $\dfrac{c^2}{2}+\dfrac{c}{x}-\dfrac{x}{c}$. **27.** $1-\dfrac{x}{2}-\dfrac{x^2}{8}$. **28.** $1-a-\dfrac{a^2}{2}$. **29.** $a-\dfrac{b}{a}-\dfrac{b^2}{2a^3}$.
30. $a+\dfrac{b}{2a}-\dfrac{b^2}{8a^3}$. **31.** $2+\dfrac{3a}{4}-\dfrac{9a^2}{64}$. **32.** $2a+\dfrac{3}{4a}-\dfrac{9}{64a^3}$.
33. $2x-\dfrac{1}{4x}-\dfrac{1}{64x^3}$. **34.** $1-y^2-\dfrac{y^4}{2}$. **35.** $a+\dfrac{b}{4a}-\dfrac{b^2}{32a^3}$.

Page 183. — **3.** 53. **4.** 63. **5.** 66. **6.** 96. **7.** 47. **8.** 41. **9.** 256. **10.** 233. **11.** 207. **12.** 266. **13.** 344. **14.** 821. **15.** 886. **16.** 881. **17.** 328. **18.** 15.3. **19.** 4.16. **20.** 24.01. **21.** 969. **22.** 2424. **23.** 3546. **24.** 5555. **25.** 1.157. **26.** .027. **27.** 32.16. **28.** 172.15. **29.** .0849. **30.** 2.2360+. **31.** 3.3166+. **32.** 3.6055+. **33.** 3.8729+. **34.** 2.6832+. **35.** 2.3021+. **36.** .2645+. **37.** .0836+. **38.** .7559+. **39.** .6123+. **40.** .7453+. **41.** .7385+. **42.** 1.1547+. **43.** .7745+. **44.** .6831+. **45.** .8164+.

Page 185. — **3.** $x+2y$. **4.** $3a+1$. **5.** $2x-3$. **6.** $3x+4$.

Page 186. — **7.** $a+\dfrac{1}{a}$. **8.** $a-4$. **9.** $2a+1$. **10.** $3x^2-2x+1$.
11. $2m^2+3m+1$. **12.** $1-a+a^2$. **13.** $m-1-\dfrac{1}{m}$. **14.** $x-y+2z$.
15. $2a-7b$. **16.** y^2-2y+3. **17.** $1-3a+4a^2$. **18.** $2c^2-3c+1$.
19. $a^2+1+\dfrac{1}{a^2}$. **20.** $x^2-4xy+4y^2$.

ANSWERS.   395

Page 187.—1. $1-2a$. **2.** $x+1$. **3.** $x+1$. **4.** $a+2b$. **5.** $x-y$.
6. $a-\dfrac{x}{y}$. **7.** $c+z$.

Page 190.—3. 42. **4.** 64. **5.** 55. **6.** 327. **7.** 507. **8.** 89.
9. 57. **10.** 63. **11.** .25. **12.** 2.34. **13.** 177. **14.** 126. **15.** 536.
16. 39.2. **17.** .04. **18.** 1.442+. **19.** .646+. **20.** 1.856+. **21.** .166+.
22. .941+. **23.** .956+. **24.** .648+. **25.** .643+.

Page 192.—17. 64. **18.** 5. **19.** 81. **20.** 4. **21.** 216. **22.** 243.
23. $\frac{27}{64}$. **24.** $\frac{27}{64}$.

Page 193. — 1. $\dfrac{3}{a^2}$. **2.** $\dfrac{4}{b^3}$. **3.** $\dfrac{1}{a^2b^3}$. **4.** $\dfrac{1}{xy^2}$. **5.** $\dfrac{6y}{a^3}$. **6.** $\dfrac{x^3}{y^2}$.
7. $\dfrac{a^2}{b^5}$. **8.** $\dfrac{4x^3}{y^7}$. **9.** $\dfrac{3x}{cd^3}$. **10.** $\dfrac{2ac}{b^n}$. **11.** $\dfrac{7b^2}{ac^3}$. **12.** $\dfrac{5b^2c^n}{a^3}$. **13.** $2xy^{-2}$.
14. a^2xy^{-3}. **15.** $3axy^4$. **16.** $cx^{-1}y^{-1}$. **17.** $abcxy^{-2}$. **18.** $5x^2y^{-1}z^{-2}$.
19. $3a^2x^3y^{-n}z^{-3}$. **20.** $2bcd^n$.

Page 194.—1. $6a^{\frac{17}{3}}$. **2.** $15x^3$. **3.** $6a^{\frac{7}{2}}$. **4.** $8b$. **5.** $10c^{\frac{14}{3}}$.
6. $7a$. **7.** $3ab^{\frac{1}{4}}$. **8.** $2a^{\frac{1}{3}}b^2$. **9.** $3ax^{\frac{1}{3}}$. **10.** $4a^{\frac{5}{3}}x^2$. **11.** $8c^{\frac{1}{5}}x$.
12. $10ad^{1-n}$. **13.** $6a^{-\frac{1}{6}}b$. **14.** $ab^{\frac{1}{4}}$. **15.** $15abc^{-\frac{1}{6}}$. **16.** $ab^{\frac{5}{4}}c$.
17. $ab^{-\frac{1}{2}}c^{\frac{3}{2}}$. **18.** $x^{\frac{1}{2}}y^{\frac{5}{3}}z^{\frac{8}{5}}$.

Page 195.—21. $a+2a^{\frac{1}{2}}b^{\frac{1}{2}}+b$. **22.** $x+x^{\frac{2}{3}}y^{\frac{1}{3}}+x^{\frac{1}{3}}y^{\frac{2}{3}}+y$. **23.** $x^{\frac{2}{3}}+x^{\frac{1}{2}}y^{\frac{1}{6}}+2x^{\frac{1}{3}}y^{\frac{1}{3}}+x^{\frac{1}{6}}y^{\frac{1}{2}}+y^{\frac{2}{3}}$. **24.** $4x+x^{\frac{5}{3}}+3x^{\frac{4}{3}}+4x^{\frac{1}{3}}+6x^{\frac{2}{3}}$. **25.** $x^{\frac{1}{2}}+x^{\frac{3}{4}}y^{\frac{1}{2}}-x^{\frac{1}{2}}y^{\frac{1}{4}}-y^{\frac{1}{2}}$. **26.** $a^{-\frac{1}{3}}b^{\frac{2}{3}}-a^{\frac{2}{3}}b^{-\frac{1}{3}}$. **27.** $x^3-x-2-x^{-1}$. **28.** $x^6y^{-6}-7+x^{-6}y^6$. **29.** $x^3y^{-2}-x^2y+x^{-1}y^2+x^3y^{-1}-x^2y^2+x^{-1}y^3-x^2y^{-1}+xy^2-x^{-2}y^3$. **30.** $-x^3y^{-3}+x^{\frac{3}{2}}y^{-\frac{3}{2}}+x^{-3}y^3+x^{-\frac{3}{2}}y^{\frac{3}{2}}$.

Page 196.—1. a^2. **2.** a^5. **3.** a^{-2}. **4.** a^4. **5.** $a^{\frac{17}{12}}$. **6.** $c^{-\frac{2}{15}}$.
7. $x^{-\frac{13}{12}}$. **8.** $z^{-\frac{5}{21}}$. **9.** $b^{\frac{1}{2}}$. **10.** $bc^{-\frac{11}{12}}$. **11.** $a^{\frac{1}{4}}x^{\frac{11}{12}}$. **12.** $x^{-\frac{27}{20}}y^{\frac{11}{20}}$.
15. $x^{\frac{1}{2}}+y^{\frac{1}{2}}$. **16.** $x^{\frac{2}{3}}+x^{\frac{1}{3}}y^{\frac{1}{3}}+y^{\frac{2}{3}}$. **17.** $a^{\frac{2}{3}}+b^{\frac{2}{3}}$. **18.** $a^{\frac{2}{3}}+2a^{\frac{1}{3}}x^{\frac{1}{3}}+x^{\frac{2}{3}}$.
19. $x^{\frac{1}{3}}-y^{\frac{1}{3}}$.

Page 197.— 20. $x^{-2}+y^{-2}$. **21.** $x^{-4}-2x^{-2}y^{-2}+y^{-4}$. **22.** $-b^{-2}+a^{-2}+a^{-1}b^{-1}$. **23.** $c^{\frac{2}{3}}-c^{\frac{1}{3}}+1$.

1. a^2. **2.** $a^{-\frac{1}{3}}$ or $\dfrac{1}{a^{\frac{1}{3}}}$. **3.** c^{-1} or $\dfrac{1}{c}$. **4.** b. **5.** $d^{\frac{1}{5}}$. **6.** $x^{-\frac{1}{2}}$ or $\dfrac{1}{x^{\frac{1}{2}}}$.
7. $x^{\frac{6}{25}}$. **8.** z^{-2} or $\dfrac{1}{z^2}$. **9.** c^{-4} or $\dfrac{1}{c^4}$. **10.** $\dfrac{c}{27}$. **11.** $\dfrac{a^2}{4}$. **12.** $\dfrac{x^{-6}}{216}$ or $\dfrac{1}{216x^6}$. **13.** $z^{-\frac{1}{5}}$ or $\dfrac{1}{z^{\frac{1}{5}}}$. **14.** $x^{-\frac{1}{2}}y^{-\frac{1}{3}}$ or $\dfrac{1}{x^{\frac{1}{2}}y^{\frac{1}{3}}}$. **15.** $\dfrac{x^{-7}}{16384}$ or $\dfrac{1}{16384x^7}$.
16. $\dfrac{125\,x^{-6}y^{-\frac{9}{2}}}{27}$ or $\dfrac{125}{27x^6y^{\frac{9}{2}}}$.

1. $a^{\frac{2}{3}}+2a^{\frac{1}{3}}b^{\frac{1}{3}}+b^{\frac{2}{3}}$. **2.** $x^{\frac{4}{3}}-2x^{\frac{2}{3}}y^{\frac{2}{3}}+y^{\frac{4}{3}}$. **3.** $c^{\frac{5}{3}}+2c^{\frac{5}{6}}x^{\frac{5}{6}}+x^{\frac{5}{3}}$.
4. $x^{\frac{4}{3}}-y^{-\frac{4}{3}}$. **5.** $1+2x^{\frac{1}{2}}y^{\frac{1}{2}}+xy-x^{-1}y-2x^{-\frac{1}{2}}y^{\frac{3}{2}}-y^2$. **6.** $x^3+2x^{\frac{9}{4}}z^{\frac{3}{4}}-2x^{\frac{3}{4}}z^{\frac{9}{4}}-z^3$.

HIGH SCHOOL ALGEBRA

Page 198.—7. $a-b$.   8. $c-2c^{\frac{1}{2}}+6c^{\frac{1}{4}}-c^{\frac{3}{4}}-4$.   9. $8a+4a^{\frac{4}{5}}b^{\frac{1}{5}}+6a^{\frac{3}{5}}b^{\frac{2}{5}}+18a^{\frac{2}{5}}-4a^{\frac{3}{5}}b-2a^{\frac{2}{5}}b^{\frac{6}{5}}-3a^{\frac{1}{5}}b^{\frac{7}{5}}-9b$.   10. $x^{3p}+y^{3p}$.   11. $7c^{\frac{2}{3}}-2c^{\frac{1}{3}}+1$.   12. $ab^{-1}+a^{-1}b$.   13. $x^{4n}+x^{3n}y^n+x^{2n}y^{2n}+x^ny^{3n}+y^{4n}$.   14. $a^{\frac{2}{3}}-a^{\frac{1}{3}}+1$.   15. $3c^{\frac{1}{2}}-2+c^{-\frac{1}{2}}$.   16. $2a^{\frac{1}{2}}+3b^{\frac{1}{3}}+c$.   17. $2x^{\frac{2}{3}}-5+x^{-\frac{1}{3}}$.   18. $x^{\frac{1}{2}}+x^{\frac{1}{4}}+1$.   19. $2x^{\frac{1}{4}}-5x^{\frac{1}{8}}y^{\frac{1}{8}}+3y^{\frac{1}{4}}$.   20. $x^{\frac{1}{3}}+4$.   21. $x^{\frac{2}{3}}-x^{\frac{1}{3}}-1$.   22. $\frac{1}{2}x-2y^{\frac{1}{2}}$.   23. $2a+1-3a^{-1}$.

Page 201.—3. $3\sqrt{5}$.   4. $5\sqrt{3}$.   5. $3\sqrt[3]{2}$.   6. $6\sqrt{3}$.   7. $8\sqrt{3}$.   8. $3\sqrt[3]{6}$.   9. $2\sqrt[3]{-7}$.   10. $5\sqrt[3]{3}$.   11. $7\sqrt[3]{-2}$.   12. $3\sqrt[3]{2}$.   13. $3x\sqrt{2}$.   14. $6a\sqrt{b}$.   15. $5xy^2\sqrt{3x}$.   16. $10a^3b\sqrt{b}$.   17. $12ax\sqrt{3ay}$.   18. $9a\sqrt{xy}$.   19. $6yz\sqrt{x}$.   20. $15xyz\sqrt{2x}$.   21. $a\sqrt{1-x}$.   22. $x\sqrt[3]{x-y^2}$.   23. $a\sqrt{a-b}$.   24. $x\sqrt[3]{x^2+y^2}$.   25. $x\sqrt{a-x}$.   26. $b\sqrt{2b+b^2y}$.   27. $b\sqrt[3]{3+bx}$.   28. $x\sqrt[3]{ax-2y}$.   29. $2\sqrt{ax-2}$.   30. $3\sqrt{a^3+2y}$.   31. $x\sqrt[3]{16+54xy^2}$.   32. $y(a-x)\sqrt[3]{(a-x)^2y}$.   33. $a^2\sqrt[3]{a-x^2}$.   34. $a(a+b)\sqrt{a}$.   35. $a(a+b)\sqrt{a^2-b^2}$.   36. $(x^2-y^2)\sqrt{x}$.   37. $x(1+y+y^2)^{\frac{1}{2}}$.   38. $4a(1+y)x^{\frac{1}{2}}$.

Page 202.—40. $\frac{1}{3}\sqrt{6}$.   41. $\frac{1}{7}\sqrt{21}$.   42. $\frac{1}{5}\sqrt{40}$.   43. $\frac{1}{3}\sqrt[3]{15}$.   44. $\frac{1}{7b}\sqrt{35ab}$.   45. $\frac{1}{3y}\sqrt{7xy}$.   46. $\frac{1}{6ab}\sqrt{30abxy}$.   47. $\frac{1}{2b}\sqrt[3]{7ab^2}$.   48. $\frac{1}{3x}\sqrt[3]{18cx^2}$.   49. $\frac{x}{5a}\sqrt{15ax}$.   50. $\frac{1}{2}\sqrt{10}$.   51. $\frac{1}{2b}\sqrt{21ab}$.   52. $\frac{x-2}{x+2}\sqrt{x^2-4}$.   53. $\frac{a+b}{a-b}\sqrt{a(a-b)}$.   54. $\sqrt[3]{x(x+y)}$.   55. $\frac{a^2-b^2}{a^2+b^2}\sqrt{x(a^2+b^2)}$.   56. $\frac{3a}{7y}(21xy)^{\frac{1}{2}}$.   57. $\frac{5x}{2}(6y^2)^{\frac{1}{3}}$.

Page 203.—2. $\sqrt{9a^2x^4}$.   3. $\sqrt{4x^2y^2}$.   4. $\sqrt{16a^4x^2y^2}$.   5. $\sqrt{25x^4y^2}$.   6. $\sqrt{9x^6y^8}$.   7. $\sqrt{25a^{-4}x}$.   8. $\sqrt{(a+b)^2}$.   9. $\sqrt{9a^4x^{-6}}$.   10. $\sqrt[3]{8a^6x^3}$.   11. $\sqrt[3]{27a^3x^9}$.   12. $\sqrt[3]{8b^{12}c^3}$.   13. $\sqrt[3]{64x^{-6}y^3}$.   14. $\sqrt[3]{27a^{\frac{3}{2}}b^6}$.   15. $\sqrt[3]{ax^{\frac{3}{4}}}$.   16. $\sqrt[3]{(a-x)^3}$.   17. $\sqrt[3]{(a+b)^{\frac{3}{2}}}$.   18. $\sqrt{4x^2(x^2+y^2)}$.   19. $\sqrt{2x(x+y)^2}$.   20. $\sqrt{9a^2(a^2-b^2)}$.   21. $\sqrt{(a+b)(a-b)^2}$.   22. $\sqrt{a(x-y)}$.   23. $\sqrt{x(x+y)^2}$.

Page 204.—2. $\sqrt[6]{1000}$, $\sqrt[6]{9}$.   3. $\sqrt[6]{a^3x^3}$, $\sqrt[6]{x^4y^2}$.   4. $\sqrt[4]{x^4y^6}$, $\sqrt[4]{xy^2}$.   5. $\sqrt[12]{a^6x^3y^3}$, $\sqrt[12]{x^4y^8z^4}$.   6. $\sqrt[10]{a^5x^5y^{2j}}$, $\sqrt[10]{a^4x^2y^4}$.   7. $\sqrt[mn]{a^mx^m}$, $\sqrt[mn]{b^ny^n}$.   8. $\sqrt[6]{a^3x^3}$, $\sqrt[6]{b^2y^2}$, $\sqrt[6]{a^3z^3}$.   9. $\sqrt[6]{a^6x^3}$, $\sqrt[6]{a^2y^4}$, $\sqrt[6]{b^3y^9}$.   10. $\sqrt[6]{\frac{8}{27}}$, $\sqrt[6]{\frac{9}{25}}$, $\sqrt[6]{8}$.   11. $\sqrt[6]{\frac{x^3}{27}}$, $\sqrt[6]{\frac{x^2}{16}}$, $\sqrt[6]{x^9}$.   12. $\sqrt[6]{(a+b)^3}$, $\sqrt[6]{(a+b)^2}$.

Page 205.—2. $15\sqrt{2}$.   3. $11\sqrt{3}$.   4. $9\sqrt[3]{4}$.   5. $56\sqrt{2}$.   6. $12\sqrt[3]{6}$.   7. $3\frac{1}{10}\sqrt{6}$.   8. $12\sqrt{6}$.   9. $1\frac{1}{3}\sqrt{3}$.   10. $1\frac{1}{4}\sqrt{2}$.   11. $44x^2y^2\sqrt{2xy}$.   12. $25\sqrt{3}$.   13. $24xy\sqrt{2}$.   14. $9\sqrt[3]{5}$.   15. $\sqrt{6}$.   16. $11\sqrt{3}$.   17. $34\sqrt{2}$.   18. $9a\sqrt{5a}$.   19. $(2a+2b-5c)\sqrt{2y}$.   20. $8\sqrt[3]{2}$.   21. $5\sqrt{3a}$.   22. $5\frac{1}{4}\sqrt{3}$.

ANSWERS.

Page 206.—23. $\left(\dfrac{1}{x}+\dfrac{1}{y}-\dfrac{1}{z}\right)\sqrt{a}$.   24. $\left(\dfrac{a^2b}{cx^2}-\dfrac{ac}{d^2x}+\dfrac{dy^2}{px}\right)\sqrt{ax}$.
25. $3\sqrt[4]{3}$.   26. $2x\sqrt{z}$.   27. $2\tfrac{3}{5}\sqrt{5}$.   28. $\dfrac{x-y+1}{x-y}\sqrt{x^2-y^2}$.
29. $(yz+x-x^2y^2z^2)\sqrt[5]{x^3yz^2}$.   30. $13\sqrt[3]{4}$.   31. $13\sqrt{2}$.   32. $20\sqrt[3]{-3}$.
33. $\tfrac{7}{12}\sqrt{3}$.   34. $6\sqrt[3]{3x}$.   35. $2y\sqrt{2}$.   36. $6xy^2\sqrt[4]{2}$.   37. $\dfrac{37}{18b}\sqrt{3a}$.
38. $9\tfrac{1}{4}\sqrt{2}$.

Page 207.—4. $12\sqrt{6}$.   5. $24\sqrt{3}$.   6. $30\sqrt{3}$.   7. $36\sqrt{2}$.   8. $9\sqrt{6}$.
9. $24\sqrt{35}$.   10. $150\sqrt{2}$.   11. $24\sqrt[3]{2}$.   12. $60\sqrt[3]{5}$.   13. $30\sqrt[3]{4}$.
14. $6\sqrt[6]{108}$.   15. $3\sqrt[6]{72}$.   16. $2\sqrt[4]{5}$.   17. $6\sqrt[6]{675}$.   18. $6\sqrt[6]{72}$.
19. $288\sqrt[6]{2}$.   20. $3ab\sqrt[3]{2a^2b}$.   21. $a\sqrt{6bc}$.   22. $6xy^2\sqrt{2x}$.
23. $x^2y^2\sqrt[6]{108x^5}$.   24. $axy\sqrt[6]{9a^2x^5y^4}$.   25. $2ab\sqrt[6]{a^4bc^3}$.   26. $36xy\sqrt{2xy}$.
27. $\tfrac{1}{2}\sqrt{3}$.

Page 208.—28. $\tfrac{2}{3}$.   29. $\tfrac{2}{9}\sqrt{6}$.   30. $\tfrac{1}{2}\sqrt[6]{2}$.   31. $\sqrt[6]{3\tfrac{2}{75}}$.   32. $\sqrt[6]{\tfrac{2.5}{6}}$.
33. $\sqrt[6]{2\tfrac{4.5}{4.8}}$.   35. $7\sqrt{2}-12$.   36. $38+23\sqrt{3}$.   37. -55.   38. $\sqrt{6}-21$.
39. $27-33\sqrt{2}$.   40. $9-2\sqrt{15}$.   41. $28+45\sqrt{35}$.   42. $16a+16\sqrt{ab}+4b$.
43. $4-243a$.   44. $a\sqrt{a}-b\sqrt{b}$.   45. $12x-2$.   46. $x^2\sqrt{3}+3x^2+2x\sqrt{3}+x+1$.

Page 209.—4. 1.   5. 6.   6. $1\tfrac{2}{3}$.   7. $1\tfrac{2}{3}\sqrt{3}$.   8. $\tfrac{3}{4}$.   9. $4\tfrac{1}{5}\sqrt{5}$.
10. $4\sqrt{2}$.   11. $3\sqrt[6]{2}$.   12. $\tfrac{1}{2}\sqrt[6]{108}$.   13. $2\sqrt[6]{5}$.   14. $\tfrac{3}{4}\sqrt[6]{x^2z^{-1}}$.   15. $\tfrac{1}{3}\sqrt[3]{36}$.
16. 3.   17. $\tfrac{1}{2}\sqrt[6]{18}$.   18. $\tfrac{3}{2}\sqrt[6]{96}$.   19. $\dfrac{3}{xy}\sqrt[6]{a^2x^5y^3}$.   20. $\dfrac{1}{a}\sqrt[6]{2a^3x}$.
21. $\sqrt[12]{32xz}$.   22. $\sqrt[15]{\dfrac{243x}{216y}}$.   23. $\tfrac{2}{3}\sqrt[12]{x^2yz^{-5}}$.   24. $\sqrt[12]{18b^{-5}c^7}$.
25. $3ax^2\sqrt[12]{axy^4}$.   26. 2.   27. $2+3\sqrt{2}$.   28. $4+3\sqrt{5}$.

Page 210.—29. $4+3\sqrt{6}$.   30. $\tfrac{4}{3}\sqrt{3}-2$.   31. $\sqrt[3]{a^2}-b\sqrt[3]{b}$.
32. $\sqrt{x}+\sqrt{y}+\sqrt{z}$.
4. $4x$.   5. $9x\sqrt[3]{4x}$.   6. $16b\sqrt[3]{9b}$.   7. $18a^2b\sqrt[3]{2b}$.   8. $216x\sqrt{x}$.
9. $24a^4\sqrt{3a}$.   10. $324ax$.   11. $64ax^4$.   12. $16+40\sqrt{x}+25x$.   13. $9-42\sqrt{x}+49x$.   14. $4\sqrt{15}+23$.   15. $7+10x\sqrt{7ax}+25ax^3$.

Page 211.—18. $1.31+$.   19. $2.59+$.   20. $2.92+$.   21. $\sqrt[4]{ax}$.
22. $\sqrt[6]{a^2x}$.   23. $\sqrt[8]{xy^2z^3}$.   24. $1.906+$.   25. $\sqrt[9]{x^2y^5}$.   26. $2.236+$.
27. $\sqrt[15]{x^8y^7}$ 28. $1.48+$.   29. $\sqrt[21]{a^5x^9y^3}$.

Page 212.—4. $\dfrac{3\sqrt{2a}}{2a}$.   5. $\sqrt[3]{2}$.   6. $\dfrac{2\sqrt{ax}}{x}$.   7. $\dfrac{5\sqrt[3]{4a}}{2a}$.
8. $\dfrac{3a\sqrt{2a}}{2}$.   9. $\dfrac{3\sqrt{5abx}}{5x}$.   10. $\dfrac{7\sqrt[3]{3z}}{3z}$.   11. $\dfrac{3a\sqrt[5]{2x^2}}{2}$.   12. $\dfrac{2z\sqrt[4]{3a^2}}{3a}$.

Page 213.—4. $\dfrac{8-2\sqrt{5}-4\sqrt{3}+\sqrt{15}}{1}$.   5. $\dfrac{6-3\sqrt{2}}{2}$.   6. $\dfrac{x^2+x\sqrt{y}}{x^2-y}$.
7. $\dfrac{c\sqrt{x}-c\sqrt{z}}{x-z}$.   8. $\dfrac{5-2\sqrt{6}}{1}$.   9. $\dfrac{x^2+2x\sqrt{z}+z}{x^2-z}$.   10. $\dfrac{a+2\sqrt{ax}+x}{a-x}$.

11. $\dfrac{x-2+2\sqrt{x+1}}{x}$.   12. $\dfrac{2x^2-1-2x\sqrt{x^2-1}}{1}$.   13. $\dfrac{2a+3-2\sqrt{a^2+3a}}{3}$.
14. $\dfrac{b+\sqrt{b^2-c^2}}{c}$.   15. $\dfrac{\sqrt{x^4-4}-x^2}{2}$.

Page 214.—17. 4.949+.   18. 9.237+.   19. 2.828+.   20. .2309+.
21. .412+.   22. .2357+.   23. 1.   24. 8.549+.   25. 1.271+.

Page 216.—4. $\sqrt{3}+\sqrt{2}$.   5. $1+\sqrt{8}$.   6. $\sqrt{3}-\sqrt{5}$.   7. $3-\sqrt{2}$.
8. $\sqrt{5a}+\sqrt{a}$.   9. $\sqrt{3}-\sqrt{2}$.   10. $2\sqrt{2}+\sqrt{7}$.   11. $5+\sqrt{3}$.
12. $\sqrt{10}+\sqrt{3}$.   13. $\sqrt{5}-\sqrt{2}$.   14. $\sqrt{6}+\sqrt{5}$.   15. $\sqrt{m+n}-\sqrt{m-n}$.
16. $2\sqrt{2}-\sqrt{7}$.   17. $3+\sqrt{5}$.   18. $\tfrac{1}{2}(\sqrt{14}+\sqrt{10})$.   19. $5+\sqrt{3}$.
20. $m+\sqrt{n}$.   21. $2\sqrt{3}+\sqrt{13}$.

1. $a^7+7a^6b+21a^5b^2+35a^4b^3+35a^3b^4+21a^2b^5+7ab^6+b^7$.   2. $\dfrac{8x^3}{125}+\dfrac{36x^2y}{175}+\dfrac{54xy^2}{245}+\dfrac{27y^3}{343}$.   3. $16a^4-96a^3b+216a^2b^2-216ab^3+81b^4$.
4. $\dfrac{x^5}{1024}+\dfrac{5x^4y}{768}+\dfrac{5x^3y^2}{288}+\dfrac{5x^2y^3}{216}+\dfrac{5xy^4}{324}+\dfrac{y^5}{243}$.   5. $243a^5+1620a^4b+4320a^3b^2+5760a^2b^3+3840ab^4+1024b^5$.   6. $\dfrac{a^3}{8}-\dfrac{a^2b}{4}+\dfrac{ab^2}{6}-\dfrac{b^3}{27}$.

7. $a^n+na^{n-1}b+\dfrac{n(n-1)}{2}a^{n-2}b^2+\dfrac{n(n-1)(n-2)}{2\times 3}a^{n-3}b^3+\dfrac{n(n-1)(n-2)(n-3)}{2\times 3\times 4}a^{n-4}b^4+\dfrac{n(n-1)(n-2)(n-3)(n-4)}{2\times 3\times 4\times 5}a^{n-5}b^5$.

8. $a^{n-2}+(n-2)a^{n-3}b+\dfrac{(n-2)(n-3)}{2}a^{n-4}b^2+\dfrac{(n-2)(n-3)(n-4)}{2\times 3}a^{n-5}b^3+\dfrac{(n-2)(n-3)(n-4)(n-5)}{2\times 3\times 4}a^{n-6}b^4$.

9. $x^r+rx^{r-1}y+\dfrac{r(r-1)}{2}x^{r-2}y^2+\dfrac{r(r-1)(r-2)}{2\times 3}x^{r-3}y^3+\dfrac{r(r-1)(r-2)(r-3)}{2\times 3\times 4}x^{r-4}y^4+\dfrac{r(r-1)(r-2)(r-3)(r-4)}{2\times 3\times 4\times 5}x^{r-5}y^5$.

10. $1+4x^2+9y^2+z^2+4x+6y+2z+12xy+4xz+6yz$.

Page 217.—11. $4\sqrt{2}-\sqrt{78}$.   12. $8\sqrt[3]{2a}+\sqrt[9]{2a^3}$.   13. $12\tfrac{1}{2}\sqrt{3}$.
14. 10.   15. $a\sqrt{3a}$.   16. $3x^{\tfrac{1}{2}}-4y^{\tfrac{2}{3}}+2$.   17. $2a^{\tfrac{2}{3}}+a-4a^{\tfrac{4}{3}}$.   18. $x^{\tfrac{2}{3}}-2x^{\tfrac{1}{3}}y^{\tfrac{1}{3}}+y^{\tfrac{2}{3}}$.   19. $3x^{\tfrac{1}{2}}-2x^{\tfrac{1}{3}}+x^{\tfrac{1}{6}}-4$.   20. x^2-x-1.   21. $x^{\tfrac{2}{5}}-2x^{\tfrac{1}{5}}+1$.
22. $x^{\tfrac{1}{6}}+x^{\tfrac{1}{12}}y^{\tfrac{1}{12}}-2y^{\tfrac{1}{6}}$.   23. $a^{\tfrac{1}{3}}x^{-\tfrac{1}{3}}+\tfrac{1}{3}a^{-\tfrac{5}{9}}x^{\tfrac{5}{3}}-\tfrac{1}{9}a^{-\tfrac{11}{9}}x^{\tfrac{11}{3}}$ + etc.   24. $2x+2\sqrt{ax}$.   25. x.   26. $4x^{\tfrac{8}{3}}-8x^{\tfrac{4}{3}}-5+10x^{-\tfrac{4}{3}}+3x^{-\tfrac{8}{3}}$.

Page 218.—27. $m-\sqrt{mn}+n$.   28. $x^2-xy\sqrt{2}+y^2$.   29. 1.
30. $y\sqrt[3]{4y}+2\sqrt[6]{4x^3y^5}+x\sqrt[3]{y}$.   31. $4+4a^2\sqrt{2x}+y+a^4(2x+y)$.   32. $147+60\sqrt{6}$.   33. $1+2x\sqrt{1-x^2}$.   34. 3.464+.   35. $a^3x\sqrt[3]{ax^2}$.   36. $a^2\sqrt[3]{a^2}$.
37. $27\sqrt{3}$.   38. $\sqrt[5]{8}$ or 1.514+.   39. $\sqrt[5]{3x\sqrt[3]{2}}$.   40. $\dfrac{9+5\sqrt{3}}{6}$.
41. $\dfrac{3\sqrt{3}+3\sqrt{2}+\sqrt{15}+\sqrt{10}}{1}$.   42. $\dfrac{a+\sqrt{ab}}{a-b}$.   43. $\dfrac{24+17\sqrt{2}}{1}$.

ANSWERS.

44. $\dfrac{3\sqrt{2}-2\sqrt{3}}{1}$. **45.** $\dfrac{8-\sqrt{42}}{1}$. **46.** $\dfrac{2a^2+2a\sqrt{a^2-x^2}-x^2}{x^2}$.

47. $\dfrac{m^2-\sqrt{m^4-1}}{1}$. **48.** 1.341. **49.** 4.081. **50.** 4.791. **51.** .269.
52. $\sqrt{5}-2$. **53.** $5-3\sqrt{2}$. **54.** $\sqrt{21}-\sqrt{3}$. **55.** $2\sqrt{3}-\sqrt{10}$.
56. $3\sqrt{5}-\sqrt{2}$. **57.** $\sqrt{12}-\sqrt{11}$.

Page 221. — **5.** 4. **6.** 32. **7.** 56. **8.** 6. **9.** 27. **10.** 12. **11.** 5.
12. 6. **13.** 9. **14.** 25. **15.** 121. **16.** 4. **17.** 64. **18.** 18. **19.** 25.
20. $5\frac{1}{3}$. **21.** 1. **22.** 27. **23.** $\frac{3}{4}$. **24.** 2. **25.** $a-1$. **26.** $\dfrac{81}{a}$. **27.** 4.
28. 64. **29.** $2\frac{1}{3}$. **30.** 25. **31.** 100.

Page 222. — **32.** $4\frac{1}{2}$. **33.** 10. **34.** $\dfrac{2ab}{b^2+1}$. **35.** $\frac{1}{3}$. **36.** $\dfrac{a(b-1)^2}{4b}$.

37. $\frac{4}{9}$. **38.** 4. **39.** $\dfrac{4a^2}{(a+1)^2}$. **40.** $\dfrac{a^2(b-1)^2}{(b+1)^2}$. **41.** $\dfrac{4ab^2}{(1+a)^2}$. **42.** $\frac{2}{5}$.
43. $\frac{9}{32}$. **44.** $2\frac{2}{5}$. **45.** 27. **46.** 9. **47.** -3. **48.** $16a$. **49.** $-b$.
50. $\pm\frac{1}{2}\sqrt{5}$. **51.** $\dfrac{a(b-c)}{2\sqrt{bc}}$. **52.** $\dfrac{a(\sqrt{b}-1)^2}{2\sqrt{b}}$.

Page 224. — **3.** ± 7. **4.** ± 2. **5.** ± 6. **6.** ± 6. **7.** ± 4. **8.** ± 5.
9. ± 10.

Page 225. — **10.** ± 5. **11.** ± 6. **12.** ± 3. **13.** ± 3. **14.** ± 1.
15. ± 8. **16.** $+\frac{1}{2}$. **17.** $\pm 3\sqrt{2}$. **18.** ± 2. **19.** $\pm a\sqrt{2}$. **20.** ± 1.
21. ± 2. **22.** $\pm\frac{1}{2}\sqrt{3}$. **23.** $\pm m\sqrt{3}$. **24.** $\pm\dfrac{a}{\sqrt{2a-1}}$. **25.** $\pm\frac{3}{2}$.
26. $\pm 4\frac{1}{2}$. **27.** ± 6. **28.** ± 1. **29.** $\pm\sqrt{2}$. **30.** $\pm\frac{2}{3}\sqrt{2}$. **31.** $\pm\sqrt{a^2+b^2}$.
32. $\pm\frac{1}{2}\sqrt{5}$. **33.** $\pm\sqrt{\dfrac{a}{a-2}}$. **34.** $\pm\sqrt{(a-2)^2-1}$.

Page 226. — **1.** $\frac{2}{3}$. **2.** 4. **3.** 8. **4.** 8. **5.** 30, ± 50. **6.** Son, 8;
Father, 32. **7.** \$150. **8.** 12 ft., 18 ft. **9.** 3, 9. **10.** 6, 8. **11.** 6, 7.

Page 227. — **12.** 8, 12. **13.** 12, 20. **14.** 8 yd. **15.** Length, 40 rd.;
Breadth, 36 rd. **16.** 8, 10. **17.** 6, 10, 16. **18.** 25 da. \$2.50 per da.
19. 9, 36. **20.** 4, 16. **21.** 7, 11.

Page 230. — **4.** 5, or -9. **5.** 3, or -9. **6.** 2, or -10. **7.** 1, or
-11. **8.** 1, or -21. **9.** 1, or -19. **10.** 1, or -25. **11.** 15, or -3.
12. 11, or -3. **13.** 17, or -3. **14.** 30, or -2. **15.** 32, or -2.
16. 2, or $-3\frac{1}{2}$. **17.** 3, or $-4\frac{1}{3}$. **18.** 3, or $-6\frac{1}{2}$. **19.** 2, or -5.
20. 4, or $-\frac{1}{2}$. **21.** 20, or 1. **22.** 2, or $-5\frac{1}{3}$. **23.** 4, or -1.

Page 233. — **5.** 1, or $-2\frac{2}{3}$. **6.** 2, or $-5\frac{1}{2}$. **7.** 4, or $-5\frac{1}{4}$. **8.** 5, or
$-4\frac{1}{5}$. **9.** 10, or $-4\frac{2}{3}$. **10.** 6, or $-4\frac{1}{4}$. **11.** 2, or $-2\frac{4}{5}$. **12.** 8, or $-7\frac{1}{4}$.
13. 6, or $-4\frac{4}{5}$. **14.** 4, or $-4\frac{2}{3}$. **15.** 8, or -2. **16.** 14, or -1.
17. 1, or -18. **18.** 12, or -1. **19.** 5, or 4. **20.** 2, or $-4\frac{1}{2}$. **21.** $2\frac{1}{3}$,
or -3. **22.** $6.229+$, or $-2.729+$. **23.** $3.525+$, or $-2.325+$. **24.** 3,
or $-2\frac{1}{3}$. **25.** 13, or $-4\frac{1}{3}$. **26.** 7, or $-1\frac{3}{7}$. **27.** $1\frac{7}{18}$, or $-1\frac{2}{3}$. **28.** 2, or
-5. **29.** 14, or -10. **30.** 3, or $-\frac{1}{2}$. **31.** 7, or $-1\frac{2}{3}$. **32.** 1, or $\frac{7}{35}$.
33. 5, or $5\frac{2}{5}$. **34.** 6, or $-1\frac{1}{3}$. **35.** 3. **36.** 4, or $-1\frac{2}{3}$. **37.** -3, or $-3\frac{2}{5}$.

Page 234. — **38.** 4, or -21. **39.** 1, or $\frac{2}{6}$. **40.** 4, or -1. **41.** $2\frac{2}{3}$.
42. 4. **43.** 3, or $\frac{3}{4}$. **44.** 5, or -3. **47.** $-b(1\pm\sqrt{2})$. **48.** $6b$, or $-2b$.

400 HIGH SCHOOL ALGEBRA.

49. $b(2 \pm \sqrt{11}$. **50.** $2a$, or $-5a$. **51.** $2b$, or $-7b$. **52.** $2c$, or $\dfrac{c}{2}$.
53. $\dfrac{3}{4a}$, or $-\dfrac{1}{3a}$. **54.** $\dfrac{a}{2}$, or $-\dfrac{7a}{2}$. **55.** a, or 1.

Page 235.—56. $\tfrac{10}{9}$, or -2. **57.** -16, or $-2a$. **58.** $\dfrac{4b-15}{4} \pm$ $\tfrac{1}{4}\sqrt{16b^2-240b+225}$. **59.** $\dfrac{b}{a}$, or $-\dfrac{a}{b}$. **60.** $\dfrac{c(c+d)}{c-d}$. **61.** $\dfrac{3b}{4}$, or $\dfrac{b}{2}$.
62. $4c$, or $-\dfrac{c}{4}$. **63.** $\dfrac{b^2}{ac}$. **64.** c, or $-\dfrac{1}{c}$. **65.** c, or $-\dfrac{c}{7}$. **66.** $\dfrac{3b}{a}$, or $\dfrac{2b}{a}$. **67.** c, or $\dfrac{1}{c}$. **68.** $1 \pm \sqrt{-2}$. **69.** $\dfrac{b}{3}$, or $\dfrac{2c}{3}$. **70.** $a-b$, or c.
71. $a+c$, or $4c$. **72.** $\dfrac{c}{c+d}$, or $\dfrac{-d}{c+d}$. **73.** $9n$, or $-n$. **74.** $\dfrac{a+b}{c}$, or $\dfrac{c}{a+b}$.
75. $2a$, or $2b\dfrac{(2a-3b)}{3a-2b}$. **76.** $\dfrac{c+d}{c-d}$, or $\dfrac{c-d}{c+d}$.

Page 237.—4. ± 2, or $\pm \sqrt{-2}$. **5.** 2, or $\sqrt[3]{-5}$. **6.** 2, or $\sqrt[3]{-4}$.
7. ± 2, or $\pm \sqrt{-2}$. **8.** ± 1, or $\pm \tfrac{1}{2}\sqrt{-3}$. **9.** 2, or $\tfrac{1}{2}\sqrt[3]{-62}$. **10.** $\pm \sqrt{2}$, or $\pm \sqrt{-6}$. **11.** 16, or 256. **12.** 4, or $\sqrt[3]{121}$. **13.** 1, or -64.
14. 625, or 256. **15.** $\sqrt[n]{-\dfrac{b}{2a} \pm \dfrac{1}{2a}\sqrt{b^2+4ac}}$. **16.** ± 1, or $\pm \sqrt{-10}$.

17. 3, or -6. **18.** ± 5, or ± 2. **19.** ± 2, or $\pm \sqrt{-7}$. **20.** 4, -3, or $\tfrac{1}{2} \pm \tfrac{1}{2}\sqrt{-43}$. **21.** 4, or -1. **22.** 4, or 20. **23.** 4, 2, or $-\tfrac{7}{2} \pm \tfrac{1}{2}\sqrt{17}$.
24. 3, 2, or $-3 \pm \sqrt{3}$. **25.** 3, -1, or $1 \pm \tfrac{1}{2}\sqrt{-10}$. **26.** 9, -2, or $\tfrac{7}{2} \pm \tfrac{1}{2}\sqrt{173}$. **27.** 3, $-4\tfrac{1}{2}$, or $-\tfrac{3}{4} \pm \tfrac{1}{4}\sqrt{-55}$.

Page 238.—28. ± 4, or $\pm \tfrac{4}{3}\sqrt{15}$. **29.** 4, or 69. **30.** 14, or 23.
31. 1, or -10. **32.** 50, or $19\tfrac{46}{81}$. **33.** ± 3, or $\pm \sqrt{-1}$. **34.** $2-b$, or $-(3+b)$. **35.** $-2 \pm \sqrt{3}$, or $-2 \pm \sqrt{-5}$. **36.** $1 \pm \sqrt{17}$, or $1 \pm \sqrt{2}$.
37. b^4-a, or $81 b^4-a$. **38.** 9, 1, or $5 \pm 2\sqrt{2}$.
2. 7, 3. **3.** 7, 20.

Page 239.—4. 28 rd., 40 rd. **5.** 20 sheep. **6.** 50 rows, 40 in a row. **7.** 10, 12. **8.** 11 persons. **9.** 6 days. **10.** 8¢ per doz. **11.** \$20.
12. 20 persons. **13.** 3 in. **14.** \$30.

Page 240.—15. A, \$1.14+per rd.; B, \$.89+per rd. A dug 43.84 rd.; B dug 56.16 rd. **16.** 6 rd. **17.** 12 yd., 24 yd. **18.** 9 gal.
19. One, 9 mi.; other, 10 mi. **20.** Silver, 2; copper, 25.

Page 244.—7. $x=6$; $y=2$. **8.** $x=2$, or 3; $y=3$, or 2. **9.** $x=5$, or $1\tfrac{1}{2}$; $y=3$, or 10. **10.** 6, or $-1\tfrac{1}{2}$; 1, or -4. **11.** 3, or -5; 2, or 6.
12. 5, or 6; 4, or $3\tfrac{1}{3}$. **13.** 5, or 10; 10, or 5. **14.** ± 6, or $\pm 4\sqrt{-2}$; ± 4, or $\pm 3\sqrt{-2}$. **15.** 5, or -2; 2, or -5. **16.** 4, or 3; 3, or 4.
17. 4, or -3; 3, or -4. **18.** 3, or 1; 1, or 3. **19.** 3, or 1; 1, or 3.
20. 3, or -1; 1, or -3. **21.** ± 6; ± 5. **22.** ± 2; ± 3. **23.** ± 7, or $\pm 5\sqrt{2}$; ± 3, or $\pm 2\sqrt{2}$. **24.** 8, or $17\tfrac{2}{3}$; 6, or $-13\tfrac{1}{3}$. **25.** 18, or $12\tfrac{1}{2}$; 3, or $-2\tfrac{1}{2}$. **26.** 2, or -46; 3, or 15. **27.** 4, or 2; 2, or 4.

Page 245.—28. 6, or 4; 4, or 6. **29.** ± 2, or $\pm \tfrac{1}{2}\sqrt{2}$; ± 3, or $\pm \tfrac{5}{2}\sqrt{2}$. **30.** 64, or 8; 8, or 64. **31.** 4, or -2; 2, or -4. **32.** 2, or 1;

ANSWERS. 401

1, or 2.   **33.** 3, or −2; 2, or −3.   **34.** 2, or 3; 3, or 2.   **35.** ±3√3, or ±4; ±√3, or ±5.   **36.** ±2, ±4√3; ±5, ±√3.   **37.** ±3, ±8; ±5.   **38.** ±5, or ±3; ±3, or ±5.   **39.** ±3, or ±2; ±2, or ±3.   **40.** 9, or 4; 4, or 9.   **41.** ±2, or ±6; ±4, or ±1.   **42.** 5, or −4; 4, or −5.   **43.** 3, or −1; 1, or −3.   **44.** 4, 2, or −2; 2, 4, or −4.   **45.** ±2; ±½.   **46.** 2, or 24⅜; 5, or −10 $\frac{1}{15}$.   **47.** ±2, or ±1; ±1, or ±2.   **48.** ±2, or ±√2; ±4, or ±3√2.   **49.** ±3, or ±2; ±2, or ±3.   **50.** 3, or 4; 4, or 3.   **51.** 6, or −102; 5, or 59.   **52.** 3, or 1; 1, or 3.   **53.** 3, or 9; 9, or 3.   **54.** ±6; ±3.

Page 246. — **55.** $\frac{a^2+b^2}{a-b}$; $\frac{2ab}{a-b}$.   **56.** $a-b, b-a$; $a+b, 2a$.

57. $a+b$; $a-b$.   **58.** 5; 4.   **59.** 4, or 3; 3, or 4.   **60.** $\frac{25a}{29}$; $\frac{34a}{29}$.

1. 2, 6.   **2.** 10, 2.   **3.** 4, 9.   **4.** 9, 11.   **5.** 36, 64.   **6.** 4, 2.   **7.** 8, 6.   **8.** 8, 6.   **9.** $\frac{a}{2} \pm \frac{1}{2}\sqrt{2b-a^2}$; $\frac{a}{2} \mp \frac{1}{2}\sqrt{2b-a^2}$.

Page 247. — **10.** 6, 4.   **11.** 48.   **12.** A, 36; B, 24.   **13.** Linen, 16 yd.; Cotton, 48 yd.   **14.** 24 rd. long, 18 rd. wide.   **15.** 49 yd., $3 per yd.   **16.** 40 children, 10 cents.   **17.** 60 cows, $40 per head.

Page 248. — **18.** 30, 20.   **19.** 8 miles.   **20.** J, 15; F, 20.   **21.** 67.   **22.** 2, 8.   **23.** $600, $1000.   **24.** 600, 504.   **25.** Fore, 12; hind, 15.

Page 249. — **26.** ½(3±√5), ½(1±√5).   **27.** A's $192; B's $224.   **28.** Length, 31; Breadth, 19.   **29.** 8 men; 12 women. Men, $3; women, $2.   **30.** Gold, 5; Silver, 4.   **31.** 18, 3.   **32.** 20 mi.

Page 253. — **2.** $x^2-x-6=0$.   **3.** $x^2-14x+48=0$.   **4.** $x^2-x-72=0$.   **5.** $x^2+17x+70=0$.   **6.** $x^2-12x-45=0$.   **7.** $x^2+\frac{x}{2}-3=0$.   **8.** $x^2-\frac{12x}{5}-\frac{9}{5}=0$.   **9.** $x^2-6\frac{1}{3}x+9\frac{1}{3}=0$.   **10.** $x^2-\frac{17}{12}x+\frac{6}{12}=0$.   **11.** $x^2+\frac{31}{6}x+\frac{35}{6}=0$.   **12.** $x^2-6x+7=0$.   **13.** $x^2-4x+1=0$.   **14.** $x^2-2x-4=0$.   **15.** $x^2-2ax+bx+a^2-ab=0$.   **16.** $x^2-2ax-bx+a^2+ab-2b^2=0$.   **17.** $x^2-2ax+a^2-a^4=0$.   **18.** $x^2-ax+\frac{a^2-b}{4}=0$.

Page 255. — **3.** −2, −1.   **4.** −3, −4.   **5.** −3, +7.   **6.** −2, +9.   **7.** −4, −2.   **8.** −4, −8.   **9.** +13, −3.   **10.** +16, −4.   **11.** +1, 1½.   **12.** 2, 1.   **13.** −3a, −a.   **14.** $\frac{-4b}{3}$, −2b.   **17.** ±√−2, or 5.   **18.** −3, or ±½√6.   **19.** ±√5, $-\frac{2}{a}$.   **20.** 3, −1, −2.   **21.** −5, 7, −3.   **22.** 3, $-\frac{3}{2} \pm \frac{1}{2}\sqrt{-27}$.

Page 266. — **2.** 4½.   **3.** 1½.   **4.** 2⅔.   **5.** ±6.   **6.** 1½.   **7.** 8.   **8.** 16.   **9.** 8.   **10.** 11.   **11.** 2, or −6.   **12.** 4.   **13.** $12, $28.   **14.** 12, 18.   **15.** 12, 20.   **16.** 200 bu., wheat; 300 bu., oats.

Page 267. — **19.** 5, 3.   **20.** 5, 2.   **21.** 8, 6.   **22.** 8, 4.   **23.** 4, 2.   **24.** 4, 3.

Page 269. — **3.** 6.   **4.** $\frac{9b^2}{a}$.   **5.** ⅘.   **6.** $\frac{2ab}{b^2+1}$.   **7.** ±½√5.   **8.** $\frac{a(\sqrt{b}-1)^2}{2\sqrt{b}}$.   **9.** $\frac{a(b-c)}{2\sqrt{bc}}$.

ALGEBRA — 26.

402   HIGH SCHOOL ALGEBRA.

Page 272.—**2.** 55. **3.** 47. **4.** 6. **5.** 3. **6.** −68. **7.** 30 a. **8.** 0.
9. 2 n−1. **10.** $1.72. **11.** 214$\frac{1}{2}$ feet.

Page 273.—**2.** 144. **3.** 108. **4.** 124. **5.** 52$\frac{1}{2}$. **6.** 80 a. **7.** 9 a+ 9 b+36 c.

Page 274.—**8.** n^2x. **9.** −12. **10.** 330 mi. **11.** 78. **12.** $3360.
13. $676. **14.** 438 mi. **15.** 1102$\frac{1}{2}$ mi. **16.** 36 mi. or 120 mi.
17. 3720 ft. **18.** 6433$\frac{1}{3}$ ft.

Page 276.—**3.** 17. **4.** 3. **5.** 13. **6.** 58. **7.** 216. **8.** −43. **9.** 192.
10. 6. **11.** −11. **12.** $\frac{1}{2}$. **13.** −$\frac{1}{15}$. **14.** 13. **15.** 1518.
2. 6, 9, 12, 15. **3.** 20, 27, 34, 41, 48, 55, 62, 69. **4.** $\frac{1}{5}$, 4$\frac{2}{5}$, 8$\frac{3}{5}$, 12$\frac{4}{5}$.
5. 243, 293, 343, 393. **6.** $\frac{19}{36}$, $\frac{5}{9}$, $\frac{7}{12}$, $\frac{11}{18}$, $\frac{23}{36}$, $\frac{2}{3}$, $\frac{25}{36}$, $\frac{13}{18}$. **7.** −8, −$\frac{52}{7}$,
−$\frac{48}{7}$, −$\frac{44}{7}$, −$\frac{40}{7}$, −$\frac{36}{7}$, −$\frac{32}{7}$, −4. **8.** $\frac{4}{5}$, $\frac{3}{5}$, $\frac{2}{5}$, $\frac{1}{5}$, −0, −$\frac{1}{5}$, −$\frac{2}{5}$, −$\frac{3}{5}$, −$\frac{4}{5}$.

Page 278.—**2.** 30. **3.** 10. **4.** 8. **5.** 3, 5, 7. **6.** 1, 3, 5. **7.** 3, 6, 9.
8. 2, 5, 8. **9.** 3, 5, 7. **10.** 1, 2, 3, 4. **11.** 3, 5, 7, 9. **12.** 2, 4, 6, 8.

Page 279.—**13.** 2, 5, 8, 11, 14. **14.** 1, 3, 5, 7, 9. **15.** 1, 4, 7, 10.
16. 234.

Page 280.—**2.** 160. **3.** 512. **4.** 2187. **5.** 512. **6.** 128 a^7.
7. 768 a^8x^8. **8.** 2^{n-1}. **9.** 3×4^{n-1}. **10.** $\frac{1}{729}$. **11.** $2187. **12.** $32000.
13. $512.

Page 281.—**14.** $196.83. **15.** $31.25. **16.** 67108864 bu.
17. $2928.20.

Page 282.—**2.** 2047. **3.** 9841. **4.** 16380. **5.** 265719. **6.** 2046 a.
7. 59048 x^2. **8.** 2(2^n−1). **9.** 3$\frac{255}{256}$. **10.** 10$\frac{1365}{2048}$. **11.** 1364. **12.** 1022.
13. 4. **14.** 12.

Page 283.—**15.** 3. **16.** $\dfrac{x^2}{x^2-1}$. **17.** $\dfrac{x^2}{x+y}$. **18.** $510.
19. $1048575. **20.** $60500. **21.** $255. **22.** 50 ft. **23.** $42949672.95.

Page 284.—**2.** 2, 4, 8, 16. **3.** 62, 124, 248. **4.** 3, 9, 27, 81, 243.
5. 6, 18, 54, 162, 486, 1458, 4374. **6.** $\frac{1}{2}$, 2, 8, 32. **7.** −6, −18, −54,
−162, −486, −1458. **8.** −$\frac{9}{4}$, $\frac{27}{16}$, −$\frac{81}{64}$, $\frac{243}{256}$. **9.** 12. **10.** 15.
11. 28. **12.** 21.

Page 285.—**13.** 8. **14.** 3$\frac{1}{3}$. **15.** 20 x^2y.

1. $a = rl - (r-1)s$. **2.** 2. **3.** −1. **4.** $r = \sqrt[n-1]{\dfrac{l}{a}}$. **5.** 4. **6.** $r = 3$;
$\frac{2}{3}$, 2, 6, 18, series. **7.** $a = \dfrac{l}{r^{n-1}}$. **8.** $a = 5$; 5, 15, 45, 135, 405, 1215,
series. **9.** $r = \dfrac{s-a}{s-l}$. **10.** $r = \frac{1}{2}$; 2, 1, $\frac{1}{2}$, $\frac{1}{4}$, $\frac{1}{8}$, $\frac{1}{16}$, $\frac{1}{32}$, series. **11.** 2.
12. $l = \dfrac{a+(r-1)s}{r}$. **13.** 1215.

Page 287.—**2.** 7. **3.** 1. **4.** $\frac{2}{3}$. **5.** 5. **6.** 1, 3, 9. **7.** 1, 2, 4.
8. 1, 3, 9, 27. **9.** 1, 2, 4, 8. **10.** 2, 4, 8. **11.** 4, 6, 9, 13$\frac{1}{2}$. **12.** 2, 6, 18.
13. 2, 6, 18.

Page 288.—**15.** $629.38+. **1.** 10 $x^{\frac{1}{2}}y - 2$. **2.** $(5-5c)x^4y^2$ +
$2x^{-n}y^{\frac{3}{2}} + 3x^2 + 2a + cy + 2x^3$.

Page 289.—**3.** 10 $cy^{-\frac{2}{3}} + 12m + 16ax - 3b$. **4.** 10 $xy^2 - xz + 10a -$
$(12-2b)x$. **5.** $a^5 - b^5$. **6.** $6x^m - 4xy^{n-2} - 9x^{m-1}y^2 + 6y^n$. **7.** $9x^{-1} -$
$4y^{\frac{4}{3}}$. **8.** $4x^{\frac{m}{n}} - 9y^{m-n}$. **9.** $x^{4n} + 2x^{2n}y^{2m} + y^{4m}$. **10.** $x^3 - x^2y + xy^2 - y^3$.
11. $x^6 - x^5y + x^4y^2 - x^3y^3 + x^2y^4 - xy^5 + y^6$. **12.** $x^{n-1} + x^{n-2}y + x^{n-3}y^2 +$

ANSWERS.

$x^{n-4}y^3+x^{n-5}y^4+x^{n-6}y^5$. **13.** $x^2+\frac{3}{4}$. **14.** $(2x+y)(2x+y)$. **15.** $(x^2+y^2)(x+y)(x-y)$. **16.** $(x-7)(x+5)$. **17.** $(x-9)(x+3)$. **18.** $(x^4+y^4)(x^2+y^2)(x+y)(x-y)$. **19.** $x-y$. **20.** $2x^2+3$. **21.** $x-3$. **22.** x^2-5x-8. **23.** $12\,a^2x^2y^3$.

Page 290.— 24. $y(x^2-y^2)$. **25.** $\dfrac{x-5}{2x+3}$. **26.** $\dfrac{x-2}{x+4}$. **27.** $\dfrac{a-b-c}{a+b-c}$. **28.** $\dfrac{3x^2}{x^2-1}$. **29.** $\dfrac{1}{x+2}$. **30.** 0. **31.** 0. **32.** $x^4-\dfrac{1}{x^4}$. **33.** $\dfrac{8\,a^2b^2}{a^4-b^4}$. **34.** x. **35.** $\dfrac{(x^2+y^2)^2}{x^4+y^4}$.

Page 291.— 36. $\dfrac{x-y-z}{x+y+z}$. **37.** $x-1$. **38.** $\sqrt{y}-\sqrt{x}$. **39.** $x^7+7\,x^6y+21\,x^5y^2+35\,x^4y^3+35\,x^3y^4+21\,x^2y^5+7\,xy^6+y^7$. **40.** $32\,a^5+240\,a^4b+720\,a^3b^2+1080\,a^2b^3+810\,ab^4+243\,b^5$. **41.** $a^{10}+10\,a^9b+45\,a^8b^2+120\,a^7b^3+210\,a^6b^4+252\,a^5b^5+210\,a^4b^6+120\,a^3b^7+45\,a^2b^8+10\,ab^9+b^{10}$. **42.** $x^3+3\,x^2y+3\,xy^2+y^3$. **43.** $(x+y)^2\sqrt[3]{xy}$. **44.** $(x-y)\sqrt{3\,z}$. **45.** $10\sqrt[3]{4}$. **46.** $x-y$. **47.** $a+2\sqrt{ab}+b$. **48.** 7. **49.** $\dfrac{2\,n}{a-b}$. **50.** a^2-b^2. **51.** $\dfrac{ad}{bc}$. **52.** 6. **53.** 12. **54.** $\dfrac{bn-am}{m-n}$.

Page 292.— 55. 5. **56.** ± 2, or $\pm\sqrt{-6}$. **57.** $\pm\sqrt{7}$. **58.** 81. **59.** 25. **60.** 4. **61.** $\frac{3}{4}$. **62.** $x=3, y=4, z=5$. **63.** $x=11\frac{7}{15}, y=-7\frac{1}{3}, z=74\frac{2}{3}$. **64.** $x=2, y=4, z=3, w=3, v=1$. **65.** $x=\dfrac{a}{c}, y=ac, z=\dfrac{c}{a}$. **66.** $x=\pm 2, y=\pm 4$. **67.** $x=2$, or 3; $y=3$, or 2. **68.** $x=4$, or 5; $y=5$, or 4. **69.** $x=9$, or 25; $y=25$, or 9. **70.** $x=4$, or -2; $y=2$, or -4. **71.** $x=\pm 5$; $y=\pm 4$. **72.** $x=\pm 3$; $y=\pm 4$. **73.** $x=2$, or 3; $y=3$, or 2. **74.** $x=2$, or 16; $y=2$, or $\frac{1}{2}$. **75.** $x=4$, or 2; $y=2$, or 4. **76.** $x=5$, or 4; $y=4$, or 5. **77.** $x=4$, or 1; $y=8$.

Page 293.— 78. $x=8$, or 6; $y=6$, or 8. **79.** $x=3$, or -2; $y=2$, or -3. **80.** 21, 28. **81.** 20 minutes past 5. **82.** 20. **83.** 9, 12. **84.** 9, 3. **85.** 17, 14, 27, 8, 33. **86.** 2, 3. **87.** 276. **88.** 50 apples, 150 pears. **89.** 42 miles. **90.** A's age, 21; B's age, 39.

Page 294.— 91. 333. **92.** 8 cents. **93.** $\frac{2}{3}$. **94.** 144 sq. yd. **95.** 40 horses. **96.** $180, $120. **97.** 357. **98.** 30 shillings. **99.** 27, 13.

Page 295.— 100. 24000 men. **101.** A to B 12 mi.; B to C 4 mi.; C to D 18 mi. **102.** $1-\sqrt{2}$. **103.** 8 persons. **104.** $27\frac{3}{11}$ minutes past 11. **105.** 40 rd. and 16 rd. **106.** 10, 8. **107.** $577.18+. **108.** Length $118.48+$ feet; breadth $88.86+$ feet. **109.** $x=18$, or 6; $y=6$, or 18.

Page 296.— 110. $x=20$, or -16; $y=16$, or -20. **111.** 7, 13, 19, 25. **112.** 2 yd. and 5 yd. **113.** $40. **114.** 1, 3, 9. **115.** $x=\pm 3, y=\pm 1$. **116.** 5 ft. and 4 ft. **117.** 2, 4, 8. **118.** ax^2+bx+c. **119.** 10, 20, 40. **120.** $1600, $400, $100.

Page 297.— 121. 38 gal. and 62 gal. **122.** 24 bales or 72 casks. **123.** 18 acres; $12 per acre. **124.** 4. **125.** 5, 3. **126.** A, 96; B, 108. **127.** 15 pieces. **128.** 2, 5, 8. **129.** B, 15 days; C, 18 days.

Page 298.— 130. $\frac{1}{2}(3\pm\sqrt{-3})$, $\frac{1}{2}(3\mp\sqrt{-3})$. **131.** $\frac{1}{4}(5\pm\sqrt{5})$, $\pm\frac{1}{2}\sqrt{5}$. **132.** A, 55 h.; B, 66 h. **133.** 6 days. **134.** 3, or -2.

135. 1, 2, 3.   **136.** $x=\pm 3$, or ± 4; $y=\pm 4$, or ± 3.   **137.** $x=\pm 2$, or ± 1; $y=\pm 1$, or ± 2.   **138.** 300 miles.

Page 300. — **2.** $5\sqrt{-1}$.   **3.** $5\sqrt{-2}$.   **4.** $20\sqrt{-1}$.   **5.** $6\sqrt{-3}$.
6. $8\sqrt{-5}$.   **7.** $3a\sqrt{-1}$.   **8.** $10b\sqrt{-1}$.   **9.** $17ax\sqrt{-1}$.   **10.** $-a\sqrt{-1}$.
11. $-2mx\sqrt{-3}$.

Page 301. — **2.** $-2\sqrt{5}$.   **3.** $-12\sqrt{6}$.   **4.** -36.   **5.** $-6a^2\sqrt{x}$.
6. 2.   **7.** $-2\sqrt{-1}$.   **8.** $-6\sqrt{6}$.   **9.** $15\sqrt{10}$.   **10.** $-30\sqrt{3}$.   **11.** $-8\sqrt{-1}$.
12. $96+12\sqrt{-1}$.   **13.** $x\sqrt{-1}+y\sqrt{-1}$.

3. $3\sqrt{2}$.   **4.** $\tfrac{2}{3}$.   **5.** $\dfrac{2a}{3}$.   **6.** $1-\sqrt{-1}$.   **7.** $1-\sqrt{-1}$.   **8.** $\tfrac{1}{2}\sqrt{3}$.

Page 302. — **9.** $2+\sqrt{-2}$.   **10.** $2a\sqrt{-x}$.   **11.** $-6\sqrt{-1}$.
12. $9\sqrt{-1}$.   **13.** $2\sqrt{-x}$.   **14.** $2\sqrt{-1}$.   **15.** $1+4\sqrt{-1}$.
16. $\sqrt{-1}$.

Page 304. — **3.** ∞.   **4.** Indeterminate, $\tfrac{0}{0}$.   **4.** $\tfrac{0}{0}$.   **5.** ∞.   **6.** $\tfrac{0}{0}$.

Page 307. — **2.** 1 and 4; 2 and 3.   **3.** 1 and 6; 2 and 5; 3 and 4.
4. 1 and 4; 5 and 3; 9 and 2; 13 and 4.

Page 308. — **7.** 272, 228.   **8.** $x=8$; $y=5$.   **9.** $x=3$; $y=7$.
10. $x=48$; $y=19$.   **11.** $x=8$; $y=3$.   **12.** $x=11$; $y=18$.   **13.** $x=37$; $y=13$.   **14.** Cows, 5, 10, 15, etc.; sheep, 83, 66, 49, etc.   **15.** Calves, 1 or 4; sheep, 27 or 8; geese, 72 or 88.

Page 309. — **16.** 63.   **17.** 31.   **18.** 59, 119.   **19.** 215.   **20.** 1147.
21. Calves, 1, 2, 3, 4, 5, etc.; sheep, 34, 32, 30, 28, 26, etc.; lambs, 5, 6, 7, 8, 9, etc.   **22.** Oats, 38; barley, 30; rye, 132.
23. Calves, 4, 9, 14, etc.; sheep, 78, 60, 42, etc.; lambs, 72, 85, 98.

Page 310. — **24.** Least number, 253.   A, 89; B, 73; C, 61; D, 26.
25. Histories, 6, 13; lexicons, 17, 7.   **26.** Calves, 11, 12, 13; sheep, 15, 10, 5; pigs, 4, 8, 12.   **27.** 1st part, 6, 12, 18, etc.; 2d part, 56, 42, 28, etc.; 3d part, 8, 16, 24, etc.   **28.** Sheep, 37, 36, 35, 34, etc.; pigs, 4, 12, 20, 28, etc.; lambs, 59, 52, 45, 38, etc.   **29.** Sheep, 5, 10, 15; turkeys, 42, 24, 6; geese, 53, 66, 79.

Page 313. — **3.** $x>7$.   **4.** $x>8$.   **5.** $x>a$.   **6.** $x>6$.   **7.** $x>3$.
8. $x>6$.   **9.** $x>1\tfrac{1}{5}$; $y<5\tfrac{2}{5}$. $x=2$; $y=3$.   **10.** $x<2\tfrac{4}{5}$; $y>7\tfrac{2}{5}$.
$x=2$; $y=9$.   **11.** $x>\tfrac{4}{7}$; $y<2\tfrac{1}{14}$. $x=1$; $y=2$.   **12.** $x<4\tfrac{6}{11}$; $y>2\tfrac{2}{11}$.
$x=4$; $y=3$.   **13.** $x>11\tfrac{7}{19}$; $y<11\tfrac{6}{19}$. $x=12$; $y=10$.   **14.** $x<5\tfrac{10}{93}$;
$y>3\tfrac{209}{217}$. $x=5$; $y=4$.   **16.** 2.   **17.** $\dfrac{a^3+b^3}{a^2+b^2}$.

Page 319. — **2.** 2.50243.   **3.** 2.45484.   **4.** 2.68664.   **5.** 2.52504.
6. 1.52634.   **7.** 0.42813.   **8.** $\overline{1}.58433$.   **9.** 3.68404.   **10.** 3.58500.
11. 3.44483.   **12.** 3.50093.   **13.** 3.27301.   **14.** 0.37014.   **15.** 0.22636.
16. 3.32323.   **17.** 3.17833.   **18.** 3.48487.   **19.** 3.35160.   **20.** 3.49914.
21. 3.47480.   **22.** 3.61679.   **23.** 3.56169.   **24.** 3.70233.   **25.** 3.57461.
26. 3.61982.   **27.** 3.54195.   **28.** 0.75724.   **29.** 1.79552.   **30.** 3.95260.
31. 3.89470.   **32.** 3.54666.   **33.** 3.67320.   **34.** 3.92236.   **35.** 3.96727.
36. 3.69854.   **37.** 3.82775.   **38.** 0.76246.   **39.** 1.79851.   **40.** 2.62726.
41. 3.88195.   **42.** 1.92133.   **43.** 0.93257.   **44.** 2.73287.   **45.** 1.89757.
46. 0.94221.   **47.** 3.79782.   **48.** $\overline{2}.04532$.   **49.** $\overline{4}.00000$.   **50.** $\overline{1}.47726$.
51. $\overline{2}.15229$.   **52.** $\overline{2}.30320$.   **53.** 3.38935.   **54.** 3.64019.   **55.** 3.93399.
56. $\overline{3}.89209$.   **57.** $\overline{2}.15229$.

ANSWERS.

Page 320.—**2.** 241.31. **3.** 153.55. **4.** 1.7040. **5.** .19339. **6.** .09652. **7.** 1528.6. **8.** 731.72. **9.** .001765. **10.** 965.06. **11.** 385.85. **12.** 8886.9. **13.** 1.6464. **14.** 531.95. **15.** .16382. **16.** 326.02. **17.** 1387.04. **18.** .04448. **19.** 63339.

Page 321.—**2.** 8.51. **3.** 87.5. **4.** 756. **5.** 74.87+. **6.** 418.2. **7.** 5.824. **8.** .000598. **9.** .0000225. **10.** .884. **11.** .3248. **12.** .48958. **13.** .00045. **14.** .0035425. **15.** 1182.19. **16.** .0120989. **17.** 182863. **18.** .00912. **19.** 16147. **20.** .000016. **21.** 120.964. **22.** 2402.7. **23.** 20.8018. **24.** .0006372. **25.** 18311.04.

Page 322.—**2.** .25. **3.** .763. **4.** 30.2. **5.** 3650. **6.** .15. **7.** 321.90+. **8.** 2.68+. **9.** 3130. **10.** 21600. **11.** 41.6. **12.** 4420. **13.** .428+. **14.** 94.27+. **15.** 10308.4+. **16.** 17. **17.** 605. **18.** 210. **19.** 6250. **20.** 5.02. **21.** 206.89+. **22.** 13498.16+.
2. 4225. **3.** 7055.9+. **4.** 421882+. **5.** 16776045+. **6.** 658500+. **7.** 50652+. **8.** 5765000+. **9.** 753568+. **10.** 1061183+. **11.** 131077639+. **12.** 3112101+. **13.** 225621+. **14.** 374543+. **15.** 199170403+. **16.** 307.535+. **17.** 273253164556+. **18.** 510095238+. **19.** 76.735+. **20.** .88473+. **21.** 44.361+. **22.** 858.49+. **23.** 28.095+. **24.** .000000274625. **25.** .0074305+. **26.** 590424+. **27.** 17.075+. **28.** 26.872+. **29.** 51684.52+.

Page 323.—**2.** 14. **3.** 16. **4.** 24. **5.** 34. **6.** 41. **7.** 63. **8.** 16. **9.** 72. **10.** 24. **11.** 2.5. **12.** 42−. **13.** 45. **14.** 3.072. **15.** 4.020+. **16.** 3.054+. **17.** 2.491+. **18.** 1.472+. **19.** 1.691+. **20.** 1.677+. **21.** 1.601+. **22.** 1.581+. **23.** 1.484+. **24.** 1.558+. **25.** 1.456+. **26.** 1.047+. **27.** .0071+. **28.** 7.962+. **29.** 9.804+.

Page 325.—**2.** \$1010. **3.** \$740.25. **4.** \$701.79. **5.** \$332.50. **6.** \$541.46. **7.** \$675.62. **8.** $4\frac{1}{2}\%$. **9.** 11.89 yr.

Page 327.—**3.** \$10515.96. **4.** \$5298.19. **5.** \$4454.59. **6.** \$8376.15. **7.** \$5383.82. **8.** \$12462.06.

Page 330.—**1.** 3024. **2.** 6720. **3.** 70. **4.** 20160. **5.** 120. **6.** 5040.
Page 331.—**7.** 495. **8.** 31. **9.** 18564. **10.** 3628800.
Page 334.—**1.** $a^5 + 5\,a^4x + 10\,a^3x^2 + 10\,a^2x^3 + 5\,ax^4 + x^5$. **2.** $a^7 - 7\,a^6x + 21\,a^5x^2 - 35\,a^4x^3 + 35\,a^3x^4 - 21\,a^2x^5 + 7\,ax^6 - x^7$. **3.** $a^8 + 4\,a^6b^3 + 6\,a^4b^6 + 4\,a^2b^9 + b^{12}$. **4.** $a^{-4} + 4\,a^{-3}c^{-2} + 6\,a^{-2}c^{-4} + 4\,a^{-1}c^{-6} + c^{-8}$. **5.** $c^{\frac{3}{2}} + 3\,cd^{-2} + 3\,c^{\frac{1}{2}}d^{-4} + d^{-6}$. **6.** $a^{\frac{3}{3}} - 4\,a^2x^{\frac{3}{4}} + 6\,a^{\frac{4}{3}}x^{\frac{3}{2}} - 4\,a^{\frac{2}{3}}x^{\frac{7}{4}} + x^3$. **7.** $x^{-\frac{5}{2}} + 5\,x^{-2}y^3 + 10\,x^{-\frac{3}{2}}y^6 + 10\,x^{-1}y^9 + 5\,x^{-\frac{1}{2}}y^{12} + y^{15}$. **8.** $\dfrac{8\,x^3}{y^3} - \dfrac{36\,x}{y} + \dfrac{54\,y}{x} - \dfrac{27\,y^3}{x^3}$.
9. $x^4 + 4\,x^3y^{\frac{1}{2}} + 6\,x^2y + 4\,xy^{\frac{3}{2}} + y^2$. **10.** $a^9 + 12\,a^{\frac{15}{2}}x^3 + 60\,a^6x^{\frac{2}{3}} + 160\,a^{\frac{9}{2}}x + 240\,a^3x^{\frac{4}{3}} + 192\,a^{\frac{3}{2}}x^{\frac{5}{3}} + 64\,x^2$. **11.** $243\,x^{10} - 135\,x^{\frac{17}{2}} + 30\,x^7 - \frac{10}{3}\,x^{\frac{11}{2}} + \dfrac{5\,x^4}{27} - \dfrac{x^{\frac{5}{2}}}{243}$. **12.** $\dfrac{16\,x^2}{y^{\frac{8}{3}}} - \dfrac{32\,x}{y^{\frac{3}{2}}} + \dfrac{24}{y^{\frac{1}{3}}} - \dfrac{8\,y^{\frac{5}{6}}}{x} + \dfrac{y^2}{x^2}$.

Page 335.—**1.** $56\,a^3b^5$. **2.** $210\,a^6x^4$. **3.** $462\,a^5c^6$. **4.** $495\,a^4b^{16}$. **5.** $-2835\,x^9$. **6.** $-10\frac{1}{2}\,a^6x^3$. **7.** $35\dfrac{x}{a}$. **8.** $-20\,a^{\frac{15}{2}}$. **9.** $-\frac{35}{81}\,x^4y^{\frac{9}{2}}$. **10.** $\frac{7}{4}\,a^{-1}x^5$. **11.** $70\,a^2x^2$. **12.** $-36\,a^{\frac{4}{3}}c^{\frac{7}{2}}$. **13.** $64\,x^3$. **14.** $-\dfrac{20\,ab}{3}\sqrt{\dfrac{ab}{3}}$.

1. $x^3 + 3\,x^2y + 3\,xy^2 + y^3 + 3\,x^2z + 6\,xyz + 3\,y^2z + 3\,xz^2 + 3\,yz^2 + z^3$. **2.** $1 + 4\,x + 2\,x^2 - 8\,x^3 - 5\,x^4 + 8\,x^5 + 2\,x^6 - 4\,x^7 + x^8$. **3.** $x^8 - 4\,x^7 + 14\,x^6 - 28\,x^5 + 49\,x^4 - 56\,x^3 + 56\,x^2 - 32\,x + 16$. **4.** $a^6 + 3\,a^5x + 6\,a^4x^2 + 7\,a^3x^3 + 6\,a^2x^4 +$

406   HIGH SCHOOL ALGEBRA.

$3\,ax^5+x^6$.   **5.** $a^3+3\,a^2x^2+3\,ax^4+x^6+3\,a^2z+6\,ax^2z+3\,x^4z+3\,a^2+6\,ax^2$ $+3\,x^4+3\,az^2+6\,az+3\,a+3\,x^2z^2+6\,x^2z+3\,x^2+z^3+3\,z^2+3\,z+1$.   **6.** $x^8-8\,x^7-20\,x^6+40\,x^5-74\,x^4-120\,x^3+108\,x^2+216\,x+81$.

Page 339.—3. $1-2\,x+2\,x^2-2\,x^3+2\,x^4$.   **4.** $1+x+3\,x^2+9\,x^3+27\,x^4$.   **5.** $\tfrac{1}{2}-\tfrac{1}{4}x+\tfrac{3}{8}x^2-\tfrac{9}{16}x^3+\tfrac{27}{32}x^4$.   **6.** $1+3\,x+4\,x^2+7\,x^3+11\,x^4$.   **7.** $1+2\,x+8\,x^2+28\,x^3+100\,x^4$.   **8.** $2-5\,x+3\,x^2+2\,x^3-5\,x^4$.   **9.** $1+3\,x+7\,x^2+15\,x^3+31\,x^4$.   **10.** $3+x-2\,x^2-3\,x^3-x^4$.   **11.** $x-x^2+0\,x^3+x^4-x^5$.   **12.** $2+3\,x+4\,x^2+5\,x^3+6\,x^4$.   **13.** $1-2\,x^2+x^4+4\,x^6-11\,x^8$.   **14.** $x+9\,x^2+32\,x^3+92\,x^4+240\,x^5$.

Page 341.—5. $\dfrac{2}{x+1}+\dfrac{1}{x-2}$.   **6.** $\dfrac{2}{x+1}+\dfrac{3}{x+2}$.   **7.** $\dfrac{1}{x-2}+\dfrac{1}{x-3}$.   **8.** $\dfrac{3}{x-2}+\dfrac{4}{x+4}$.   **9.** $\dfrac{5}{x-5}+\dfrac{4}{x-3}$.   **10.** $\dfrac{46}{5(x-3)}-\dfrac{37}{5(2x-1)}$.   **11.** $\dfrac{1}{x}+\dfrac{1}{x+1}+\dfrac{3}{x-2}$.   **12.** $\dfrac{1}{x}+\dfrac{1}{x-1}+\dfrac{1}{x-2}$.   **13.** $\dfrac{7}{(x-2)^2}+\dfrac{3}{x-2}$.   **14.** $\dfrac{6}{(x-1)^3}+\dfrac{11}{(x-1)^2}+\dfrac{5}{x-1}$.

Page 343.—2. $x=y-y^2+y^3-y^4+\cdots$.   **3.** $x=y-3\,y^2+13\,y^3-67\,y^4+\cdots$.   **4.** $x=y+\tfrac{1}{2}y^2+\tfrac{1}{6}y^3+\tfrac{1}{24}y^4+\cdots$.   **5.** $x=\tfrac{1}{2}y-\tfrac{3}{8}y^2+\tfrac{5}{16}y^3-\tfrac{35}{128}y^4+\cdots$.   **6.** $x=\tfrac{1}{2}y-\tfrac{1}{2}y^2+\tfrac{5}{8}y^3-\tfrac{7}{8}y^4+\cdots$.

Page 344.—8. $.454631-$.   **9.** $.274649+$.

Page 345.—2. $m=x^2$; $n=x$.   **3.** $m=2\,x^2$; $n=x$.   **4.** $m=3\,x^2$; $n=2\,x$.   **5.** $m=2\,x^2$; $n=3\,x$.

Page 347.—2. $\dfrac{1+2\,x}{1-x-x^2}$.   **3.** $\dfrac{1+2\,x}{1-x-2\,x^2}$.   **4.** $\dfrac{1-x}{1-3\,x-2\,x^2}$.   **5.** $\dfrac{1+x}{1-2\,x+x^2}$.

Page 351.—2. $a^{\tfrac{1}{2}}+\tfrac{1}{2}a^{-\tfrac{1}{2}}b-\tfrac{1}{8}a^{-\tfrac{3}{2}}b^2+\tfrac{1}{16}a^{-\tfrac{5}{2}}b^3-\cdots$.   **3.** $a^{\tfrac{2}{3}}+\tfrac{2}{3}a^{-\tfrac{1}{3}}b-\tfrac{1}{9}a^{-\tfrac{4}{3}}b^2+\tfrac{4}{81}a^{-\tfrac{7}{3}}b^3-\cdots$.   **4.** $a^{-2}-2\,a^{-3}b+3\,a^{-4}b^2-4\,a^{-5}b^3+\cdots$.   **5.** $a^{-\tfrac{1}{2}}-\tfrac{1}{2}a^{-\tfrac{3}{2}}b+\tfrac{3}{8}a^{-\tfrac{5}{2}}b^2-\tfrac{5}{16}a^{-\tfrac{7}{2}}b^3+\cdots$.   **6.** $a^{-3}+3\,a^{-4}b+6\,a^{-5}b^2+10\,a^{-6}b^3+\cdots$.   **7.** $a^{\tfrac{3}{4}}-\tfrac{3}{4}a^{-\tfrac{1}{4}}b-\tfrac{3}{32}a^{-\tfrac{5}{4}}b^2-\tfrac{15}{384}a^{-\tfrac{9}{4}}b^3+\cdots$.   **8.** $a^{-\tfrac{1}{4}}-\tfrac{1}{4}a^{-\tfrac{5}{4}}b+\tfrac{5}{32}a^{-\tfrac{9}{4}}b^2-\tfrac{15}{128}a^{-\tfrac{13}{4}}b^3+\cdots$.   **9.** $a^{-\tfrac{3}{2}}-\tfrac{3}{2}a^{-\tfrac{5}{2}}b+\tfrac{15}{8}a^{-\tfrac{7}{2}}b^2-\tfrac{35}{16}a^{-\tfrac{9}{2}}b^3+\cdots$.   **10.** $1+5\,x+5\,x^2-\tfrac{5}{3}x^3+\cdots$.   **11.** $\sqrt[4]{8\,a^3}+\dfrac{9\,b}{4\sqrt[4]{2\,a}}-\dfrac{27\,b^2}{64\sqrt[4]{2\,a^5}}+\dfrac{135\,b^3}{512\sqrt[4]{4\,a^9}}$.   **12.** $1-\tfrac{1}{2}x-\tfrac{3}{8}x^2-\tfrac{7}{16}x^3+\cdots$.   **13.** $a^{-\tfrac{3}{5}}+\tfrac{3}{5}a^{-\tfrac{8}{5}}b+\tfrac{12}{25}a^{-\tfrac{13}{5}}b^2+\tfrac{52}{125}a^{-\tfrac{18}{5}}b^3-\cdots$.   **15.** $2.609+$.   **16.** $4.242+$.   **17.** $5.099+$.   **18.** $3.072+$.   **19.** $3.036+$.   **20.** $3.0049+$.

Page 354.—2. $-1, 3$.   **3.** $1, 2$.   **4.** $-3, 2$.   **5.** $8, 2$.   **6.** $-\tfrac{1}{2}\pm\tfrac{1}{2}\sqrt{-39}$.

Page 356.—1. $x^3-9\,x^2+26\,x-24=0$.   **2.** $x^3-9\,x^2+23\,x-15=0$.   **3.** $x^3+2\,x^2-5\,x-6=0$.   **4.** $x^4-2\,x^3-13\,x^2+38\,x-24=0$.   **5.** $x^3-4\tfrac{1}{2}x^2+5\,x-1\tfrac{1}{2}=0$.   **6.** $x^4+\tfrac{7}{6}x^3-6\,x^2-\tfrac{7}{6}x+1=0$.   **7.** $1, 2, 3, 4$.   **8.** $2, -1$.

Page 357.—1. Two positive and one negative.   **2.** Three positive.   **3.** Two positive and one negative.   **4.** One positive and one negative.   **5.** One positive and one negative.   **6.** Two positive and two negative.

ANSWERS.

Page 359.—**2.** $y^5-3y^4+2y^2+3y+6=0$. **3.** $y^6-3y^4-4y^2-7y+13=0$. **4.** $y^5-2y^4-5y^3-3y-4=0$.
1. $y^3+36y-432=0$. **2.** $y^3+6y^2-81=0$. **3.** $y^3-6y^2-243=0$.
4. $y^3-27y-54=0$.

Page 360.—**1.** $y^4-y^3+2y-288=0$. **2.** $y^3-14y^2+11y-75=0$.
3. $y^3-18y^2+27y-120=0$. **4.** $y^3-11y^2-12y+31=0$. **5.** $y^4-12y^3+150y^2-13500y+23625=0$. **6.** $y^4-65y^3+1170y^2-153000y+162000=0$. **7.** $y^4+4y^3-75y^2+500y=0$. **8.** $y^5-36y^3+13824y-5971968=0$.

Page 363.—**1.** $1-9+26-24$. **2.** $1+4-18-76-55$. **3.** $1-6+11-6$.

Page 364.—**3.** $x^4-3x^2+5x-6=0$. **4.** $x^3-30x^2+225x-491=0$.
5. $y^5+50y^4+1005y^3+10150y^2+51497y+104965=0$. **6.** $y^4+9y^3+12y^2-14y=0$. **7.** $y^4+43y^3+687y^2+4847y+12453=0$.

Page 365.—**8.** $x^5+10x^4+77x^3+302x^2+430x+153=0$. **9.** $y^4-10y^3+35y^2-50y+24=0$. **10.** (1) $y^5+6y^4+11y^3-19y-15=0$. (2) $y^4+48y^3+244y^2+432y+291=0$. (3) $x^5+10x^4+40x^3+80x^2+80x+31=0$. (4) $y^3+6y^2-68y-62=0$.

Page 366.—**4.** 4, 5, 1. **5.** $\frac{3}{2}$, 3, $-\frac{8}{3}$. **6.** 5, -6, $-2 \pm \sqrt{6}$. **7.** -6, 8, 4 -1. **8.** $\frac{5}{3}$, $\frac{5}{2}$. **9.** 2, 3, 4, 7. **10.** 7, 9, 14, 18.

Page 367.—**1.** 1. **2.** $b(b-1)$. **3.** $\dfrac{2a(2b^2-5)}{4a-3b}$. **4.** 13. **5.** 8. **6.** 4.
7. 1. **8.** 4. **9.** $\frac{1}{3}a$. **10.** 5, or $-\frac{5}{4}$. **11.** $\dfrac{2ac}{c^2+1}$. **12.** $\dfrac{b}{(\sqrt{a}-2\sqrt{b})^2}$.
13. $b\left\{1-\left(\dfrac{2\sqrt{c}}{1+c}\right)^4\right\}$. **14.** $\dfrac{b^2-4a^2}{4a}$.

Page 368.—**15.** 3. **16.** 1, or $1\frac{1}{2} \pm 2\sqrt{15}$. **17.** 4. **18.** $\frac{1}{2}$, or $\frac{1}{10}$.
19. $10\sqrt{-1}$. **20.** $x=4$; $y=2$; $z=6$; $u=5$. **21.** $x=3$; $y=2$; $z=4$; $u=-2$; $v=-3$. **22.** $x=\pm\dfrac{a(b-c)}{2\sqrt{bc}}$. **23.** $x=4$. **24.** $x=\pm\dfrac{a(n-1)}{\sqrt{2n-1}}$.
25. $x=\pm 4.54+$. **26.** $x=\pm\sqrt{\dfrac{a^2(c-d)^2-b^2(c+d)^2}{2(c^2+d^2)}}$.

Page 369.—**27.** $x=\dfrac{2}{\sqrt{a^2+4}}$. **28.** $x=\pm\frac{1}{2}$. **29.** $x=\pm 3$, or ± 2; $y=\pm 2$, or ± 3. **30.** $x=a$, or b; $y=b$, or a. **31.** $x=2$; $y=2$. **32.** $x=5$; $y=3$. **33.** $x=3$, or 6; $y=6$, or 3. **34.** $x=4$, or $(\frac{1}{108})^2$; $y=9$, or 324.
35. $x=4$, or 1; $y=8$. **36.** $x=4$, or -2; $y=2$, or -4. **37.** $x=2$; $y=6$. **38.** $x=4$, or 16; $y=5$, or -7. **39.** $x=3$, or -7; $y=2$, or $\frac{4}{7}$.
40. $x=3$, or $-\frac{11}{3}$; $y=1$, or $-\frac{7}{3}$. **41.** $x=5$; $y=3$. **42.** $x=2$, or 4; $y=4$, or 2. **43.** $x=3$, or 2; $y=2$, or 3. **44.** $x=2$, or $\frac{3}{5}$; $y=2$, or -5.
45. $x=4$, or 2; $y=2$, or 4. **46.** $x=3$; $y=2$. **47.** $x=8$, or 7; $y=7$, or 8. **48.** $x=6$, or 9; $y=4$, or 1.

Page 370.—**49.** $x=\pm 8$; $y=\pm 6$. **50.** $x=5$, or 4; $y=4$, or 5.
51. $x=1$, or $1\frac{7}{17}$; $y=2$, or $-\frac{1}{17}$. **52.** $x=9$, or 4; $y=4$, or 9. **53.** $x=4$, or 2; $y=2$, or 4. **54.** $x=\pm 4$, or ± 3; $y=\pm 3$, or ± 4. **55.** $x=12$; $y=6$. **56.** $x=4$; $y=8$. **57.** $x=3$, or 2; $y=2$, or 3. **58.** $x=2$, or 1; $y=1$, or 2. **59.** $x=\pm 4$; $y=\pm 5$. **60.** $x=2$, or 3; $y=3$, or 2.
61. $x=4$, or 6; $y=6$, or 4. **62.** $x=3$, or 2; $y=2$, or 3. **63.** $x=8$, or 6; $y=6$, or 8. **64.** $x=4$; $y=5$; $z=6$. **65.** $x=6$; $y=8$; $z=5$.

Page 371. — **66.** $x = \dfrac{a}{2} + \dfrac{1}{2}\sqrt{\dfrac{4b-a^3}{3a}}$; $y = \dfrac{a}{2} - \dfrac{1}{2}\sqrt{\dfrac{4b-a^3}{3a}}$. **67.** $x = \dfrac{a^2}{a+b}$; $y = \dfrac{b^2}{a+b}$. **68.** $x = \dfrac{\sqrt{2b-a} \pm \sqrt{a}}{2\sqrt[6]{2b-a}}$; $y = \dfrac{\sqrt{2b-a} \mp \sqrt{a}}{2\sqrt[6]{2b-a}}$.
69. $\dfrac{11}{21} + \dfrac{23}{42} + \dfrac{4}{7} + \dfrac{25}{42} + \dfrac{13}{21} + \dfrac{9}{14}$. **70.** $3, 1, \tfrac{1}{3}$. **71.** $\tfrac{3}{35}$. **72.** 10. **73.** $a^{\frac{1}{2}} - \tfrac{1}{2} a^{-\frac{1}{2}} x - \tfrac{1}{8} a^{-\frac{3}{2}} x^2 - \tfrac{1}{16} a^{-\frac{5}{2}} x^3 - \tfrac{5}{128} a^{-\frac{7}{2}} x^4$. **74.** $a^{3n+3} + (n+1)a^{3n} b^2 + \dfrac{n(n+1)}{2} a^{3n-3} b^4 + \dfrac{n(n+1)(n-1)}{6} a^{3n-6} b^6 + \dfrac{n(n+1)(n-1)(n-2)}{24} a^{3n-9} b^8 + $.
75. $1 + 6x + 9x^2 - 4x^3 - 9x^4 + 6x^5 - x^6$. **76.** $3360\, a^{-6} x^2$. **77.** $3.3437+$. **78.** $1 + 2x + 8x^2 + 28x^3 + 100x^4$. **79.** $\dfrac{3}{x^4+1} + \dfrac{4}{x^4-1}$. **80.** $\dfrac{2}{x} + \dfrac{1}{x+1} + \dfrac{2}{x+2}$. **83.** $3 - \sqrt{2},\ 3 + \sqrt{2}$. **84.** $x^3 + 5x^2 + 2x - 8 = 0$.

Page 372. — **85.** $x = 4$; $y = 12$; $z = 4$. **86.** 376, 214. **87.** $28\tfrac{164}{181}$, $260\tfrac{28}{181}$, $37\tfrac{170}{181}$. **88.** \$80. **89.** 12 inches. **90.** Length, 20 ft.; breadth, 16 ft.; height, 14 ft. **91.** $656\tfrac{1}{10}$ ft. nearly. **92.** Base, 15 ft.; hyp., 39 ft.; perp., 36 ft. **93.** 39, 93.

Page 373. — **94.** $\dfrac{ab}{a+b}$. **95.** 25.4 in. +. **96.** 4, 13, 16. **97.** Length, 16 ft.; breadth, 2 ft. **98.** Length, diminished 102 ft.; breadth, increased 114 ft. **99.** A, 36; B, 30. **100.** $\tfrac{1}{2}(3 \pm \sqrt{5})$; $\tfrac{1}{2}(1 \pm \sqrt{5})$. **101.** $\pm \tfrac{1}{2}\sqrt{5}$; $\tfrac{1}{4}(5 \pm \sqrt{5})$.

Page 374. — **102.** \$500, \$320. **103.** 100 feet. **104.** 7 hours; 5 hours. **105.** 79 days; number of men alive, 28. **106.** 40 ft. long; 25 ft. wide. **107.** 38, 62.

Page 375. — **108.** 25 miles from Baltimore. **109.** Length, 90 yd.; breadth, 60 yd. **110.** 6 days; 26 men. **111.** Length, 30 yd.; breadth, 19 yd.; height of wall, 4 yd. **112.** .655. **113.** 1.895.

English Grammar

Conklin's English Grammar and Composition
    Cloth, 12mo. 296 pages . . . . . . 60 cents
    A complete graded course in Grammar and Composition for Intermediate or Grammar School grades.

Harvey's Elementary Grammar and Composition—Revised
    Cloth, 12mo. 160 pages . . . . . . 42 cents

Harvey's Practical English Grammar—Revised
    Cloth, 12mo. 272 pages . . . . . 65 cents
    A practical and systematic course in language study, including Language Lessons, Composition and English Grammar.

Maxwell's First Book in English
    Cloth, 12mo. 176 pages . . . . . . 40 cents

Maxwell's Introductory Lessons in English Grammar
    Cloth, 12mo. 172 pages . . . . . . 40 cents
    A brief graded course in English for Elementary and Grammar grades.

Metcalf's Elementary English
    Cloth, 12mo. 200 pages . . . . . . 40 cents

Metcalf's English Grammar
    Cloth, 12mo. 288 pages . . . . . . 60 cents
    A logical and progressive series based on the inductive method. Adapted for use in graded or ungraded schools.

Swinton's New English Grammar
    Cloth, 12mo. 256 pages . . . . . 56 cents
    A working class book for the study and practice of English.

FOR ADVANCED CLASSES

Baskervill and Sewell's English Grammar
    Cloth, 12mo. 349 pages . . . . . 90 cents
    An advanced English Grammar based on the actual use of the language. For use in High School, Academy and College classes.

Maxwell's Advanced Lessons in English Grammar
    Cloth, 12mo. 334 pages . . . . . 60 cents
    For use in higher Grammar classes and High Schools.

Lyte's English Grammar and Composition
    Cloth, 12mo. 270 pages . . . . . 65 cents

Copies of any of the above books will be sent prepaid to any address, on receipt of the price, by the Publishers:

American Book Company

New York   •   Cincinnati   •   Chicago

Arithmetics

SCHOOL ARITHMETICS

Appletons' First Lessons in Arithmetic	$0.36
Numbers Applied	.75
Milne's Elements of Arithmetic	.30
Standard Arithmetic	.65
White's First Book of Arithmetic	.30
New Complete Arithmetic	.65
Ray's New Elementary Arithmetic	.35
New Practical Arithmetic	.50
New Higher Arithmetic	.85
Robinson's New Rudiments of Arithmetic	.30
New Practical Arithmetic	.65
New Higher Arithmetic	1.00
Bailey's American Comprehensive Arithmetic	.65
Moore's Grammar School Arithmetic	.60

MENTAL ARITHMETICS

Bailey's American Mental Arithmetic	.35
Davies's Intellectual Arithmetic	.25
Dubbs's Complete Mental Arithmetic	.35
French's Mental Arithmetic	.36
Milne's Mental Arithmetic	.35
Ray's New Intellectual Arithmetic	.25
Robinson's New Intellectual Arithmetic	.35

AIDS TO THE STUDY OF ARITHMETIC

Baird's Graded Work in Arithmetic.	
First Book. Numbers to 20	.18
Second Book. Numbers to 100	.18
Third Book. Numbers to 1,000,000	.20
Fourth Book. Intermediate Grades	.20
Dubbs's Arithmetical Problems, Teachers' Edition	1.00
Dubbs's Arithmetical Problems, Pupils' Edition, 2 Parts, each	.25
White's Oral Lessons in Number	.60
Kirk and Sabin's Oral Arithmetic. Parts I. and II. Each	.25

Copies of any of these books will be sent, prepaid, to any address on receipt of the price by the Publishers:

American Book Company

NEW YORK ♦ CINCINNATI ♦ CHICAGO

Supplementary Reading

JOHONNOT'S HISTORICAL READERS

SIX BOOKS. 12MO. ILLUSTRATED.

Grandfather's Stories. 140 pages	27 cents
Stories of Heroic Deeds. 150 pages	30 cents
Stories of Our Country. 207 pages	40 cents
Stories of Other Lands. 232 pages	40 cents
Stories of the Olden Time. 254 pages	54 cents
Ten Great Events in History. 264 pages	54 cents

JOHONNOT'S NATURAL HISTORY READERS

SIX BOOKS. 12MO. ILLUSTRATED.

Book of Cats and Dogs. 96 pages	17 cents
Friends in Feathers and Fur. 140 pages	30 cents
Neighbors with Wings and Fins. 229 pages . . .	40 cents
Some Curious Flyers, Creepers and Swimmers. 224 pages	40 cents
Neighbors with Claws and Hoofs. 256 pages . . .	54 cents
Glimpses of the Animate World. 414 pages . . .	$1.00

These books are admirably adapted for use as supplementary readers. Each series contains a full course of graded lessons for reading from instructive topics, written in a style that is of the most fascinating interest to children and young people, while training them to habits of observation and storing their minds with valuable information. Each book is fully illustrated in an artistic and attractive manner.

Copies of any of the above books will be sent prepaid to any address, on receipt of the price, by the Publishers:

American Book Company

New York   ♦   Cincinnati   ♦   Chicago

Books for Teachers

FOR THE STUDY OF PEDAGOGY

Calkins's Manual of Object Teaching	$1.25
Hailmann's History of Pedagogy	.60
Hewett's Pedagogy for Young Teachers	.85
How to Teach (Kiddle, Harrison, and Calkins)	1.00
King's School Interests and Duties	1.00
Krüsi's Life and Work of Pestalozzi	1.20
Mann's School Recreations and Amusements	1.00
Page's Theory and Practice of Teaching	1.00
Palmer's Science of Education	1.00
Payne's School Supervision	1.00
Payne's Contributions to the Science of Education	1.25
Sheldon's Lessons on Objects	1.20
Shoup's History and Science of Education	1.00
Swett's Methods of Teaching	1.00
White's Elements of Pedagogy	1.00
White's School Management	1.00

FOR THE STUDY OF PSYCHOLOGY

Halleck's Psychology and Psychic Culture	1.25
Hewett's Psychology for Young Teachers	.85
Putnam's Elementary Psychology	.90
Roark's Psychology in Education	1.00

FOR THE TEACHER'S DESK

Schaeffer's Bible Readings for Schools	.35
Eclectic Manual of Methods	.60
Swett's Questions for Written Examination	.72
Appletons' How to Teach Writing	.50
Morris's Physical Education	1.00
Smart's Manual of School Gymnastics	.30
White's Oral Lessons in Number	.60
Dubbs's Arithmetical Problems. Teachers' Edition	1.00
Doerner's Treasury of General Knowledge. Part I.	.50
The Same. Part II.	.65
Webster's Academic Dictionary. New Edition.	1.50

Any of the above books sent, prepaid, on receipt of the price by the Publishers:

American Book Company

NEW YORK ♦ CINCINNATI ♦ CHICAGO

For the Study of Literature

Matthews' Introduction to the Study of American Literature
By BRANDER MATTHEWS, Professor of Literature in Columbia University. Cloth, 12mo, 256 pages . . . $1.00

A text-book of literature on an original plan, admirably designed to guide, to supplement, and to stimulate the student's reading of American authors.

Watkins's American Literature (Literature Primer Series).
By MILDRED CABELL WATKINS.
Flexible cloth, 18mo, 224 pages . . . . . 35 cents

A text-book of American Literature adapted to the comprehension of pupils in common and graded schools.

Brooke's English Literature (Literature Primer Series).
By the Rev. STOPFORD BROOKE, M.A. New edition, revised and corrected. Flexible cloth, 18mo, 240 pages . . 35 cents

Equally valuable as a class-book for schools or as a book of reference for general readers.

Smith's Studies in English Literature
By M. W. SMITH, A.M. Cloth, 12mo, 427 pages . . $1.20

Containing complete selections from Chaucer, Spenser, Shakespeare, Bacon and Milton, with a History of English Literature to the death of Dryden in 1700.

Cathcart's Literary Reader
By GEORGE R. CATHCART. Cloth, leather back, 12mo, 541 pages $1.15

A manual of English literature containing typical selections from the best British and American authors, with biographical and critical sketches, portraits and fac-simile autographs.

Anderson's Study of English Words
By J. M. ANDERSON. Cloth, 12mo, 118 pages . . 40 cents

A summary of the most important facts of the English language, with special reference to the growth of English words.

Koopman's Mastery of Books
By H. L. KOOPMAN, Librarian of Brown University. Cloth, 12mo, 214 pages . . . . . . 90 cents

A guide to the selection of the best books for reading and reference.

Copies of any of the above books will be sent, prepaid, to any address on receipt of the price by the Publishers:

American Book Company

NEW YORK ♦ CINCINNATI ♦ CHICAGO

Composition and Rhetoric

Butler's School English

 Cloth, 12mo, 272 pages . . . . 75 cents

 A brief, concise and thoroughly practical manual for use in connection with the written English work of secondary schools. It has been prepared specially to secure definite results in the study of English, by showing the pupil how to review, criticise, and improve his own writing.

Quackenbos's Practical Rhetoric

 Cloth, 12mo, 477 pages $1.00

 This book develops, in a perfectly natural manner, the laws and principles which underlie rhetorical art, and then shows their use and application in the different processes and kinds of composition. It is clear, simple, and logical in its treatment throughout, original in its departure from technical rules and traditions, copiously illustrated with examples for practice, and calculated to awaken interest and enthusiasm in the study. A large part of the book is devoted to instruction and practice in actual composition work, in which the pupil is encouraged to follow and apply genuine laboratory methods.

Waddy's Elements of Composition and Rhetoric

 Cloth, 12mo, 416 pages $1.00

 A complete course in Composition and Rhetoric, with copious exercises in both criticism and construction. It is inductive in method, lucid in style, orderly in arrangement, and clear and comprehensive in treatment. Sufficiently elementary for the lower grades of high school classes and complete enough for all secondary schools.

Copies of the above books will be sent prepaid to any address, on receipt of the price, by the Publishers:

American Book Company

New York   •   Cincinnati   •   Chicago

Important New Books

Crockett's Plane and Spherical Trigonometry

By C. W. CROCKETT, C.E., Professor of Mathematics and Astronomy in Rensselaer Polytechnic Institute, Troy, New York. With Tables. Cloth, 8vo. 310 pages . . $1.25
The Same. Without Tables 1.00
Logarithmic and Trigonometric Tables (separate) . . 1.00

A clear analytic treatment of the elements of Plane and Spherical Trigonometry and their practical applications to Surveying, Geodesy, and Astronomy, with convenient and accurate "five place" tables for the use of the student, engineer, and surveyor. Designed for High Schools, Colleges, and Technical Institutions.

Raymond's Plane Surveying

By W. G. RAYMOND, C.E., Member American Society of Civil Engineers, Professor of Geodesy, Road Engineering, and Topographical Drawing in Rensselaer Polytechnic Institute.
Cloth, 8vo. 485 pages. With Tables and Illustrations . $3.00

A modern text-book for the study and practice of Land, Topographical, Hydrographical, and Mine Surveying. Special attention is given to such practical subjects as system in office work, to labor-saving devices, to coördinate methods, and to the explanation of difficulties encountered by young surveyors. The appendix contains a large number of original problems, and a full set of tables for class and field work.

Todd's New Astronomy

By DAVID P. TODD, M.A., Ph.D., Professor of Astronomy and Director of the Observatory, Amherst College.
Cloth, 12mo. 500 pages. Illustrated $1.30

A new Astronomy designed for classes pursuing the study in High Schools, Academies, and other Preparatory Schools. The treatment throughout is simple, clear, scientific, and deeply interesting. The illustrations include sketches from the author's laboratory and expeditions, and numerous reproductions from astronomical photographs.

Copies of the above books will be sent, prepaid, to any address on receipt of the price by the Publishers:

American Book Company

New York   ✦   Cincinnati   ✦   Chicago

A New Astronomy

BY

DAVID P. TODD, M.A., Ph.D.

Professor of Astronomy and Director of the Observatory, Amherst College.

Cloth, 12mo, 480 pages. Illustrated - - Price, $1.30

This book is designed for classes pursuing the study in High Schools, Academies, and Colleges. The author's long experience as a director in astronomical observatories and in teaching the subject has given him unusual qualifications and advantages for preparing an ideal text-book.

The noteworthy feature which distinguishes this from other text-books on Astronomy is the practical way in which the subjects treated are enforced by laboratory experiments and methods. In this the author follows the principle that Astronomy is preëminently a science of observation and should be so taught.

By placing more importance on the physical than on the mathematical facts of Astronomy the author has made every page of the book deeply interesting to the student and the general reader. The treatment of the planets and other heavenly bodies and of the law of universal gravitation is unusually full, clear, and illuminative. The marvelous discoveries of Astronomy in recent years, and the latest advances in methods of teaching the science, are all represented.

The illustrations are an important feature of the book. Many of them are so ingeniously devised that they explain at a glance what pages of mere description could not make clear.

Copies of Todd's New Astronomy will be sent, prepaid, to any address on receipt of the price by the Publishers:

American Book Company

NEW YORK • CINCINNATI • CHICAGO